Henry Thompson

The Diseases of the Prostate

Their Pathology and Treatment

Henry Thompson

The Diseases of the Prostate
Their Pathology and Treatment

ISBN/EAN: 9783744692182

Printed in Europe, USA, Canada, Australia, Japan

Cover: Foto ©berggeist007 / pixelio.de

More available books at **www.hansebooks.com**

THE

DISEASES OF THE PROSTATE,

THEIR

PATHOLOGY AND TREATMENT;

COMPRISING THE SECOND EDITION OF "THE ENLARGED PROSTATE,"
AND A DISSERTATION "ON THE HEALTHY AND MORBID
ANATOMY OF THE PROSTATE GLAND,"

TO WHICH

THE JACKSONIAN PRIZE, FOR THE YEAR 1860, WAS AWARDED BY
THE ROYAL COLLEGE OF SURGEONS OF ENGLAND.

BY

HENRY THOMPSON, F.R.C.S.,

SURGEON EXTRAORDINARY TO H.M. THE KING OF THE BELGIANS;
SURGEON TO UNIVERSITY COLLEGE HOSPITAL; CONSULTING SURGEON TO THE ST. MARYLEBONE
INFIRMARY; FELLOW OF UNIVERSITY COLLEGE; HONORARY CORRESPONDING
MEMBER OF THE SOCIÉTÉ DE CHIRURGIE OF PARIS, ETC., ETC.

LONDON:
JOHN CHURCHILL, NEW BURLINGTON STREET.
MDCCCLXI.

TO

JAMES SYME, F.R.S.E.,

PROFESSOR OF CLINICAL SURGERY IN THE UNIVERSITY OF EDINBURGH,
ETC. ETC.

My dear Mr. Syme,

Not the least among the many sources of pleasure which writing this book has afforded me, is the opportunity of inscribing your name on its first page, in token of my gratitude to you for innumerable acts of personal kindness, no less than of my admiration of that skill and judgment, as a surgeon and as a clinical teacher, which have rendered your name illustrious.

Assuring you that I feel your acceptance of this trifling acknowledgment to be another addition to the many obligations you have conferred on me,

I beg to remain,

My dear Mr. Syme,

Your sincere and grateful friend,

HENRY THOMPSON.

PREFACE TO THE FIRST EDITION.

HAVING during the last few years enjoyed considerable opportunities for the study of Prostatic disease, I have aimed at embodying, in the following pages, the observations which a careful and laborious prosecution of it has led me to make. I should not have ventured to do so, had the results altogether coincided with those obtained by previous inquirers. The views here maintained of the Anatomy of the Healthy Prostate, but particularly of the organ in its most common deviation from the normal state, viz. when the subject of senile enlargement, differ materially from those which have been commonly held. The conclusions I have arrived at are based on extended anatomical researches, embracing an examination of not less than seventy original dissections, forming preparations now in my own possession, in addition to such observation of the contents of our metropolitan museums as I have been able to make. The data from which such conclusions were drawn have been appended, so far as it was possible to do so, that the scientific enquirer may form his own opinions respecting them. The points to which I desire especially to request his attention may be briefly stated as follows:—

The assignment of the "third" or "middle lobe," as a separate anatomical portion of the Prostate, to the abnormal history of the organ;—discussed in the first chapter.

The analogy between the enlargements and tumours of the Prostate and those of the Uterus;—discussed in the second chapter (fifth of Second edition).

An examination of the alleged causes of enlargement of the Prostate, resulting in new views of this subject;—in the third chapter (seventh of Second edition).

The effects of enlarged Prostate in relation to the function of micturition;—considered in the fifth chapter (ninth of Second edition).

The researches in relation to Malignant and Tubercular disease of the Prostate;—in the ninth and tenth chapters (fourteenth and fifteenth of Second edition).

The consideration of "the bar at the neck of the bladder;"—in chapter the twelfth (seventeenth of Second edition).

Besides these, I have treated at length the subject of Diagnosis and Treatment of enlargement, and of the various complications which arise in connection with it, perhaps, I may venture to say, more fully than any preceding author.

And lastly, I have devoted a long chapter to a consideration of that important, but not uncommon complication of enlarged Prostate, stone in the bladder; and especially of the best modes of successfully applying Lithotrity as a means for its removal. I venture to hope that in discussing thus fully the question of Treatment, whether in relation to the simple or the complicated forms of this common complaint, some hints will be found which may prove useful in the varied emergencies of practice. I shall feel abundantly rewarded should some of my professional brethren discover any such fruit as the result of my labours.

16, Wimpole Street, Cavendish Square,
  London, Nov. 1857.

PREFACE TO THE SECOND EDITION.

The Council of the Royal College of Surgeons of England announced in 1859, as the subject for the Jacksonian prize, "The Healthy and Morbid Anatomy of the Prostate Gland."

In the spring of the present year, this prize was awarded to an Essay by myself. At the same time, the first edition of my previous work, "The Enlarged Prostate," had become exhausted; I thought it desirable, therefore, to incorporate, in a Second edition, the greater part of the Essay, and to present the whole under a more comprehensive title, viz., "The Diseases of the Prostate, their Pathology and Treatment."

Considerable additions to the original volume have thus been made;—first, to that part which relates to the general and minute anatomy of the prostate;—secondly, to the various sections describing the morbid anatomy in every condition in which deviation from health has been observed, for which purpose more numerous data than any before obtained have been made available;—thirdly, those portions which are devoted to the subject of Treatment, in its various departments, have been augmented; while much of the old matter has been re-arranged, and some re-written, with a view to render the work more complete and useful. Lastly, a number of new plates, illustrative of important points in pathology, have been added: for the pains bestowed in producing these and for the successful results attained, my best acknowledgments are due to the skill and admirable fidelity of the artist, Mr. Clérevaux.

16, Wimpole Street, Cavendish Square,
London, Sept. 1861.

CONTENTS.

PART I.—THE ANATOMY OF THE PROSTATE.

CHAP. PAGE
I. The Topographical and Structural Anatomy of the Prostate 1
II. Facts relative to Weight, Size, and Morbid condition obtained by dissecting the Prostate . . . 43

PART II.—THE DISEASES OF THE PROSTATE.

III. Inflammation of the Prostate, acute and chronic . 50
IV. Suppuration of the Prostate—Abscess, acute and chronic 67
V. Hypertrophy of the Prostate; its anatomical characters 79
VI. Tumours and Outgrowths of the Prostate . . 101
VII. The Causes of Hypertrophy of the Prostate . . 122
VIII. The Symptoms of Hypertrophied Prostate . . 139
IX. The Effects of Enlarged Prostate in relation to the function of Micturition, Retention, Incontinence, Engorgement, and Overflow 151
X. The Diagnosis of Prostatic and other Obstructions at the Neck of the Bladder 161
XI. The Treatment of Prostatic Enlargement, from Hypertrophy and Simple Tumour 174
XII. The Treatment of Retention of Urine from Enlarged Prostate 228
XIII. Atrophy of the Prostate 256
XIV. Cancer of the Prostate 262
XV. Tubercular Disease of the Prostate 283
XVI. Cysts or Cavities in the Prostate 288
XVII. The Bar at the Neck of the Bladder . . . 293
XVIII. Prostatic Concretions and Calculi 312
XIX. On the relation between Hypertrophied Prostate and Stone in the Bladder 329

EXPLANATION OF PLATE I.

A healthy Prostate, from a man aged 35 years. It was dissected, as described, pages 6-7, and was accurately drawn while in the fresh state.

The organ lies with its posterior or rectal surface downwards. The portion contiguous to the bladder is that which is nearest to the eye; the internal meatus being seen above, and the ejaculatory ducts in their depression below.

Referred to, page 7.

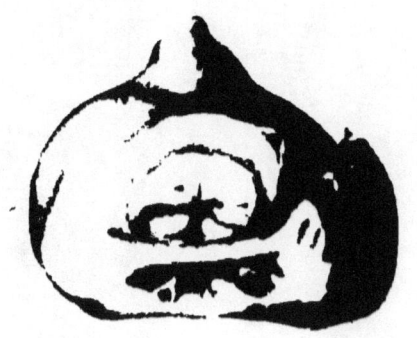

Cvp Clearevance del &hte

Pl. II.

EXPLANATION OF PLATE II.

The upper figure represents a section in the median line of a healthy Prostate, of its natural size, accurately drawn from a fresh average specimen, taken from a subject aged 54 years: weight of prostate, 4 drachms 48 grains. The utricle is laid open, as well as the ejaculatory duct of the left side marked a. This is seen to traverse the wall of the utricle, being slit up to show its situation there. The right duct, previously in contact, is removed with the right half of the organ. Both ducts lie closely together in the centre of the median portion of the organ, and there is no sign of the existence of any natural separation of a part of it in this place, to indicate a "third" or "middle lobe" as a distinct formation. The situation of the glandular structures, and the direction of their ducts may be traced, as described, page 33.

The lower figure is drawn from a specimen exhibiting enlargement in a very early stage. It was taken from a subject aged 62 years: weight of prostate, 6 drachms 5 grains. The portion lying above the duct is protruding upwards. A small fibrous tumour was found in each lateral lobe, and one of them may be seen projecting a little into the urethra, and rendering the mucous membrane slightly prominent (indicated in the figure by an asterisk). A few concretions are seen in the mouths of ducts around the verumontanum. In this case the *right* ejaculatory duct was opened; the left, therefore, lies beyond it, and the utricle is not exposed as in the preceding case.

EXPLANATION OF PLATE III.

GENERAL HYPERTROPHY OF THE PROSTATE, AND TUMOURS.

Part of the bladder, and the prostate from a. man aged 65 years.

The lateral lobes are about equally hypertrophied. The enlargement at the neck of the bladder was constituted by three or four small tumours originating in the median portion. Obstruction to the outflow of urine during life was very considerable.

Referred to at pages 91, 159, and 169.

Plate III

vp Clerevaux del & lith   W West imp

Cyp Clerevaux del. & lith.

EXPLANATION OF PLATE IV.

HYPERTROPHY OF PROSTATE, AND TUMOURS.

The bladder and prostate of a man aged 76 years, drawn when fresh, and now forming a preparation in the possession of the author.

a, a, indicate a tumour embedded in the interior part of the organ, divided into nearly two equal hemispheres by an incision into the urethra. At the neck of the bladder another eminence is seen, which produced almost total obstruction to the outflow of urine; it consisted of a spheroidal tumour, carrying before it a thin layer of prostatic tissue.

b, shows a sac communicating with the proper cavity of the bladder by a small opening, in which a slender bougie is placed.

Referred to at pages 91, 159-60.

EXPLANATION OF PLATE V.

HYPERTROPHY OF THE PROSTATE.

Very considerable hypertrophy of the prostate; in which a large barrier exists at the neck of the bladder, preventing the outflow of urine, and forming a formidable obstacle to ordinary catheterism.

Referred to at page 240.

From a preparation in the Museum of University College Hospital, London.

Cyp. Clairvaux, del et lith.　　　　　　　　　　W Wess, ith.

Cyp. Clérevaux, del. et lith. W. West, imp.

EXPLANATION OF PLATE VI.

HYPERTROPHY; SIDE VIEW.

This preparation shows extremely well, by a side-view, the influence upon the direction of the urethra which considerable hypertrophy of the prostate very commonly occasions, and the obstruction to catheterism which results.

Referred to at page 240.

From a preparation in the Museum of University College Hospital, London.

EXPLANATION OF PLATE VII.

ENORMOUS HYPERTROPHY OF THE PROSTATE.

This drawing represents one of the most remarkable examples of hypertrophy of the prostate on record, for the amount of enlargement the organ attained. It was removed from a gentleman between 70 and 80 years of age, who was frequently under the care of my colleague, Mr. Quain, of University College Hospital, to whom I am indebted for the interesting fact that the patient usually possessed the faculty of emptying his bladder to the last, only occasionally requiring the aid of the catheter.

Referred to at pages 85 and 93.

From a preparation in the Museum of University College Hospital, London.

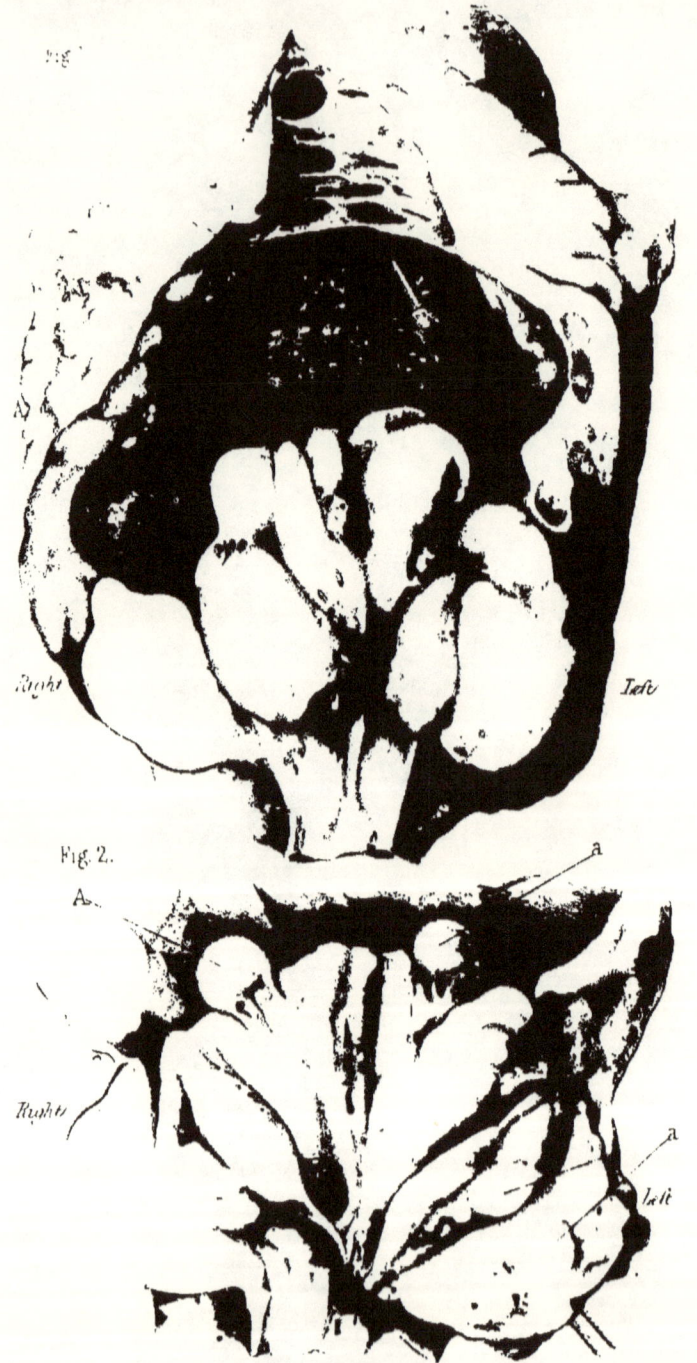

Fig. 2.

EXPLANATION OF PLATE VIII.

FIBROUS TUMOURS OF THE PROSTATE.

Enlargement of prostate from fibrous tumours; bladder diseased and sacculated. (From Case No. III., pp. 119 20.

Fig. 1 represents a section through anterior part of bladder and prostate. The enlargement of the prostate is seen obstructing the vesical neck and projecting into the cavity of the bladder. A indicates a rounded eminence occupied by one of the tumours, which is dissected out in fig. 2, and denoted there by the same letter.

Fig. 2. Some of the prostatic tumours are dissected out (A and *a*). Others are divided by transverse section; one is well seen, marked *a* at the lower corner of the figure, near to the word "left."

Referred to at pages 81, and 119-20.

EXPLANATION OF PLATE IX.

HYPERTROPHY OF THE PROSTATE FORMING A LARGE BAR AT THE NECK OF THE BLADDER.

General hypertrophy of all the parts of the prostate, from a patient aged 74 years, and now forming a preparation in the possession of the author. The lateral lobes blended so intimately with the intervening median portion, that a bank or bar nearly an inch high was formed at right angles with the course of the urethra in the prostate. A depression is seen in which the catheter was probably caught during life.

a, indicates a considerable cavity filled with numerous dark concretions.

Referred to at pages 169 and 210.

Plate IX

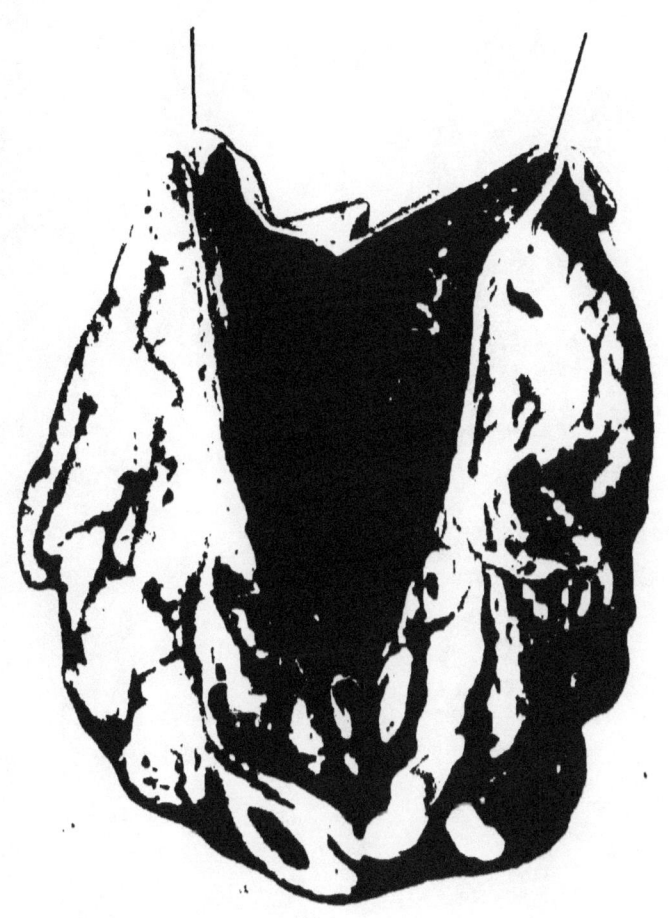

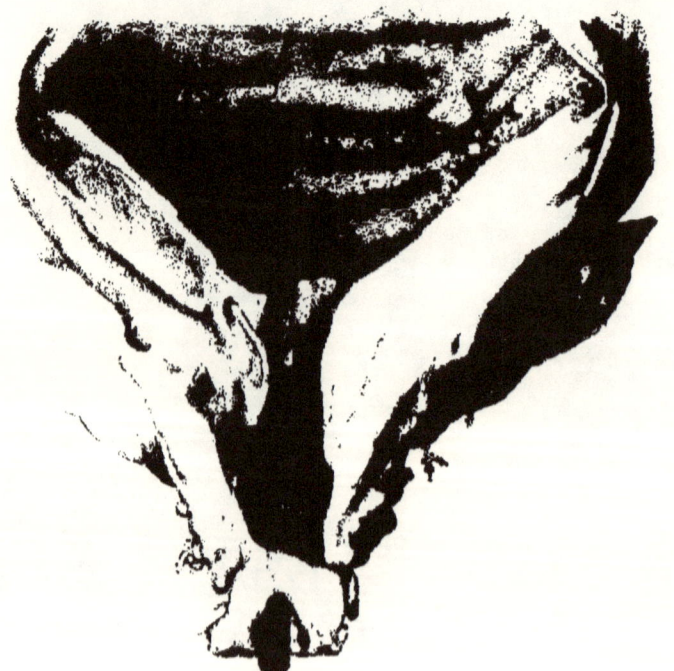

Gyp Clérevaux del et lith.   W West, imp

EXPLANATION OF PLATE X.

TWO SPECIMENS OF BAR AT THE NECK OF THE BLADDER.

The upper figure represents a preparation in the Museum of the Royal College of Surgeons, London, and I am indebted to the Council for permission to represent it here.

It is a well-marked specimen of prostatic bar, and is thus described in the Hunterian Catalogue: "Prep. No. 2488. Part of the bladder with an enlarged prostate gland. The enlargement has taken place chiefly in the middle lobe, which is raised in a broad thick transverse ridge or bar behind and below the vesical orifice of the urethra. The anterior surface of this ridge forms nearly a right angle with the rest of the prostatic part of the urethra."

The lower figure is that of a preparation in the Museum of University College, presenting a good example of bar produced, not by enlarged prostate, no such affection existing, but by hypertrophy of the muscular and fibrous structures at the neck of the bladder, induced by their action to overcome obstruction in the form of long-standing stricture of the urethra. See pp. 302-3.

Referred to at pages 169, 240, 298, and 303.

EXPLANATION OF PLATE XI.

ABSCESS OF THE PROSTATE.

The bladder, prostate, and a part of the urethra, drawn while fresh from a subject who died with abscess of the prostate. The parts form a preparation now in possession of the author. The case is reported at length, pp. 74-76.

The sac of the abscess formed a cavity capable of containing ten or twelve drachms of fluid. It undermined the mucous membrane of the urethra, opening into the canal by an aperture the size of a florin, situated in the upper part; thus the floor of the urethra alone remained, forming a kind of bridge through the cavity, which extended below, above, and on either side of it. This cavity is bounded by the capsule of the prostate, nearly all the substance of the organ having disappeared. Passing through the cavity is the right ejaculatory duct, found to be dissected out entire.

Referred to at pp. 76, 77.

Pl XI

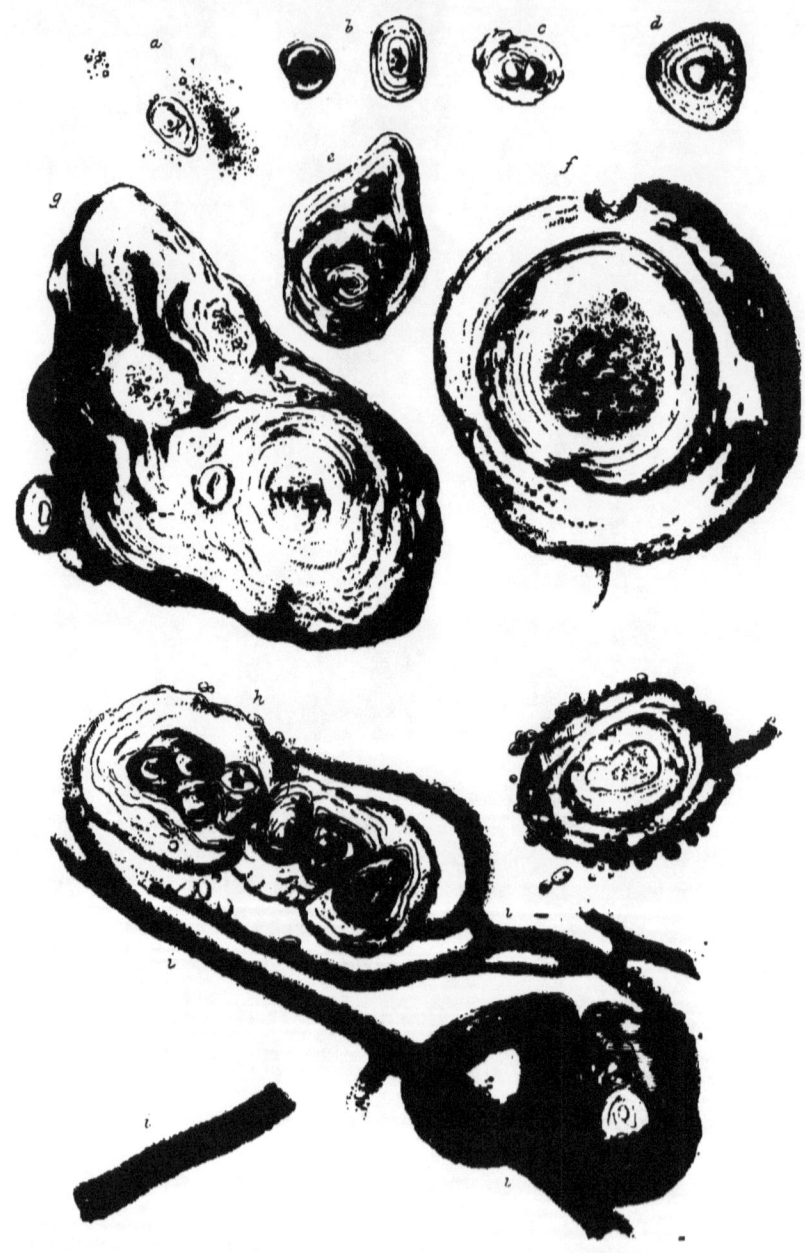

EXPLANATION OF PLATE XII.

PROSTATIC CONCRETIONS AND THEIR FORMATION.

Prostatic concretions of various sizes, ducts loaded with "yellow bodies," &c.; all drawn under a power of 150 diameters. (See page 313, *et seq.*)

a, b, c, and *d* represent small concretions, apparently in an early stage of formation.

e, f, g, and *h* are larger; most of them appear to be made up of smaller concretions adhering together.

i, i, i, represent ducts loaded with yellow granules of various sizes, "yellow bodies;" this appearance is very frequently observed in the neighbourhood of concretions in an advanced stage of formation, p. 317.

Referred to at pages 313 and 316-17.

EXPLANATION OF PLATE XIII.

PROSTATIC CONCRETIONS AND THEIR FORMATION.

Fig. 1. A concretion embedded in prostatic tissue. It is surrounded by yellow granules; and a vessel stuffed with the same matter is seen.

Fig. 2. Small vessels stuffed with yellow granules, and larger bodies among them. Many of the latter lie free on the section, p. 317.

Fig. 3. An aggregation of numerous small concretions forming one large one. Prostatic epithelium seen free in the field; some appears to contain granules of a yellow colour.

Fig. 4. A large duct cut transversely. Many of these ducts are lined with epithelium containing yellow granules. A concretion lying in the duct. A circular arrangement of fibres seen around the opening, p. 318.

Fig. 5. Fluid expressed from the prostate; it contains similar epithelium, free yellow granules and concretions, p. 316.

Fig. 6. Fluid from vesiculæ seminales, containing numerous yellow granules, p. 322. All the above are represented as seen by a power of 150 diameters.

Referred to at pages 316-17 and 321-2.

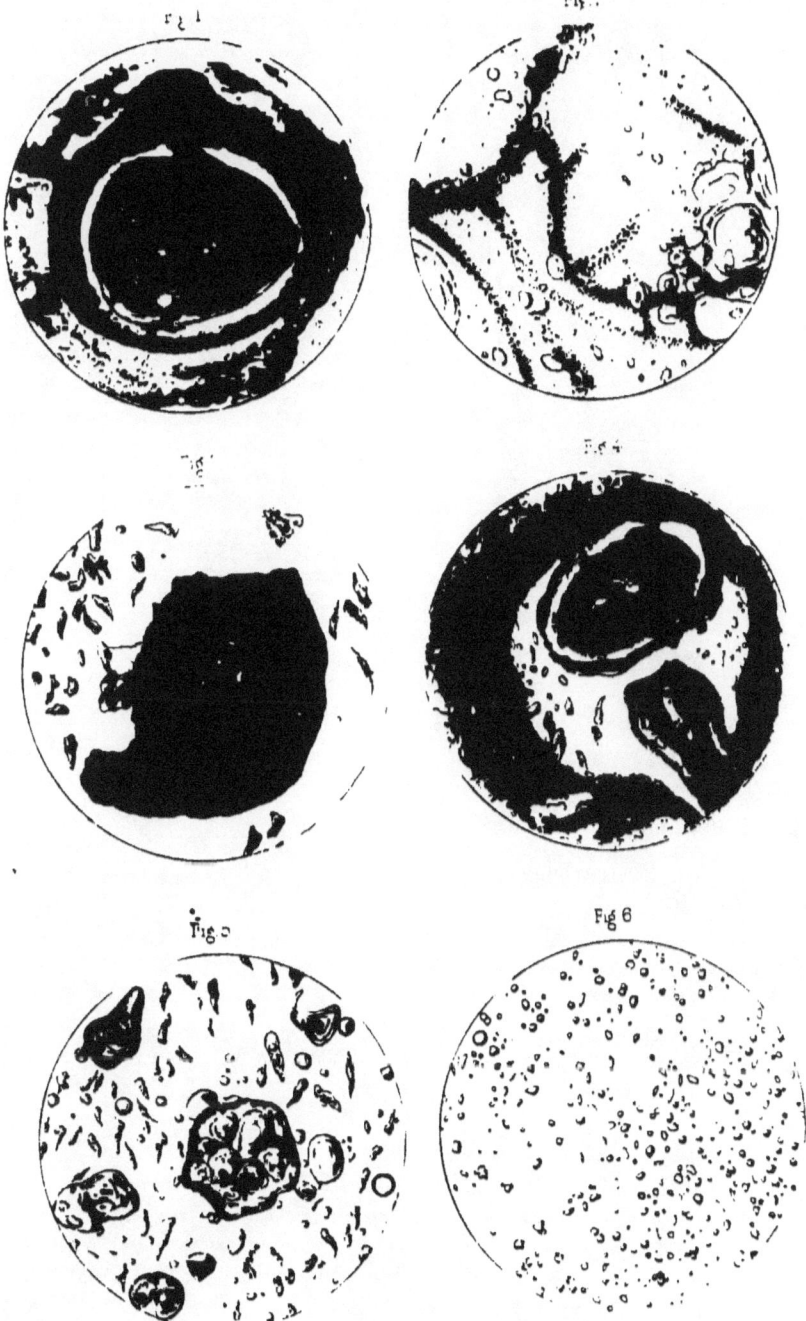

THE ANATOMY AND DISEASES OF THE PROSTATE.

PART I.—THE ANATOMY OF THE PROSTATE.

CHAPTER I.

THE TOPOGRAPHICAL AND STRUCTURAL ANATOMY OF THE PROSTATE.

The Prostate as an Independent Organ.—Mode of Dissecting.—Limits of.—Form.—Size.—Measurements.—Weight.—External Relations.—Vessels and Nerves.—Anatomical Conformation; Lobes.—History of the "Third Lobe."—Prostatic Urethra, its Course and Character.—The Utricle. — Ejaculatory Ducts.—MINUTE ANATOMY.—Its Fibrous Elements.—Its Glandular Elements.—Relative Proportions of.—Prostate Gland in Young Subjects.—Its Weight, Size, and Relations.

IN entering upon a study of the Pathology of the Prostate, there appears to exist a peculiar necessity for observing accurately its normal structure, conformation, and anatomical relations. The most common, as well as important, morbid states to which this organ is subject, either consist in, or are associated with, deviations from the natural size and form, as well as from the natural disposition of the structures entering into its composition. It is apparent that its anatomical relations also must, of necessity, be deranged, when the deviation from size in the direction of enlargement becomes considerable, as it not unfrequently does. Further, while the normal dimensions of the organ have been approximately determined, it is certain that the results which have been hitherto obtained are not based upon observations sufficiently numerous and extended, definitively to settle the question, inasmuch as very different weights and measurements are named in describing it by various authors. And lastly, our knowledge of its structure, to say nothing of its physiological uses, is, perhaps, less exact and defined, or, at all events, has been so until late years, than other topics in its history which are even less important to the student of disease. I propose, therefore, first to consider its normal anatomy in detail.

In entering on this subject, it may be admitted that clearly defined marks are wanting in the structure of that part of the human organism, to which the term "prostate gland" is commonly applied, to indicate with exact precision its anatomical boundaries in every direction, or even to substantiate an undisputed title to its recognition as an independent organ, in that sense in which the word organ is very commonly understood. Forming a small portion of a large and important apparatus, which is continuous in structure throughout, it nevertheless exhibits peculiar characters in the arrangement and combination of its component tissues; a special appellation may be therefore regarded as appropriate and necessary for anatomical purposes. In like manner, a large and continuous division of the intestinal tube is recognised by anatomists as two separate portions, under the names of jejunum and ileum, although no physical boundary line indicates the limits of their proximal extremities; specialities of structure and function characterise each of these two portions, but those of the one glide insensibly into the other.

Another important circumstance supports the custom of regarding the prostate as a separate organ, viz., the existence of such a structure, or of some analogous one, as an accessory to other sexual organs, throughout a very large portion of the animal series, not only in the Vertebrata, but in the Invertebrata also. And lastly, there is the consideration, of no little weight, which is supplied by the fact that the prostate is subject to morbid changes, which, as will be seen hereafter, are peculiar to it, and which do not occur in any adjacent or related structure.

These remarks appear to be necessary at the outset, to explain that, although the organ which is the subject of examination here, does not possess characters so independent as those which belong to such glands as the liver, the kidney, or the pancreas, it nevertheless exhibits characters of a nature so peculiar and distinguishing, that we feel compelled to assign to it the possession of an individuality and integrity which may, perhaps, be somewhat overlooked, if we study it mainly or solely from that point of view which regards it only as a fractional portion of some greater whole.

The name of "prostate," as applied to the part under consideration, is supposed to have been originated by anatomists, from the fact of its *standing before*, or anterior to, the bladder or vesiculæ seminales, in the supine position of the subject; its name of "gland," from the glandular structures which form a considerable part of its component tissues.

The prostate gland of the adult is commonly described as resembling a full-grown chestnut in size and form; sometimes it is compared in the last-named character to the ace of hearts, the small extremity being directed downwards and forwards, and the base upwards and backwards, in the erect position of the body. Assuming for the present that there is a certain amount of truth in these analogies, we shall next inquire what is the nature of its relations with the bladder on the one hand, and with the urethra on the other, before proceeding to determine accurately its limits, size, and form.

When the bladder, prostate, and vesiculæ seminales, as well as the urethra for two or three inches of its course anterior to the prostate, are removed from the body and are fairly isolated from the neighbouring parts by dissection, the prostate appears as a mass having the form of a short truncated cone flattened between its pubic and rectal aspects, the base of which surrounds the neck of the bladder and projects somewhat below it, while its blunted apex ends at the fascial partition which stretches across the angular interval between the pubic bones, and is known as the posterior layer of the deep perineal fascia. The bladder should be slightly distended with tow, and the parts above-named should be properly secured to facilitate the dissection, with the posterior or rectal aspect upward, from which it is understood that the rectum itself has been carefully removed, without removing more of the fascia than is necessary in the operation.

First, will be seen, the vesiculæ seminales and vasa deferentia bound closely by a dense fascia to the posterior border of the prostate and base of the bladder adjacent, and requiring some careful dissection to isolate them fairly. Each vas deferens has a vesicula seminalis on its outer side, and approaches obliquely the median line as it courses forward to the base of the prostate, and just before entering, it joins the vesicular duct to form the common or ejaculatory duct. These two vessels should be carefully cleaned and traced, when they will be seen to perforate the mass in a deep central interlobular depression or notch, and to enter it in the middle line side by side. A layer of fascia, a portion of the recto-vesical fascia, may be now dissected from the inferior surface and sides of the gland, for which it forms an enveloping sheath, and several venous sinuses will be encountered lying between it and the proper capsule of the prostate, especially along its lateral borders, where those vessels are found large and generally filled with coagula, particularly in elderly subjects; and sometimes in the latter containing also large phlebolithes. The proper capsule, which cannot be regarded as a mere offshoot from any adjacent fascia, but is a special envelope belonging to the prostate itself, although thin is firm in texture, and defines clearly the form and limits of the prostate here. Proceeding with the dissection by turning upwards a little the peritoneum covering the bladder, if we next clean the external layer or longitudinal muscles of that viscus and trace them with care, some bands of the paler and less superficial fibres will be found inserted into the base of the prostate above the point where the ejaculatory ducts enter. These should next be divided, and the longitudinal layer be turned aside from the base of the bladder to the right and left of the middle line as well as the mutual interlacements of the different vesical coats permit, and the inner or circular layer comes into view. These, which are very pale in colour, should be defined at the base and neck of the bladder, where they are chiefly aggregated, being very thin and scattered above, since by tracing them forwards they will be found to be continuous with muscular fibres similarly arranged, and forming a considerable portion of the prostate itself; and in order to do this, the proper capsule must be partially removed, with the more superficial as well as the lateral portions of the prostate which do not strictly belong to the circular system of fibres, but which contain a large proportion of glandular structure intermingled with interlacing muscular and fibrous tissues, and which will be described when we consider the minute anatomy of the organ. Into this outer portion also, some fibrous prolongations of the capsule proceed, and minute vessels enter the prostatic substance with them. In order to trace the circular fibres of the bladder through the prostate, a prolonged and delicate dissection is necessary.

With care and patience it is, however, quite demonstrable that complete continuity of structure exists to a considerable extent between the circular fibres of the bladder and the constituent fibres of the prostate, and that they are arranged around the tube of the urethra in the manner described. By carefully continuing the dissection, this annular series of fibres may be traced along that canal, diminishing in volume as we proceed in the anterior direction, as far as to the bulb of the urethra, where they stop, having extended over the whole of the membranous portion. In this latter situation they have dwindled down to a layer, from one-half to one-third of a line in thickness, although quite distinct. If next we make a longitudinal incision in the axis of the prostatic urethra through these fibres with great care, we may turn them aside and lay bare a longitudinal layer of delicate pale muscular fibres which lie immediately outside the mucous membrane which forms that canal. These are the muscular fibres which surround the urethra throughout great part of its course, and which have been described in somewhat different terms by Hancock, Hogg, and Kölliker. Continuity of tissue is thus demonstrated between the circularly-disposed fibres of the bladder, those of the prostate, and those enveloping the membranous portion of the urethra.

In order to present a clear and succinct view of the order in which the structures constituting the prostate are arranged, and particularly with regard to the urethra, which passes through it, we shall simply name them in the reverse order to that already pursued, viz., from within outwards.

Firstly.—The mucous membrane of the urethra.

Secondly.—A delicate layer of longitudinally-disposed pale or unstriped muscular fibres, mingled with a good deal of connective tissue and some elastic fibres; this layer forming part of a system of longitudinal fibres, underlying and surrounding in greater or less quantity the whole urethral canal.

Thirdly.— A circularly-disposed layer of pale or unstriped muscular fibres, of considerable thickness posteriorly, where they become continuous with the circular fibres of the bladder, and becoming thinner as they approach the membranous portion, over which they proceed and then terminate. This layer contains also connective and elastic fibres like the preceding.

Fourthly.—Beyond this lies the greater part of the glandular structure properly so called, which intermingles with, and is supported by, a considerable proportion of tissue, composed in part of pale muscular fibre, and in part of connective and fibrous tissues, which, interlacing together, constitute the rest of the organ. It is this composite structure which, disposed in masses chiefly in the lateral direction on either side, gives form and character to the organ.

Fifthly.—The enveloping fibrous capsule.

To what precise portion of the mass between the bladder on the one hand, and the urethra, lying anteriorly, on the other, shall we assign the limits of the prostate? In front, the posterior or deep layer of the perineal fascia may well be regarded as the boundary line. Behind, we have no such limiting distinction, but we may proceed as follows. Taking a fresh specimen, in which the bladder and prostate have been carefully separated from the adjacent veins and fascia, and in which the vesiculœ seminales and ejaculatory ducts have been separated and traced into the interlobular notch, the basic extremities of the lateral lobes being defined, and division having been also made in front at the perineal fascia, we should lay open the bladder, and, with a pair of scissors, cut away the walls of the bladder closely round the neck or vesical orifice of the urethra. In doing so we shall divide the circular fibres freely, and some of the external longitudinal fibres, as well as the mucous membrane and submucous tissues around the internal meatus, particularly those forming the uvula or luette vesicale: the remaining portion may then be accepted as a fair specimen of the prostate gland in an isolated condition.

With this it is easy to deal in respect of external physical characters.

First, as to form. A slightly truncated cone, flattened anteroposteriorly; short, so that the diameter of the base exceeds by a fourth or fifth the measurement of the longitudinal axis, the base notched or indented for the entry of ducts, gives as fair a representation as a few words can supply.

A drawing faithfully representing such a specimen in the supine position, and dissected in the manner described, not an ideal figure, will furnish a better conception. (See Plate 1.)

Observing it more closely, the anterior surface appears generally convex. I have scarcely ever found any trace of a depression in the line of the long axis, said, in anatomical works, to exist and to correspond with the track of the urethra; this has been carefully, but unsuccessfully, sought in numerous dissections. The posterior or rectal surface is smooth and also rather convex, but here the course of the urethra is denoted by a shallow median depression, while the unyielding masses of the lateral lobes are seen, one on each side, and are still more readily recognised by pressure with the finger; two slight lines of depression are also seen converging at the basic part of this surface, which indicate the tracks of the ejaculatory ducts. At the base are seen the internal meatus of the urethra, and below it the interlobular notch and the funnel-shaped opening by which the ejaculatory ducts enter. It should be observed that the urethra perforates at a point a little in advance of the base itself, approaching, in fact, to the upper surface, so that the portion of prostate below the urethra exceeds by about one-third or one-fourth the length of the portion which lies above. On each side of the base, the rounded posterior limit of a lateral lobe projects, a little backwards and outwards, leaving a notch between. The projection of a "median portion," or "third lobe," if present, is seen between the urethral opening and that for the two ejaculatory ducts beneath.

MEASUREMENTS.—The transverse diameter is almost always the greatest, exceeding the antero-posterior by a fifth or a sixth. These relations vary very much. Sometimes the organ has an appearance as if it had been compressed from before backwards. It is much less common to find the transverse measurement decreased.

In a series of 50 adult prostates which I made the subject of a laborious investigation, and presented to the Medical and Chirurgical Society,* 33 were healthy. Their measurements and weights are furnished in the following table.

* In relation to this investigation, and the method in which it was pursued, it may be as well to explain, for the purpose of saving further allusion to the subject, that the various statements made, in this and other chapters, are founded upon a dissection of upwards of 60 specimens of the prostate, removed by myself from the dead body, of which 50, preserved in spirit, were exhibited at the Medical and Chirurgical Society, Feb. 10, 1857, in illustration of a paper entitled "Some Observations on the Anatomy and Pathology of the Adult Prostate," and published in the fortieth volume of the Transactions. The method of examination in each case was as follows:—The organ was cleanly dissected out from the adjacent parts, and carefully removed. At the neck of the bladder the muscular and other fibrous structures which surround the vesical orifice of the urethra were pared away pretty closely, but some very small portions of these were necessarily left, which could not be regarded as, strictly speaking, prostatic tissue. Absolute definition is almost, if not quite, impossible. I was careful, however, to be so exact that any portions of adventitious tissue remaining would be so small as not to invalidate the conclusions drawn. The anterior limit of the organ, although not completely defined, is yet very nearly so; no difficulty, therefore, arises here.

THIRTY-THREE NORMAL PROSTATES.

No.	Age.	Weight.	Length.	Breadth.	Thickness.
		drs. grs.	inches	inches.	inches.
1.	70	4 48	1·4	1·45	·85
2.	42	3 37	1·3	1·4	·6
3.	47	4 57	1·8	1·7	·65
4.	85	4 44	1·25	1·55	·95
5.	47	5 33	1·4	1·7	·9
6.	63	4 35	1·35	1·7	·7
7.	50	4 34	1·5	1·7	·65
8.	54	4 8	1·25	1·8	·75
9.	90	4 58	1·25	1·85	·85
10.	52	4 13	1·5	1·75	·6
11.	54	4 37	1·5	1·75	·7
12.	66	4 27	1·4	1·5	·65
13.	63	4 3	1·35	1·5	·65
14.	70	4 2	1·4	1·6	·55
15.	54	4 50	1·5	1·7	·7
16.	70	4 55	1·4	1·8	·7
17.	66	4 56	1·4	1·7	·75
18.	74	4 4	1·3	1·6	·7
19.	61	4 16	1·3	1·6	·8
20.	56	3 50	1·25	1·75	·75
21.	21	3 34	1·3	1·4	·7
22.	40	4 30	1·55	1·6	·75
23.	64	5 2	1·4	1·75	·9
24.	50	5 20	1·55	1·75	·8
25.	73	4 48	1·6	1·8	·6
26.	50	4 46	1·4	1·9	·55
27.	46	5 4	1·5	2·0	·55
28.	66	4 35	1·5	1·75	·6
29.	66	4 6	1·5	1·65	·6
30.	65	4 24	1·4	2·0	·6
31.	55	5 30	1·3	1·8	·85
32.	54	4 48	1·4	1·75	·7
33.	61	4 54	1·3	1·75	·75

An analysis of the measurements produces the following results:—

From base to apex the measurements ranged between . . . . 1·3 and 1·8 inch.
But the measurement most commonly met with was . . . . 1·4 do.
The greatest transverse, near base, ranged between . . . . 1·4 and 2·0 do.
But the measurement most commonly met with was . . . 1·75 do.

MEASUREMENTS OF PROSTATE.

The point of extreme thickness when
measured ranged between . . ·55 and ·95 inch.
But the measurement most commonly
met with was about . . . ·7 do.

The measurements of 20 healthy specimens which I have since examined, and which are now the property of the Royal College of Surgeons, correspond very nearly to the above.

The measurement in inches may then be expressed (in average terms) as follows:—

From base to apex . . . . . $1\frac{1}{4}$ to $1\frac{1}{2}$ inch.
Greatest transverse diameter, about . $1\frac{3}{4}$ do.
Greatest thickness, about . . . . $\frac{5}{8}$ to $\frac{7}{8}$ do.
(See fig. 1.)

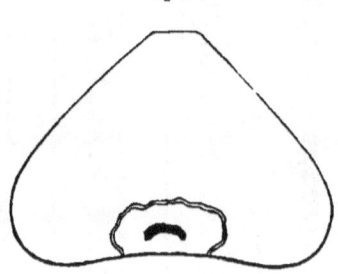
Fig. 1.

These are smaller than those given by Dupuytren, who represented, in connection with the subject of Bilateral Lithotomy, the prostate as measuring twenty to twenty-four lines transversely, and ten to twelve in thickness; and on these estimates his calculations for that operation were based.* They correspond, however, very nearly with those by Deschamps,† Senn,‡ Dr. Gross,§ and Dr. Hodgson,|| all of whom have made considerable practical researches with a view to the question of size.

The important result to be derived from measurement, as regards Lateral Lithotomy, is the length of a line directed downwards and outwards from the centre of the urethra (which may be

* Mémoire sur l'Opération de le Pierre. Paris, 1836. p. 21.
† Traité Historique et Dogmatique de l'Opération de la Taille. Paris, 1796. Vol. i. pp. 39, 40.
‡ Traité d'Anatomie Chirurgicale. Paris, 1838. Vol. ii. pp. 327-330. Par J. F. Malgaigne.
§ Diseases of the Bladder, &c. Phil., 1855. 2nd ed. p. 69.
|| The Prostate Gland, and its Enlargement in Old Age. London, 1856. p. 34.

regarded as corresponding in the operation with the bottom of the groove in the staff) to the outer border of the organ at its vesical extremity. This line may be considered as falling midway between the horizontal and vertical planes, forming with each, therefore, an angle of 45°, when the patient lies in the position for lithotomy. It may be accurately deduced from the form and measurements given above, and has been verified by numerous actual sections of the organ.

Average measurements of healthy prostate in direction described (see fig. 2) . . . $\frac{7}{8}$ of an inch.
Ditto, of small prostate, weighing under 4 drachms . . . . $\frac{3}{4}$ of an inch.

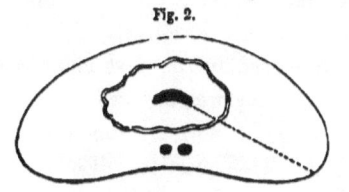

Fig. 2.

This, it will be understood, represents the distance to the extreme border; the extent, therefore, to which an incision should be carried in this situation, may be approximately determined therefrom.

WEIGHT.—Various weights are assigned by the principal writers on this subject; their estimates ranging between 4 and 8 drachms. No exact limit in this matter can be affirmed, but such only as are approximative. Probably from 4 to 6 drachms may be considered as a range within which most normal prostates will be found. The table just given, shows 3 drachms 34 grains as the lightest, and 5 drachms 33 grains as the heaviest specimen in the healthy series. The average is 4 drachms 38 grains: the prevailing weight also corresponds very closely with the average, so that the usual weight of a healthy adult prostate may be estimated at about 4½ to 4¾ drachms. Dr. Messer, of the Royal Naval Hospital, Greenwich, to whose labours I shall have again to refer, has recently dissected 100 prostates, all taken from subjects at and over the age of 60 years.* His observations led

* Report on the Condition of the Prostate in Old Age. By J. C. Messer, M.D. Trans. Med. Chir. Soc. vol. xliii.

him to divide the preparations into three classes, according to their weight, as follows:—

The first class, containing all those weighing less than 4 drachms, which he considered abnormally small.

The second class, of those weighing 4 and under 6 drachms, which he considered normal. This contained 45 preparations, weighing from 4 to 6 drachms. Average weight, 4 drachms 57 grains: ages from 60 to 94 years.

The third class, of those weighing 6 drachms and upwards, which he considered abnormally large.

His result, viz., an average of 4 drachms 57 grains to the healthy gland, from 45 preparations, comes very near my own. I think, however, that this division, though approximatively correct, is not absolutely so, and that several examples of the organ placed in the first class were not necessarily atrophied, because they weighed less than 4 drachms; and their admission into the normal class would have diminished the average weight. This question will be considered hereafter.

EXTERNAL RELATIONS.—Having thus defined the limits of the prostate gland, we may next enter upon its external relations, or topographical anatomy, before dissecting the organ itself, and considering its internal arrangement and structure.

In the erect position of the body, the adult prostate is placed just below and behind the lower border of the pubic symphysis, that is, the summit of the pubic arch, a distance of about three-eighths to five-eighths of an inch intervening between the main body of the prostate and the symphysis: and this is occupied by fascial, cellular, and muscular tissues. Its base, which, as already explained, surrounds the neck of the bladder, is the uppermost portion: the entire organ is directed somewhat obliquely downwards, and a little forwards, so that the apex is the lowest portion, the median axis corresponding with the median line of the body, on each side of which is placed a lateral lobe, the anterior face directed towards the pubes, the posterior towards the rectum.

The prostate is maintained in this situation by various structures, yet not in a condition altogether fixed; since a slight degree of mobility enables it to give way before the pressure, either of a distended bladder, or of a loaded rectum ; or, it may be, from that of a finger introduced within the intestine. This condition is provided for by several attachments.

Firstly, its connection above with the neck of the bladder, which has already been described.

Secondly, it is attached below at its apex to the posterior layer of the deep perineal fascia in the following manner : This fascia, which is equally to be regarded as the pelvic fascia, of which it is really a portion, descends from the back of the pubic bones, to the lower border of which and of the adjacent ischia it is attached, closes in the pubic arch, and becomes applied to the inferior surfaces of the levatores ani. During its course, and at about an inch below the pubic symphysis, opposite the apex of the prostate, a sheath is prolonged from it which envelopes closely the organ, supporting between its meshes the veins which run on either side and in front of the prostate, and sending numerous communicating bands to the proper capsule of the prostate itself. A continuation of the same fascia is prolonged to cover the vesiculæ seminales and bind them firmly to the base, or upper and posterior part of the prostate, as well as to the bladder.

Besides these, there are the ligamentous and muscular attachments of the prostate, the latter of which contribute perhaps in some measure to its mobility.

First, the anterior true ligaments of the bladder. These are constituted by the recto-vesical portion of the pelvic fascia, the anterior part of which passes from the posterior aspect of each pubic bone at its lower border, as a strong whitish band, to the anterior surface of the prostate on its way to the bladder, where it becomes continuous with the fibrous structures which surround the neck of that viscus, and belong to its muscular apparatus. They have acquired also the name of pubo-prostatic ligaments. The ligament of each side corresponds with its fellow, and there is a groove or depression between the two, where the fascia sinks to the apex of the prostate, and becomes continuous with its sheath there.

The muscular attachments of the prostate are the levatores prostatæ. Each muscle of this pair arises from an oblique line on the posterior surface of the pubic bone, its most anterior fibres descending to meet those of its fellow just in front of and below the apex of the prostate, its middle and posterior fibres continuing to be inserted along the lateral borders; in this manner the organ is, in a measure, suspended in its place.

The posterior surface of the prostate is somewhat convex in its character, and is closely attached to the anterior wall of the rectum; only a small quantity of thin fascia, continuous with its sheath from the recto-vesical fascia, existing between the two. It is applied to that portion of the rectum which is known as its middle third, lying in the anterior concavity of the bowel here, but as this turns downwards through its lower third to open at the anus, there is a slight interval between the bowel and the apex; the line of the urethra and the line of the bowel diverging considerably as each pursues its course. The outline and position of the prostate may be readily made out by the finger, which, when carried through the sphincter ani (the hand being in the supine position), feels first immediately above it, supposing the subject examined to lie on its back, the posterior extremity of the bulb and the membranous urethra; then, as it proceeds, the apex of the prostate; and lastly its body, composed of two lateral divisions, widening outwards to its base, which is also definable if the bladder is empty, but not so, or with less ease, when it is distended with fluid.

VESSELS AND NERVES SUPPLYING THE PROSTATE.—The inferior vesical artery furnishes a branch, vesico-prostatic, which passes to the side of the prostate, and is the chief source of supply; this vessel divides into smaller branches in front of the gland which supplies it, and anastomoses with others from the corresponding vessel of the opposite side. A smaller supply is derived also by small and unnamed branches from the internal pudic artery, and from the middle hæmorrhoidal branch of the inferior vesical artery.

Such is the usual mode of vascular supply to the prostate; but there is also an exceptional and unusual arrangement of the vessels, which, occasionally occurring, it is no less necessary to be acquainted with than with the usual mode of arterial distribution supplied to, or otherwise affecting the prostate.

When the pudic artery is small or defective, and fails in one or two, or, it may be, even in three of its named branches, another vessel supplies its place, and usually derives its origin from the trunk of the pudic, just before it makes its transit through the sacro-sciatic foramen. This vessel is called the "accessory pudic," having been thus named by Mr. Quain, who has described the deviation referred to, and has given drawings of it in his work on the Arteries, pointing out, moreover, an important practical fact connected with the anatomy of the prostate which is involved by this deviation.*

* Mr. Quain describes this vessel as follows:—"The course (of the Accessory Pudic) within the pelvis towards the prostate gland, differs in different cases according to the place of origin. Most frequently the artery proceeds forward near the lower part of the urinary bladder : it lay on the side of that organ in the body from which figure 5 in plate 63 has been drawn ; and when placed on the anterior part of the pelvis arising from the obturator or epigastric, it descends immediately behind the body of the pubes.

"In passing by the prostate and urethra—and it is here that the exact situation of this artery is of serious concern to the practical surgeon—the accessory pudic lies on the upper part of the gland, or, it may be, for a short space likewise on the posterior margin ; and then proceeding forward above the membranous part of the urethra, it reaches the perinæum and divides into the terminal branches" . . . . "I have not seen the accessory pudic artery approach the side of the prostate in any case but one, and of this a drawing is given in plate 63, fig. 4."

"BRANCHES.—The accessory pudic has in some bodies the course of the vesico-prostatic, and being substituted for this artery, or more properly, perhaps, an extension of it, furnishes branches to the same organs in the pelvis." The Anatomy of the Arteries of the Human Body. By Richard Quain, F.R.S. Page 443. See in connection with this subject plates 63, 64, and 65, and the explanatory letter-press.

Mr. Spence, of Edinburgh, describes cases in which the artery ordinarily supplying the prostate has not divided into branches in the usual way, but has pursued its course to the apex of the prostate, and has thus become the source of dangerous hæmorrhage in lateral lithotomy.*

Numerous Veins pass in front and along the lateral borders of the prostate, into which the smaller vessels of the prostate itself enter; the large veins in question commence in the dorsal vein of the penis, which is a vein of considerable magnitude. The dorsal vein penetrates the deep perineal fascia about half an inch below the arch of the pubes. On arriving at the prostate it divides into two branches, each of which passes by one side of the gland and the neck of the bladder, where it unites and anastomoses with the vesical veins, which are numerous and aggregated about the neck and base of the bladder. The name of prostatic venous plexus has in consequence been given to the veins associated in this situation, and they are prone to be large, and sometimes tortuous or varicose, especially in elderly subjects. From these sources the blood is returned towards the heart through the internal iliac veins.†

Some minute vessels, chiefly venous, may be traced beneath the mucous membrane of the prostatic urethra, for the most part taking parallel lines, on either side of the verumontanum; these, however, are much more obvious in the membranous and bulbous portions of the urethra than in the prostatic.

The Lymphatics of the prostate enter lymphatic ducts, accompanying the veins under the fascial sheath. They are distributed on the surface of the proper capsule, and their efferent ducts carry their contents to the lymphatics associated with the iliac vessels. A few lymphatics also are found under the urethral mucous membrane.

* Edin. Med Journal, vol. i. p. 157, 1841.
† This condition is well depicted in the work on the arteries just named. See Plate 65, figs. 2 and 3, the latter representing the varicose and tortuous condition alluded to.

The Nerves supplying the prostate and adjacent parts are of considerable size, and constitute the prostatic plexus of the sympathetic, which is a prolongation downwards from the inferior hypogastric plexus. As the nerves pass forward to the penis, they may be traced between the levator ani on each side and the gland, to which they supply filaments in their course.

ANATOMICAL CONFORMATION OF THE PROSTATE—LOBES.

It has been customary to regard the prostate as presenting, on anatomical analysis, several divisions, naturally indicated by conformation, but subordinate to a general continuity of structure throughout.

These are as follows:—Two " lateral lobes," which are symmetrical ; and a " middle " or " third lobe," between them behind the urethra. Besides these there are portions of prostatic substance, which unite the lateral lobes before and behind the urethra, to which special names have been applied ; for example *the isthmus* (to the posterior uniting portion) ; or, *the anterior and posterior isthmus* (to the anterior and posterior uniting portions) ; and to the same part, *the anterior and posterior commissure*. These shall be considered separately.

I. THE LATERAL LOBES.—Not only does the conformation of the prostate at every age, from childhood to adult life, indicate that it is mainly divisible into two lateral portions, but the history of its early development corroborates the same view. Up to about the fourth month of foetal life, the organ exists as two independent lobes, which become united first posteriorly and then anteriorly during the fifth month, so as to present the complete form soon after the middle of intra-uterine existence ; the size, however, is extremely small, and the general contour is less angular than that which it subsequently assumes.

The propriety of accepting these two divisions may be held as undoubted. Each lobe may be regarded as nearly ovoid in form. The two may be described as lying side by side, as applied or very closely approximating at their apices, as diverging a little at their bases and posterior aspect, and as slightly separated by the canal of the urethra, which covers with mucous membrane and delicate sub-mucous tissues their convex surfaces, which are contiguous to each other in the median line. The anterior borders of the lateral lobes may be regarded either as coalescing or perhaps more correctly as united by a thin intervening portion of muscular and connective tissue (see page 39), but in any case it may be denominated the anterior commissure or isthmus. The posterior borders have a larger interval between them, occupied consequently by a larger interposing mass of substance, which has next to be considered. These two lobes have rounded posterior extremities, which, projecting somewhat, present an angular interval or notch between them, and this it is which gives the conventional heart-shape to the entire organ.

II. THE "THIRD" OR "MIDDLE" LOBE.—Lying between the posterior borders of the lateral lobes is a stratum of tissue uniting them throughout their length. As it approaches the base, this stratum becomes thicker and sometimes has a rounded form, as if it were a distinct isolated and independent formation, perforated about the middle by the ejaculatory ducts. It is that moiety of the stratum which lies nearer the apex to which has been applied the term "isthmus," or "posterior commissure," while the thicker part, situated at the base, has received the appellation of "third or middle lobe," and its existence as an independent lobe is regarded by most authorities in this country as a fact in normal anatomy. This, however, I am compelled to call in question, and shall give the reasons for doing so in detail. The idea appears to have originated with Sir Everard Home, who accorded to the part in question the title of "third" or "middle lobe," after five examinations of the organ by dissection, performed by Sir Benjamin —then Mr.—Brodie, and he announced the result as the discovery of "a middle lobe," to the Royal Society, Feb. 20, 1806; the inquiry having been first instituted, to use the words of his paper, only two months before. "Previous to this investigation," says Sir E. Home, "it was not known to me that any distinct portion of the prostate gland was situated between the vasa deferentia and the bladder. These ducts were considered to pass in the sulcus between the two posterior portions, in close contact with the body of the gland."* Notwithstanding this remark, the vasa deferentia had been described as perforating the posterior portion of the prostate, by several of the anatomical authorities of the time, as E. Home afterwards learned, a fact admitted in his first volume on "Diseases of the Prostate," published five years subsequently, viz., in 1811. Respecting the five dissections spoken of, the author says, " the appearance was not exactly the same in any two of them." The first preparation appears to have been from the person of an elderly patient, " who had died in consequence of this part *being diseased:* the nipple-like process was very prominent." In the next, " there was *no apparent glandular substance*" at all in the spot indicated. All that is said of the third is, that "there was a lobe blended laterally with the sides of the prostate gland;" but that there was really no distinct portion marked off as a lobe is clear, from the importance attached to such a formation being detected in the two subsequent cases. The most distinct appearance was found in two subjects of 24 and 25 years of age respectively; and upon the condition of the organs in these (and not in five) his account seems to have been based, and from these two cases the existence of a law was thus hastily deduced. Whatever may have been their condition (and it must be admitted that, at all events, no change similar to that of senile enlargement was likely to have been present in either), the existence of a distinctly-marked portion of a spheroidal form, situate between the ejaculatory ducts and the verumontanum, at the inner meatus of the urethra, is so rare an occurrence as to mark an unusual state, whether congenital or acquired. Had the researches extended to a large number of bodies, it is certain that these cases must

---

* Philosophical Transactions, 1806, Paper viii. "An Account of a Small Lobe of the Human Prostate Gland, which has not before been taken notice of by Anatomists." By Everard Home, F.R.S.

have appeared to be exceptional, as my own inquiries have proved such to be. Any appearance of a lobe in this situation must be regarded as belonging, not to Normal, but to Morbid Anatomy, a slight development of it being usually attended with some signs of obstruction to the function of micturition. And so far from being overlooked, it is in this manner that it has been regarded, after very careful examination, by the earliest labourers in Pathological Anatomy, who have left to us the records of their observations. Thus Morgagni, in that vast collection of cases which forms his work, "De Sedibus et Causis Morborum," refers to it in several places as to a morbid growth causing retention of urine; in one of which he describes it, from his own dissection, in the following remarkably precise terms: "A roundish protuberance, of the bigness of a small grape, covered over with the internal coat of the bladder. What this protuberance was I readily supposed: and by forcing the knife into it, I cut through this and the contiguous prostate gland at the same time, lengthways, and showed that it was of the same nature with the gland; that it was very evidently continuous from it, and that there was no doubt but, if it had grown out to a greater degree, it must have been a very considerable impediment to the discharge of urine."* In another case he describes the same appearance, pronouncing it "beyond a doubt an excrescence of the prostate gland."† He quotes several similar cases from Valsalva, Thomas Bartholin, of Padua, and Valisneri, who indeed speaks of an enlargement, "as it were, a kind of lobe, from the glandular substance" (of the prostate) "which rose up within the bladder, of the shape and size of a walnut, and not on the anterior part, but on that which lies adjacent to the intestinum rectum." He enumerates others, mostly from the "Sepulchretum" of Bonetus,‡ besides alluding

* De Sedibus et Causis Morborum per Anatomen indagatis. By J. B. Morgagni. 2 vols. folio. Venice, 1761. Letter xli. article 18.
† Idem. Letter xxxvii. article 31.
‡ The Sepulchretum, sive Anatomia Practica. By Theophilus Bonetus, M.D. Lyons, 1700. Book iii. sects. 24, 25.

prospectively to another case which subsequently appeared in the succeeding letter of his own work;* and then makes the following generalizations from the whole.

"If you attentively examine those examples which I have pointed out," .... "you will observe that they were all from old men; and in like manner, if you examine all my observations in which there was the beginning of an excrescence, you will find that this was found to grow out in the very middle of the internal and upper circumference of the gland, posteriorly; but whether all these things happened by chance, or otherwise, future observations will show."†

Some considerable time after this that future was realized; Morgagni returned to the investigation of this subject, and the cause of his doing so is of great interest in relation to the present inquiry. It appears that "a celebrated anatomist" of the time (Morgagni refers, without doubt, to Lieutaud, although he omits all mention of his name, as he states to be his invariable custom when he proves a contemporary to be in error), declared that an eminence at the neck of the bladder was not a morbid growth, but a small part quite natural, and common to all bodies, calling it the "uvula vesicæ." Morgagni therefore devotes great part of the sixty-sixth and the whole of the seventieth letters, to the refutation of this view. In the former, stating that, during forty-four years as anatomical professor, he had most carefully dissected at Padua sixty or seventy bodies, and found it only in four; that he had made vivisection of a dog for the express purpose of seeking it there, but in vain, and that he had decided that it could be nothing but "a morbid excrescence of the prostate gland appearing in old men," .... "not rare, but not so frequent." That Valsalva, Pohlius, and his friend Santorini, regarded it in the same light, the latter presenting it in a drawing as a body

* De Sedibus, &c. Letter xlii. article 11.
† Idem. Letter xli. article 19. See the English translation by Dr. B. Alexander. London, 1769.

"prominent in diseased bladders,"* besides referring to it as "a circumstance which is diseased and unfrequent, and does not deserve to be exhibited as perpetual and constant, to the great detriment and misleading of younger practitioners."* The seventieth letter Morgagni devotes to the purpose of giving the result of his forty-fifth year of teaching anatomy, in relation to this very subject, stating that he had dissected in public five subjects, and in none of them, although he had sought most carefully, was there any trace of this "roundish protuberance," or "uvula."

John Hunter similarly regarded it, after independent examinations of the organ, stating that "a small portion of it (the prostate) which lies behind the very beginning of the urethra, swells forward like a point, as it were, into the bladder, acting like a valve, to the mouth of the urethra."†

It seems remarkable, considering the very slight grounds upon which the existence of a distinct third lobe as a normal and ordinary constituent of the prostate, were affirmed by Sir Everard Home, that it should have been so generally received without question by English anatomists to the present day. Its existence is denied by most French observers. Cruveilhier expresses the general opinion when he says that the ejaculatory ducts, being received into a groove or channel in the substance of the prostate, a portion of variable size is indicated by them, but that it has no title to be called a lobe. It is not, he says, an isolated piece, and should be called "the median portion."‡

The following is the result I have obtained after carefully prosecuting numerous dissections of this part.

* Observationes Anatomicæ. By Jo. Dom. Santorini. Venice, 1724. Chapter x., "De Virorum Naturalibus," in explanation of tabula 2, fig. 2, sects. 20 and 22, pp. 201-205.

† A Treatise on the Venereal Disease. By John Hunter. 2nd edit. London, 1788. p. 170.

‡ Anat. Path. du Corps Humain, livr. xvii. p. 3. Paris, 1835-42. Par J. Cruveilhier.—See also Traité d'Anat. Path. Générale. Tome 3eme, p. 46. Paris, 1856. By the same. Langenbeck also takes the same view.—Neue. Bibl., b. i. p. 360. Hanover, 1818.

First.—I cannot find in healthy bodies, below 50 years of age, any formation in the situation described, capable of being recognized as a distinct "third" or "middle" lobe, and am compelled to conclude that any marked prominence there, which appears to possess independent characters (as regards size or form), must be considered abnormal or morbid.

Secondly.—There is unquestionably a thick uniting stratum of tissue between the lateral lobes, which is sometimes slightly thicker in the middle line, at the vesical or basic end, than at its borders, where it becomes blended with those lobes. It is at this thickest part, three or four lines below its urethral aspect, that there exists a perforation for the entry of the ejaculatory ducts. But in many cases this thickening in the median line does not exist; and it is to be remembered that the part in question immediately underlies the uvula, which is not a portion of the prostate, but a prominence caused by crossing interlacement of muscular fibres from the inner coat of the bladder, and sometimes prone to be unduly developed. If this fact is not borne in mind, a confounding of the two structures may arise, and the error of regarding them as one will lead the observer to attribute inaccurately a middle lobe to the prostate; and I think it has not unfrequently happened that the existence of a distinct middle lobe has been affirmed after simple inspection of the interior of the bladder, without further examination, which alone can determine, in all cases, whether a small projection at the neck is due to hypertrophy of the tissues of the uvula, or of the prostate, or to a small tumour embedded in the organ at this point.

But, thirdly.—The posterior commissural part in question does possess a specific character, which distinguishes it from other parts—a character which appears to have been less prominently regarded, perhaps, than it has deserved to be in the discussion of its title to the name of third lobe, and which appears to entitle it to some distinguishing appellation. Moreover, this character seems to be connected with that tendency to enlarge, which this part of the organ undoubtedly possesses, and which will be fully discussed hereafter. It is this, that the portion in question certainly contains a larger proportion of glandular structure than most parts of the entire organ. Thin slices from this portion placed under the microscope and compared with slices from other parts demonstrate this very clearly.

Considering, then, all the terms which have been proposed to designate the part in question, there does not appear to be sufficient ground for continuing the term "lobe." On the contrary, it appears desirable to adopt that which the French anatomists have employed, viz., "the median portion," since it is sufficiently accurate, and it is perhaps not wise, without ample justification, to alter a term already used in that modern language, which in scientific literature is perhaps, at the present day, the most widely understood. If, retaining the terms "lateral lobes" as universally accepted, we apply to the part anterior to the urethra (standing position of the body) the term "anterior commissure,"—a part which is about one inch in length; and the term "posterior commissure" to that which corresponds to it for the same length behind the urethra, it leaves that thicker portion of the organ which lies behind the verumontanum, which is penetrated by the ejaculatory ducts, which is largely perforated by glandular structures, and which is prone to great increase of size in age, to be designated by the term, "median portion," conformably to the practice already pointed out. This part, it may be added, it has been proposed by Mercier of Paris to call the "portion susmontanale," as indicating its position behind the verumontanum. The above reasons for adopting the simple and hitherto better-known term of "median portion," apply, perhaps, equally to this proposal also, although there is no other objection to its adoption by any who prefer it, as it indicates pretty clearly the part intended by the writer.

COURSE OF THE URETHRA THROUGH THE PROSTATE.—Commencing at the neck of the bladder, the urethra continues through the prostate, in the erect position, downwards and slightly forwards. The great masses of the organ lie on either side of the canal; a stratum of prostatic tissue lies in front, and another behind it. It has been a disputed question whether the anterior or posterior stratum be the larger. Most observers state that, in the normal condition, the larger portion is found behind the urethra. Thus, Mr. Adams, in his article on the prostate, in the "Cyclopædia of Anatomy and Physiology," as well as in his published work, states that "two-thirds of the gland are below this canal" (the urethra).* There is, however, one important authority who denies, and very positively, this assertion. M. Mercier long ago asserted that the anterior stratum is larger and thicker than the posterior, and he states that, after renewed examination, he is assured of the accuracy of his view.† Dr. Hodgson, who has very carefully studied the anatomy of the prostate, writes as follows respecting this point:—"It is generally stated that the urethra lies at a distance of two lines from the upper and four lines from the lower surface of the gland. These measurements are by no means constant; according to my observation, they are pretty regular in the prostates of young persons of eighteen to twenty-two years of age, while in older individuals the amount of glandular tissue above the urethra equals, and occasionally rather exceeds, that below the canal."‡

In order to examine more completely this question, I removed ten healthy adult prostates, and have made a very careful section from the anterior or pubic aspect to the posterior or rectal one. From these preparations, now in the College of Surgeons, the accuracy of the following report can be tested.

The ages of the ten subjects ranged between 35 to 57 years; all may be regarded as non-hypertrophied, by the weight test; the lowest being 4 drachms 10 grains, and the highest, 5 drachms 36 grains.

In seven, the area of cut surface displayed by a median section from pubic to rectal aspect, was about equal, above and below

* Cyclop. Vol. iv. p. 146. Anatomy and Diseases of the Prostate Gland. 2nd ed. 1853. p. 1.
† Récherches Anatomiques. Paris, 1841. p. 30. Recently, Gazette Hebdomadaire. 1860. p. 131.
‡ The Prostate Gland. London, 1856. p. 22.

the urethra (supposing the organ removed and lying with its pubic surface upwards). In two, the area of cut surface displayed was rather larger below than above; in one the converse condition was found.

But it appears to me, judging from the conformation of the specimens, and not from the cut surfaces merely, that there is a rather larger mass of prostatic substance below than above the urethra, in all these specimens except two or three.

THE PROSTATIC URETHRA.—Supposing a specimen of the adult bladder and about 3 inches of the adjoining urethra, including the prostate gland and other underlying structures, to have been removed from the body, and to be placed on a table with the pubic aspect uppermost, the urethra is to be exposed by dividing all the tissue lying, in that position, above it, and by opening out the sides of the incision. The following parts are then to be observed in that part of the urethra which lies under consideration.

The vesical boundary of the prostatic urethra is the *uvula vesicæ*, seen as a slightly-rounded prominence of the floor of the neck of the bladder. From this point forward the canal gradually expands a little and presents a general indication of hollowing or depression of the floor, broken by a central longitudinal elevation, the verumontanum. This eminence commences almost imperceptibly in a faint whitish line directly in front of the uvula, and gradually rising for 6 or 8 lines reaches then its highest point (about the ⅛ of an inch), having formed a longitudinal ridge in the central line of the urethra. It widens and becomes rounded at the highest point, and then somewhat suddenly diminishing in size as it advances forward, exhibits on the anterior slope an opening which will be presently examined in detail; the ridge, gradually lessening, leaves the prostatic and enters the membranous urethra. The depression of the prostatic urethra, "*sinus prostaticus*," is thus divided into two lateral furrows, one on each side of the verumontanum; in these may be observed numerous little orifices, mouths of the prostatic ducts, which when examined with a common magnifying glass of low power, may be counted without difficulty. Not less than 20 or 24, sometimes 30, may in this manner be distinguished. They are ranged chiefly on each side of the eminence of the verumontanum, but a few may be seen both before and behind it.

The length of the prostatic urethra varies from an inch and an eighth to an inch and a quarter. Its diameter is usually represented by a fixed measurement; but as the surfaces lie in apposition, except when distended by a functional act, it is difficult to represent accurately the calibre. The walls and surrounding structures are extensible, and the calibre of the canal is consequently variable, corresponding within certain limits to the amount of pressure exerted. Without unduly extending them, the diameter, opposite the crest of the verumontanum, may be regarded as amounting to about $\frac{5}{18}$ to $\frac{6}{18}$ of an inch; but at each extremity of the prostatic limit the diameter, calculated in the same way, is rather less.

The whole of the urethra thus examined presents a surface of mucous membrane, which is covered with cylindrical epithelium, always arranged in two layers, a superficial and a deeper layer. The membrane has when fresh a pale pinkish-yellow tint, the structures which underlie it being pale, unstriped, muscular, connective, and elastic tissues, with fewer minute blood-vessels than are found in the membranous and bulbous portions of the urethra, where a redder tinge consequently exists. These submucous structures are composed chiefly of the pale, unstriped, muscular fibres, intermixed with a small proportion only of the white fibrous and elastic tissues. The muscular fibres are arranged in the longitudinal axis of the urethra, and form a thin but somewhat dense and strong layer immediately beneath the mucous membrane throughout the prostate; they continue to have a similar relation throughout the canal, although in smaller quantity than in this situation. Posteriorly they are continuous with the muscular layer immediately underlying the mucous membrane of the bladder. It is by a development of these longitudinal muscular fibres of the urethra that the verumontanum is formed; they divide to define the mouth of the utricle and admit the orifices of the prostatic ducts. These cavities are lined with a delicate membrane and epithelium (cylindrical) like that of the prostatic urethra itself.

The precise mode of arrangement of the minute blood-vessels beneath the prostatic mucous membrane is as follows. From the neck of the bladder to the most elevated portion of the verumontanum very few of them are to be seen; and the mucous membrane is of a yellowish tint from their paucity in this situation. In front of it, however, numerous fine vessels course forward, side by side, for the most part in the axis of the urethra, but with a slight oblique divergence outwards, right and left of the pale line of the verumontanum, prolonged into the membranous portion. The mucous membrane is therefore redder in tint anterior to the eminence; and on examining it with a magnifying-glass, it is seen to be due to the minute and longitudinally branching vessels just described.

The Utricle, or vesicula prostatica, is a small sac already referred to as opening on the anterior aspect of the verumontanum. It is somewhat oval in form, measuring ordinarily from two and a-half to four lines in length, and from two to three lines at its greatest breadth, although occasionally found both larger and smaller. Its posterior extremity is the widest, and the anterior is the narrowest, portion. It is obliquely placed, the long diameter being directed towards, or a little below, the median portion, and in the course of the ejaculatory ducts which lie contained in the substance of its lateral walls. The external orifice or mouth of the utricle is sometimes extremely small, and it has been occasionally seen altogether wanting; the orifices of the two ejaculatory ducts were then found opening on the anterior of the verumontanum.

The sac of the utricle is made up of mucous membrane, covered with cylindrical epithelium, and of a submucous tissue attaching it to some bands of white fibrous tissue and pale muscular fibre, by which it insensibly unites with the mass of structures around. It contains mucous follicles which appear sometimes to secrete a dark reddish-brown and jelly-like material; at least, such is found sometimes nearly filling the sac.

When in a quiescent state, the sides of the prostatic urethra are pretty closely applied to each other, and the mucous membrane then lies in very small delicate longitudinal plaits, with a very shallow groove between each. A transverse section shows the canal to have a form generally approximating to triangular the apex being upwards.

COURSE OF THE COMMON EJACULATORY DUCTS THROUGH THE PROSTATE GLAND.—The common duct, formed by the junction of a vas deferens and of a duct of the vesicula seminalis, enters, in close proximity with its fellow of the opposite side, at the base of the prostate, about three or four lines below (the prostate removed and placed for examination as last described) the opening which transmits the urethra. There is a well-marked depression, in form pyramidal with the apex directed inwards, in the base of the organ, which indeed corresponds with the apex of the notch separating the two lateral lobes. This depression penetrates the substance of the posterior portion or middle lobe, about three-fourths of the mass lying between the ducts and the urethra, the remaining one-fourth forming a thin layer between the ducts and the under-surface of the prostate. This proportion varies in different subjects; in all, however, the greater part of the mass lies between the ducts and the urethra. A thin stratum of tissue, $\frac{1}{15}$ to $\frac{1}{20}$ of an inch in thickness, separates the two ducts, as they perforate the organ side by side. They then pass directly to the utricle, and continue their course embedded in the walls of that cavity, one on each side; their little slit-like orifices opening on the anterior part of the verumontanum, at the mouth of the utricle itself. By carefully making a longitudinal section of the utricle, exactly in the median line, the course of the duct may be easily traced.

The ejaculatory duct, from its place of entry at the base of the prostate to the orifice at the utricle is about six lines in length. It is composed of a mucous lining, or inner coat, some pale muscular fibres surrounding it, with connective tissue intervening; the coats are much thinner and less firm than those of the vas deferens. The muscular fibres are arranged longitudinally in the axis of the duct, and are continued into those which underlie the urethral mucous membrane, close to the mouth of the utricle, and which have been already described.

THE MINUTE ANATOMY OF THE PROSTATE.

COMPONENT TISSUES AND THEIR ARRANGEMENTS.—It has been already seen that the pale or unstriped muscular fibre enters largely into the composition of the prostate. This has been affirmed by numerous independent observers, among the earliest of which was Dr. Handfield Jones, whose statement of the fact was first recorded in a paper in the *Medical Gazette*, August, 1847.

In the following year, Kölliker, the well-known German histologist, now Professor of Anatomy and Physiology in the University of Würtzburg, published an account in the *Zeitschrift für Wissenschaft* (Leipsic), of his examination of the prostate, in which he declared that the larger portion of the organ was constituted by the pale muscular tissue, and that the smaller portion only consisted of glandular tissue. More recently, Professor Ellis, of University College, has closely investigated the same subject, and has come to a similar conclusion. His paper, communicating the results he arrived at, is published in the Thirty-ninth Volume of the Transactions of the Medical and Chirurgical Society, 1856), and is illustrated by drawings of his dissections.*

* "The prostate is essentially a muscular body, consisting of circular or orbicular involuntary fibres, with one large central hole for the passage of the urethra; and another smaller oblique opening, directed upwards below the former, for the transmission of the common ejaculatory seminal ducts to the central urinary canal. The few longitudinal fibres on the upper surface of the prostate, which are derived from the external layer of the bladder, can scarcely be said to form part of that body.

"Its circular fibres are directly continuous behind, without any separation, with the circular fibres of the bladder ; and in front a thin stratum, about one-thirtieth of an inch thick, is prolonged forwards from it around the membranous part of

Kölliker's latest account is to be found in the "Manual of Human Microscopic Anatomy," published during the current year (1860) by Parker, London. It confirms his own original description, as well as the views of the observer above-mentioned.*

A description has already been given of the anatomical arrangement (p. 6) of the fibres which immediately underlie the mucous membrane of the prostatic urethra, a layer of intermingled muscular, connective, and elastic fibres, lying in the direction of the axis of the urethra. And it has been shown that these are surrounded by a layer of fibres transversely or circularly disposed round the axis, and consequently disposed at right angles to the preceding. And lastly, that this layer is thicker than the first-named, especially at its vesical extremity, where it becomes continuous with the circular fibres of the bladder itself. The other tissues which enter into the construction of the organ, and which I have subjected to an extended and careful examination in various ways, will now be considered in detail.

First of all, then, the two layers of muscular and other fibres just enumerated are perforated by many gland-ducts, from the gland-structures presently to be described. These gland-ducts, passing in a concentric direction, open in the sinus on either side of the verumontanum in the urethra. Next, among the outer or circularly-disposed layer of fibres, may be found a few small

the urethra, so as to separate this tube from the surrounding voluntary constrictor muscle. . . . . Within and quite distinct from the circular fibres, lies the tube of the urethra, incased by its submucous layer of longitudinal fibres. Towards the lower and outer aspects, the fibres are less firmly applied together, especially where the vessels enter; and they appear to be superadded to those which join the coat of the bladder.

"As only so small a portion of the prostate is glandular, the propriety of calling that body a gland is rendered doubtful; for the small secreting glands contained in it are but appendages of the mucous membrane, which project amongst the muscular fibres in the same way as the other glands of the urethra extend into the surrounding submucous tissues." — Trans. Med. Chir. Soc. vol. xxxix. pp. 330-1.

* "The prostate, according to my own observations, which are confirmed by V. Ellis, and partly also by Jarjavay, is a very muscular organ, the glandular substance scarcely constituting more than one-third, or the half of the entire mass."—*et seq.* pp. 439-40.

and simple gland-crypts, attached to excretory ducts in their course to the urethra. These small glands are, for the most part, situated under the urethra, near to where it joins the neck of the bladder, behind and not in front of the verumontanum.

The structure of that part of the mass which lies without and beyond these two layers of muscular fibre, and which forms the lateral lobes, is that portion which mainly gives distinctive form and character to the prostate—that, in fact, which constitutes it an independent and a male organ.

We find here a combination of muscular and allied tissues, *i.e.* the mixed pale, muscular, connective, and elastic fibres (the first-named predominating), and of the glandular structure.

The muscular tissue has no longer any very defined form of arrangement. Bands of it appear to take their rise from the outermost of the circular fibres, and to pursue a direction outwards towards the capsule. In doing so, they proceed somewhat irregularly, as if interwoven loosely with each other, and thus they leave numerous interstices. Kölliker says, that these bands or bundles radiate outwards, in all directions, from the sides of the verumontanum. I cannot distinguish anything quite so precise as a general radiation; certainly not a regular radiation without interlacement: but perhaps he does not mean this. There is a general diverging of these bands from the centre to the circumference; but there appear also to be some bands having a different direction, so as to cross and coalesce with the others, at angles more or less acute—bands which do not partake much of the diverging or radiating character, but which interlace among the elements of gland-structure next to be described, which bind them together, surrounding and packing, so to speak, the lobes of that structure. In the circumferential parts of the prostate, the muscular and allied fibres are partially continued into the capsule of the gland itself, into the structure of which they enter (see page 39).

The interstices alluded to above, are occupied by the glandular structures, which, in the lateral portions of the organ, form a large proportion of the prostatic substance, and give it a particoloured appearance to the naked eye. When a longitudinal section is made of a lateral lobe they are easily distinguished by their yellowish colouring, and by this character their situation, line of direction, and their proportion in relation to the fibrous stroma may be approximately estimated when sections of the organ are made. An antero-posterior section in the median line is well represented in Plate I., fig. 1. The agglomeration of glands which occupies the situation of the "median portion," the ejaculatory duct beneath it, and, next, the glands of the lower and posterior parts are easily distinguished, the line of their divergence outwards, backwards, and downwards from the verumontanum being marked by the gland tissue above-mentioned; their lower margins, somewhat rounded, may be seen reaching to within about a line of the circumference of the organ, this portion being the capsule and associated fibrous tissue.

The glands themselves are, by their structure, somewhat peculiar to, and characteristic of, the prostate. They are classified with the multilobular or compound racemose glands, of which the salivary and pancreatic are typical examples; but they differ somewhat from these. After carefully examining them, chiefly by unravelling fresh specimens under the microscope, as well as by obtaining fine sections of hardened preparations, the following appears to be the arrangement of the prostatic gland-structures.

It will be desirable, for the sake of precision, to name the terms employed to denote them, commencing with the extreme terminal portions of the gland-structures, and proceeding in order to their excretory orifices, viz., 1, vesicles or crypts; 2, follicles; 3, ducts; and, 4, lobules.

1. By vesicles or crypts are denoted the smallest recesses of all; cavities of a partially spheroidal form, and varying in size, *during the prime of life*, from $\frac{1}{300}$ to $\frac{1}{700}$ of an inch. Sometimes only one or two of these exist at the end of a minor duct, but more commonly several are associated, the size and number being very variable, to form a group, having a central cavity

common to all or some of them. The vesicles are more or less spheroidal or ovoid in form; sometimes they appear as mere cup-shaped depressions in the surrounding tissue, while others are even flask-shaped or pedunculated. The size given above is the range usually met with during early adult life. In old age, and, indeed, in middle life, they become larger, and appear as if dilated.

2. The cavities, more or less composite, and formed by a group of vesicles in the manner just referred to, may be termed follicles, and their size is also very variable. Ordinarily, they reach about the $\frac{1}{120}$ or $\frac{1}{100}$ of an inch in diameter. As age advances they become larger, often reaching three or even five times that size. Each follicle has either a duct of its own, as most commonly happens, or it opens at once into a duct common to several others, but leading to the single excretory duct of the lobule to which it belongs. It is peculiar to the prostatic glands that the follicles are of very varying size, and are irregularly placed upon the duct. They are not crowded upon it so as to form a compact mass, as in other racemose glands, but are scattered upon it with considerable and irregular intervals. Hence the tissue is more loosely aggregated, and it is on this account more difficult to isolate, trace, and observe with accuracy.

3. The minor ducts are extremely numerous. Each follicle, or small collection of vesicles, has its duct, which is often long, and pursues a straight course to the main trunks, and these, uniting, finally form a single excretory duct from the entire lobule. The smaller ducts lie nearly parallel to each other, and slightly converge as they approach the principal duct. Their walls are easily seen by section under the microscope to be made up of fibres circularly disposed round the axis of the tube.

4. The term *lobule* is held to imply any aggregation of these crypts which, by means of subordinate ducts, is united with the one excretory duct of that series, which then passes on, unjoined by the duct from any other lobule, to the urethra. Each lobule is a complete and independent glandular structure, like a sweat-

gland, and is equally entitled with the latter to the term "gland." The custom, however, of applying that term to the entire organ, "prostate gland," renders it preferable to use the term lobule in the case before us. All these different parts are lined with epithelium of different characters, as may be verified with great ease.

First, the walls of the vesicles and crypts are covered with an extremely regular epithelium, the cells of which show a disposition to be ovoid, but apparently become polygonal under the influence of lateral pressure. The epithelial cells are very adherent to each other, and may be removed in masses of convex, or even of tubular form.

As regards the cells themselves, each contains granular contents, and a nucleus a little more elongated, in proportion to its size, than the cell itself. There is no well-marked nucleolus. Viewed in a mass, the cells have a yellow or tawny tint, derived from the granules in their interior, though one cell or a single layer will not generally exhibit it. Their size is from the $\frac{1}{2500}$ to the $\frac{1}{3000}$ of an inch.

In the minor or subordinate ducts, the lining cells are prismatic, a fact which I am perfectly satisfied of, although some descriptions represent them as spheroidal only. In the largest ducts the same is found, and the epithelium, I believe, after much observation, to be ciliated.

The vesicles themselves are sometimes empty, occasionally they are packed with epithelial cells. Sometimes they contain the liquor prostaticus, and not infrequently some may be seen filled with a yellowish, semi-solid, jelly-like, transparent matter, containing the epithelial cells, although sometimes without them and homogeneous. Sometimes they contain the concretions which are so familiar to all who have examined the prostate, and which appear to be always present in the adult organ. (See Chapter XVIII., and Plates XII. and XIII.)

The ultimate gland vesicles are held in place and united together into follicular masses, and these again are aggregated into gland-lobules, chiefly by means of connective tissue mingled with a small proportion of pale muscular fibre. Considerable intervals, filled by these tissues, are found between many of the secondary portions of the lobule. The true gland-structures may, however, be isolated, or nearly so, with great care and patience, by dissecting them under water with a pair of needles.

The connecting tissues just alluded to as surrounding the gland elements, support also a fine but rich network of capillary blood-vessels, ramifying closely upon the outside of the vesicles. The vascular supply is considerable, judging from the amount and freedom of capillary intercommunication.

The gland-lobules, each made up as above described of an assemblage of vesicles and follicles, are about 40 or 50 in number. Each sends its excretory duct to open in the urethra, near to the verumontanum. Sometimes two such independent ducts will open by a common orifice; sometimes, also, they do not open directly into the urethra at the point where they arrive at it, but continue a submucous course until they arrive at the usual place for opening above named. Sometimes they traverse a considerable distance in this manner, occasionally as much as three-eighths or even half an inch.

The middle-sized and largest ducts have well-marked fibrous walls; they are composed of both longitudinally and circularly disposed fibres of the connective tissue, mingled with a small proportion of the pale muscular fibre. Elastic fibres are not found here.

A question naturally arises, and one which we believe has never been satisfactorily answered, inasmuch as different observers make very different statements respecting it, viz., Is every part of the prostate pervaded by glandular structures; and if so, are they distributed in equal proportion throughout?

There are two ways of examining the organ with a view to obtain a satisfactory answer. The first is, by carefully making small slices of it by means of Valentin's knife, examining each part of the prostate in turn, and submitting the slices so obtained to the microscope. Another, which is, perhaps, preferable, consists in making an incision with a perfectly clean scalpel into each part of the organ successively (or corresponding parts of several organs, to speak more accurately, as several are requisite for the examination of the question), taking great care not to press with the fingers near to the part in which the incision is made, lest fluids should be forced into the opening from adjacent sources; then to remove successively from the exposed surface at various parts of the section, minute portions of tissue with a scissors, or knife and forceps; finally, to submit each portion separately to a microscope with a good quarter-inch object-glass. Nothing is to be added except pure water, and after an examination of the untouched portion, its tissue is to be unravelled under the lens with a couple of fine needles.

What will be seen is the following:—

a. Supposing that gland-structures *are* present.

First, the drop of distilled water in which the specimen is placed before attempting to dissect it, or to cover it with thin glass (if this is employed), becomes at once slightly milky-looking if gland-tissue be present—an appearance never seen if a portion is examined in which no gland-tissue is present. This is found, on putting it under the microscope, to be due to the rapid floating out in great abundance of the epithelial cells. On further examining it by transmitted light, the mass is seen to be made up of parallel, soft, pale, muscular fibres, interlaced with much connective tissue. Floating around in the clear field, as well as occupying spaces in the mass itself, are numerous nucleated epithelial cells (already described, page 35). On adding dilute acetic acid, the soft fibres almost disappear, looking softer still, and semitransparent, while numerous rod-shaped nuclei come into view, indicating, by their linear arrangement, the direction of the muscular fibres. Some few lines of elastic fibre are also now manifest, crossing the other fibres transversely. And now the epithelial cells are seen more distinctly where they occupy spaces in the mass, lining small crypts, or the smallest ducts. They are clearer because the surrounding structures are less distinct, not because they themselves are much affected by the acetic acid; (their external contour becomes a little more defined, while the nucleus is a little fainter than before.) They are now seen to be grouped together, sometimes in compact masses of laminar, or of cylindrical or spheroidal form; these are portions of the gland-tissue loaded with the secreting cell-structure. Sometimes the ducts and crypts are empty, and then is seen a stratum of the same cells regularly applied and closely adhering, side by side, to form the lining of the cavity. It is not difficult to remove a portion of this by means of fine needles, and to examine it separately; partially cylindrical and even branched portions may be thus obtained.

If we press out a little of the liquor prostaticus, and examine it under the same power, we shall find an abundance of the same cells, as well as the prismatic cells from the ducts; and acetic acid exercises the same very slight action upon them as described above.

b. What is seen if a part is examined in which no glandular structures are present.

In this case we find the soft muscular fibres and the connective and elastic tissues—in short, the stroma* of the organ, exhibiting the same appearances, in water and in dilute acetic acid, as have been described above; but now no nucleated cells of any kind are seen. The tissue is to be well unravelled; but none of these cells, either free in the field, or occupying channels and interstices among the fibres, can be discovered. We are quite sure there is no gland-structure in the portion from which such results are obtained.

Now, in the direct median section of the anterior part of the organ, in front of the urethra—in other words, through the anterior isthmus or commissure—I have never found, and after repeated

---

* The term stroma will be used in future for brevity's sake, to indicate the structure, composed, as frequently pointed out, of three elementary tissues; and to distinguish it from the glandular structures which it surrounds and supports.

examinations, any glandular elements, provided the section is made strictly in the median line. Advancing about the eighth of an inch on either side of that line, in some instances, at a smaller distance from it, secreting cells and small crypts make their appearance, and with them often small concretions. In every other part of the prostate the cell-structure is to be found, but it is most abundant in the outer and posterior parts of the lateral lobes, and in the centre of the "middle lobe" or "median portion."

It appears, then, that the true gland-structure from one lateral lobe does not meet that from the other in the middle line anteriorly, *i. e.* in the anterior isthmus, and that an interval, occupied by simpler structure, and varying in amount in different subjects, exists there; it is worthy of remark also, that in that interval I have traced a few voluntary muscular fibres reaching down nearly to the urethra. A section removed from this spot and unravelled will frequently show some striped fibres, although in small proportion. This want of continuity of the gland-structures in front, seems to indicate a persistence of the original bilobed condition well marked in fœtal life, and obtaining for the organ the plural name of "prostates" among the older anatomists. But gland-elements are found in abundance in the middle line, behind the urethra.

The proper capsule of the prostate closely covers in the whole organ except at the base and apex, where the urethra enters and issues, and the enveloping structure is there continuous with the adjacent parts respectively, as described at page 6. This capsule is distinct from the sheath given to the prostate by the recto-vesical fascia, and is an integral portion of the organ, not to be peeled off or separated, otherwise than by cutting. It is composed of densely-applied fibrous tissues, viz., the pale muscular fibre, the connective tissue, and a little elastic tissue, and bands of these enter the substance of the prostate itself in great abundance. In fact, the capsule is continuous in structure with the stroma of the organ; much interlacing of fibre exists, and minute vessels accompany the entering fibres from without, and anastomose freely with the rest of its vascular system.

THE PROSTATE IN SUBJECTS UNDER ADULT AGE.—In a very early stage of fœtal development, after the rectum has become separated from the bladder, this latter and the genito-urinary organs in front, exist in the form of a channel, known by the name of the "sinus uro-genitalis." Into this, in the male, enter the ureters and vasa deferentia, the latter, at first by one opening, and subsequently the vesiculæ seminales are developed from a portion of the sinus uro-genitalis adjacent to the opening; and each vesicula becoming connected with its neighbouring vas, two independent orifices appear in the place of the single one, and between them the utricle is developed from an original portion of the sinus itself remaining. Around this are aggregated the prostatic glands and fibrous structures, in two distinct masses or lobes, visible as such during the fourth month of intra-uterine life. During the succeeding month an intermediate portion unites them behind (rectal aspect) in the situation of the median portion, and of the commissural part below, to which latter the name of "isthmus" or "posterior commissure" has been already applied.

At birth the organ is very small, and it continues comparatively so until the time of puberty. According to the researches of Dr. Gross, of Louisville, it weighs only thirteen grains at the time of birth: his report on the size, form, and weight, of the prostate, from that period up to adult age is very complete, and may be given here.

"*The prostate at birth.* Width, at base, 4 lines; a little above middle, 5 lines; at apex, 2 lines; length along the middle, 4 lines, and at the edge, 4¾; thickness at base, 2 lines; at middle, 3¼, and at apex, 1¼. Weight, 13 grains.

"*The prostate at 4 years.* Breadth at base, 6 lines; just above the middle, 7; and at the apex, 2½; length along the middle, 6 lines; and 7 lines at the margin; thickness at base, 2¾ lines; at the middle, 4; and at apex, 2. Weight, 23 grains.

"*The prostate at 12 years.* Width, 8¼ lines, at base; 9½ above the middle, and 3 at apex; length along the middle, 8 lines, and 8½ at the edge; thickness at base, 3; middle, 4½; and at apex, 2¾. Weight, 43 grains.

"*The prostate at 14 years.* Width at base, 11 lines; at middle, 9½; at apex,

4 ; length along the middle, 8 lines, and 10 at margin ; thickness, 3½ at base, 5 at middle, and 3 at apex. Weight, 58 grains.
"*The prostate at 20 years.* Breadth at base, 14 lines; at middle, 16; at apex, 5¼; length along middle, 15 lines, and at edge, 16; thickness at base, 8 lines; middle, 10; and apex, 5¼. Weight, 4 drachms and 1 scruple." *

Mr. H. Bell, in an inaugural thesis, which he published in Paris, made observations, which have been copied by Malgaigne in his "Traité d'Anatomie Chirurgicale,"† resulting from a dissection of upwards of forty subjects ranging between two and fifteen years of age. They are as follows:—

Ages.	Diamètre Transverse. mm.	Rayon Postérieur Oblique. mm.	Rayon Postérieur Direct. mm.	Rayon Antérieur Direct. mm.
2 à 4 ans.	12·40 à 13·5	4·5 à 5·	2·25	1·
5 à 10 ,,	13·5 à 17·	5· à 7·	4·5 à 5·6	1·
10 à 12 ,,	16· à 19·	6· à 8·	4·5 à 5·6	2·25 à 3·4
12 à 15 ,,	19. à 22·	8·	4·5 à 5·6	3·4.

I dissected one specimen at twelve years, which forms a preparation at the College of Surgeons, and the weight coincides very closely with that of one at the same age in Dr. Gross's series.

At twelve years ; weighs 40 grains.
Length . . . . . an inch
Breadth . . . . . three-quarters
Thickness . . . . three-eighths.

We must not overlook the researches of Deschamps made on a great number of bodies. In his "Traité Historique et Dogmatique de la Taille"‡ he reports as follows:—

1. "In subjects from three to eight years of age, the anterior thickness of the prostate" (anterior to the urethra) " is a line and three-quarters; its posterior part two lines and a half; and its lateral parts three and a half lines." . . 2. "In subjects from eight to sixteen years, the thickness of the anterior part is about two lines; of the posterior part three lines; of the lateral parts four or five lines."

The following general considerations are those of importance in reference to this subject.

* Gross on the Urinary Organs, 2nd edition, p. 70.
† Malgaigne's Traité D'Anat. Chir., &c. Paris, 1859. Tome ii. pp. 483-4.
‡ Paris, 1 96 Vol. i. pp. 39-40.

The position of the prostate in children differs from that in the adult subject. It is more vertically placed in the pelvis than in that of the latter. The bladder has a corresponding position, its lower fundus is less depressed, less sessile on the rectum (so to speak) than it becomes later in life. The peritoneum comes very near to its base; but as the fundus is developed in later life, the peritoneum becomes further removed, and a well-marked space, or portion of the bladder (carefully noted both by the anatomist and the surgeon), remains uncovered by it beyond the prostate.

The form of the gland is more rounded in childhood, and has less of the distinctive characters and outline which mark the adult organ and its lobes.

Its consistence is soft, the capsule is easily torn, the glandular structures are simple, and slightly developed, being, apparently, little more than simple follicles, tubular and crypt-like, and ducts.

CHAPTER II.

FACTS RELATIVE TO WEIGHT, SIZE, AND MORBID CONDITION, OBTAINED BY DISSECTING THE PROSTATE.

Dissections by the Author.—By Dr. Messer.—In all, 194 Examples of the Prostate, presented in the Form of a Table.

At the conclusion of the foregoing portion of this work, and therefore occupying a position intermediate between that which relates to the normal anatomy of the prostate, and that which relates to its diseases, appears to be the most appropriate spot for placing the data from which many of the conclusions given respecting both subjects have been drawn. These consist in researches by dissection which are tabulated below. I have made numerous other dissections for the same purpose; but these having necessarily resulted in destruction of the organ are not recorded here; every one of those reported here as my own having been preserved and exhibited as wet preparations.

A TABULAR VIEW OF THE FACTS OBSERVED IN 194 DISSECTIONS OF THE PROSTATE.

1st Series.—Thirty prostates removed and carefully dissected by the author, from the bodies of individuals at and above the age of 60 years, as they consecutively appeared in the dead-house of a large metropolitan institution, and exhibited at the Royal Medical and Chirurgical Society in 1856, together with twenty others, for the purpose of showing that hypertrophy was the exceptional and not the prevailing condition of the prostate in advanced age. Trans. vol. xl.

2nd Series.—One hundred prostates, similarly treated by Dr. Messer, at the Royal Naval Hospital, Greenwich; the account of which appeared at the Royal Medical and Chirurgical Society in 1860. Trans. vol. xliii.

3rd Series.—Thirty-four prostates, presented to the Royal College of Surgeons, by the author, in illustration of the Essay on the Healthy and Morbid Anatomy of the Prostate, which obtained the Jacksonian prize for the year 1860. These were removed by Drs. Fisher and Davis, of the Royal Naval Hospital, Greenwich, and transmitted to the author, by whom they were dissected and examined.

In treating the above 164 examples of prostates, from subjects at and above the age of 60, no selection whatever has been made; the object having been to obtain them from average lives, occurring in that class of the community which is met with in such institutions.

4th Series.—Twenty prostates, from persons mostly of middle age, all being below 60. These were also dissected and exhibited by the author, at the Medical and Chirurgical Society, in 1856; and they furnish part of the data from which the size and weight of the healthy prostate were deduced.

5th Series.—Ten prostates, removed by the author from healthy subjects of middle age—35 to 57 years inclusive—not furnished consecutively from any one source, and therefore not classed with the foregoing series. They were examined by division in the middle line, to elucidate an anatomical question discussed at page 25.

1st Series.—Nos. 1 to 30.

No.	Age.	Weight.		Remarks.	Length.	Breadth.	Thickness.
		drs.	grs.		inch.	inch.	inch.
1	79	4	48	Healthy	1·4	1·45	·85
2	85	4	44	Ditto	1·25	1·55	·95
3	63	4	35	Ditto	1·35	1·7	·7
4	90	4	58	Ditto	1·25	1·85	·85
5	66	4	27	Ditto	1·4	1·5	·65
6	63	4	3	Ditto	1·35	1·5	·65
7	79	4	2	Ditto	1·4	1·6	·55
8	70	4	55	Ditto	1·4	1·8	·7
9	66	4	56	Ditto	1·4	1·7	·75
10	74	4	4	Ditto	1·3	1·6	·7
11	61	4	16	Ditto	1·3	1·6	·8
12	64	5	2	Ditto	1·4	1·75	·9
13	73	4	48	Ditto	1·6	1·8	·6
14	66	4	35	Ditto	1·5	1·75	·6
15	66	4	6	Ditto	1·5	1·65	·6
16	65	4	24	Ditto	1·4	2·0	·6
17	61	4	54	Ditto	1·3	1·75	·75
18	71	6	40	Hypertrophied; affecting the whole organ pretty equally	1·65	1·9	·9
19	72	5	18	Left lobe larger than right; not hypertrophied	1·3	1·7	·9
20	62	7	15	Hypertrophied, drawn at Plate II., showing commencing enlargement of Median portion	1·8	1·7	·85
21	74	12	30	Hypertrophied: full of tumours. Engraved at Plate VIII.	2·25	2·4	1·75
22	73	7	15	Hypertrophied	1·5	2·0	·8
23	61	7	25	Ditto	1·5	1·9	1·0
24	64	6	20	Ditto, small tumours	1·5	1·9	·75
25	75	9	50	Ditto, full of small tumours	1·7	2·1	1·0

No.	Age.	Weight (drs. grs.)		Remarks.	Length. (inch.)	Breadth. (inch.)	Thickness. (inch.)
26	79	4	36	No hypertrophy; but full of small tumours	1·3	1·75	·8
27	65	6	18	Small tumour, developing in median portion	1·5	1·7	1·0
28	62	6	5	Ditto, and in lateral lobes	1·4	1·9	1·0
29	73	2	58	Atrophied	1·1	1·8	·6
30	60	2	57	Ditto	1·2	1·45	·55

2ND SERIES.—NOS. 31 TO 130.

No.	Age.	Weight (drs. grs.)		Remarks.
31	76	3	35	Tumours present
32	76	3	20	Ditto
33	71	3	50	Ditto
34	87	3	30	Ditto
35	79	2	30	Abscess
36	80	3	5	Median portion enlarged
37	67	3	20	
38	69	2	40	
39	85	2	40	
40	77	3	40	
41	73	3	20	
42	82	3	20	
43	71	3	40	
44	75	3	50	
45	76	3	40	
46	67	2	20	
47	70	3	20	
48	75	3	20	
49	83	2	5	
50	74	3	0	
51	81	5	0	Tumours present
52	82	5	5	Ditto
53	68	5	50	Ditto
54	78	5	30	Ditto
55	82	5	0	Ditto
56	71	5	30	Ditto
57	69	5	0	Ditto
58	87	5	10	Ditto
59	85	4	30	Ditto
60	81	6	0	Ditto
61	71	4	30	Ditto
62	76	5	30	Ditto
63	82	5	50	Little enlargement of median portion
64	85	5	30	Ditto
65	72	4	30	Abscess
66	80	4	20	Ditto

Dr. Messer has placed these, from 31 to 50 inclusive, as prostates under 4 drachms, and therefore to be regarded as atrophied. I think there is scarcely evidence for so considering them; and can regard only Nos. 35, 36, 38, 39, 46, 49, and 50 as *unquestionably* atrophied, judging from the single fact of Weight. See Chap. XIII., on "Atrophy of the Prostate."

TABULAR VIEW OF FACTS

No.	Age	Weight (drs)	Weight (grs)	Remarks		
67	67	5	40	Healthy		
68	78	5	0	Ditto		
69	76	4	0	Ditto		
70	76	4	40	Ditto		
71	66	5	20	Ditto		
72	84	5	0	Ditto		
73	85	4	30	Ditto		
74	74	4	40	Ditto		
75	94	5	0	Ditto		
76	75	4	30	Ditto		
77	61	4	20	Ditto		
78	81	5	20	Ditto		
79	76	5	30	Ditto		
80	84	5	20	Ditto		
81	81	4	20	Ditto		
82	76	4	10	Ditto		
83	64	4	0	Ditto		
84	81	5	0	Ditto		
85	84	4	40	Ditto		
86	82	4	50	Ditto		
87	75	5	15	Ditto		
88	60	5	0	Ditto		
89	73	4	30	Ditto		
90	71	5	30	Ditto		
91	73	5	0	Ditto		
92	68	4	0	Ditto		
93	70	4	0	Ditto		
94	63	5	0	Ditto		
95	77	6	0	Ditto		
96	72	28	0	Hypertrophied. All lobes enlarged.		Tumours present.
97	77	9	20	Ditto	ditto	ditto
98	77	6	20	Ditto	ditto	ditto
99	63	14	0	Ditto	ditto	ditto
100	80	19	30	Ditto	ditto	ditto
101	86	7	20	Ditto	ditto	ditto
102	87	7	30	Ditto	ditto	ditto
103	75	16	0	Ditto	ditto	ditto
104	78	33	0	Ditto	ditto	ditto
105	71	10	0	Ditto	ditto	ditto
106	70	10	0	Ditto	ditto	ditto
107	76	10	0	Ditto	ditto	ditto
108	72	17	0	Ditto	ditto	But no tumours.
109	79	12	0	Ditto	ditto	abscess.
110	81	6	35	Ditto	ditto	tubercle.
111	79	30	0	Ditto	ditto	Tumours present.
112	76	24	0	Ditto	ditto	ditto
113	78	30	0	Ditto	Lateral lobes chiefly enlarged.	ditto
114	76	48	0	Ditto	ditto	ditto
115	74	25	0	Ditto	ditto	ditto
116	62	6	15	Ditto	ditto	ditto
117	74	7	10	Ditto	ditto	ditto
118	78	7	5	Ditto	ditto	ditto
119	77	6	15	Ditto	ditto	ditto
120	64	14	0	Ditto	ditto	ditto

No.	Age.	Weight.		Remarks.
		drs.	grs.	
121	84	26	30	Hypertrophied. Lateral lobes chiefly enlarged. Tumours [present.
122	75	7	0	Ditto ditto ditto
123	71	9	30	Ditto ditto ditto
124	77	9	30	Ditto ditto ditto
125	60	14	0	Ditto ditto Tumours and abscess.
126	81	12	0	Ditto ditto No tumours.
127	67	17	0	Ditto left lateral and middle lobes only enlarged ditto
128	74	9	10	Ditto right lateral lobe chiefly do.—tumours and abscess
129	81	8	30	Ditto left lateral lobe only do. { general hypertrophy—abscess.
130	80	8	30	Ditto middle lobe chiefly do. tumours present.

3RD SERIES.—Nos. 131 TO 164.

No.	Age	drs.	grs.	Remarks
131	85	5	40	Healthy
132	76	6	0	Ditto
133	83	5	25	Ditto
134	79	4	45	Ditto
135	80	6	0	Ditto
136	75	6	0	Ditto
137	82	6	10	Ditto
138	84	6	40	Ditto
139	61	6	20	Ditto
140	64	5	0	Ditto
141	68	6	10	Ditto
142	65	5	30	Ditto
143	62	5	10	Ditto
144	80	6	15	Ditto
145	87	3	50	Ditto
146	93	4	7	Ditto
147	80	3	18	Ditto
148	75	5	25	Ditto
149	74	4	46	Ditto
150	90	2	50	Atrophied—died of carbuncle.
151	78	2	45	Atrophied—died of phthisis.
152	80	10.	0	Hypertrophied.
153	81	8	10	Ditto
154	83	9	15	Ditto
155	80	8	30	Ditto
156	79	9	6	Ditto
157	84	18	0	Ditto
158	73	10	18	Ditto
159	80	7	6	Ditto
160	77	8	50	Ditto
161	75	7	30	Ditto
162	76	7	50	Ditto
163	78	11	17	Ditto
164	72	12	45	Ditto

Of these thirteen specimens:—

In three, hypertrophy affected the whole organ pretty equally.

In three, it affected the lateral lobes, mainly and equally.

In four, it affected both lobes; but the right more than the left.

In two, it affected both lobes; but the left more than the right.

In one, it affected all parts; but the median portion chiefly.

Five were the subjects of tumour.

TABULAR VIEW OF FACTS, ETC.

4TH SERIES.—Nos. 165 TO 184.

No.	Age.	Weight.		Remarks.	Length.	Breadth.	Thickness.
		drs.	grs.		inch.	inch.	inch.
165	42	3	37	Healthy	1·3	1·4	·6
166	47	4	57	Ditto	1·8	1·7	·65
167	47	5	33	Ditto	1·4	1·7	·9
168	50	4	34	Ditto	1·5	1·7	·65
169	54	4	8	Ditto	1·25	1·8	·75
170	52	4	13	Ditto	1·5	1·75	·6
171	54	4	37	Ditto	1·5	1·75	·7
172	54	4	50	Ditto	1·5	1·7	·7
173	56	3	56	Ditto	1·25	1·75	·75
174	21	3	34	Ditto	1·3	1·4	·7
175	40	4	30	Ditto	1·55	1·6	·75
176	50	5	20	Ditto	1·55	1·75	·8
177	50	4	46	Ditto	1·4	1·9	·55
178	46	5	4	Ditto	1·5	2·0	·55
179	55	5	30	Ditto	1·3	1·8	·85
180	54	4	48	Ditto	1·4	1·75	·7
181	56	5	22	Small tumours, not hypertrophied	1·55	1·75	·75
182	56	6	0	Slight general enlargement	1·3	1·8	1·0
183	50	5	50	Small tumour developed in median portion	1·3	1·8	1·0
184	21	2	46	Atrophy — emaciation from phthisis	1·25	1·3	·55

5TH SERIES.—Nos. 185 TO 194.

No.	Age.	Weight.		Remarks.
		drs.	grs.	
185	35	4	30	Healthy.
186	39	4	36	Ditto
187	41	4	50	Ditto
188	44	4	10	Ditto
189	45	5	6	Ditto
190	45	4	40	Ditto See page 25.
191	51	4	56	Ditto
192	52	4	50	Ditto
193	53	5	36	Ditto
194	57	5	0	Ditto

THE ANATOMY AND DISEASES OF THE PROSTATE.

PART II.—THE DISEASES OF THE PROSTATE.

The diseased conditions to which the prostate is liable will be considered in the order presented below. All the deviations from health are enumerated which an extended study of morbid anatomy, as well as that of the signs and symptoms manifested during life, indicate as liable to affect the organ.

Inflammation of the Prostate : Prostatitis.
        „   „   Acute.
        „   „   Chronic.
Secondary results of the inflammatory process:—
        Suppuration.
        Diffuse.
        Abscess.
        Ulceration.
Hypertrophy of the Prostate.
Simple Tumours and Outgrowths.
Atrophy.
Cancer of the Prostate.
Tubercle of the Prostate.
Cystic Disease (?).
Calculi of the Prostate.

CHAPTER III.

INFLAMMATION OF THE PROSTATE: ACUTE AND CHRONIC.

Acute Prostatitis. — Causes. — Morbid Anatomy. — Symptoms. — Treatment. — Chronic Prostatitis.—Causes.—Morbid Anatomy.—Symptoms.—Treatment, in its various Forms and Stages.

ACUTE INFLAMMATION OF THE PROSTATE. — This is by no means a common affection, if regarded as distinct and unassociated with inflammation of the urethra or bladder. When the latter organ is inflamed, the prostate appears sometimes to suffer, although in a secondary manner and degree. But in obedience to the common law which seems to apply to tracts of mucous membrane generally, inflammation appears commonly to travel from the external to the deeper parts. Accordingly, an attack of urethritis, involving the external inch or two of the urethra, may spread inwards and fix itself, as I believe is not unfrequently the case, upon that portion of the canal which is most largely surrounded by vascular tissue, viz. the bulbous portion. Hence, probably, the origin of stricture-formation affecting that locality especially. But it may proceed, as an exceptional occurrence, more deeply still, and then the prostate becomes the subject of inflammatory action. Such is the most common mode by which this organ is involved. It may not be out of place to observe, that in the respiratory tract the same line of march from the external to the internal parts is observed. A catarrh, for example, is the first sign of inflammation in the mucous membrane of this region, the action may gradually spread to the throat, larynx, bronchi, and so on to the lung-substance itself. The converse order of things we do not observe; and after the same manner also in the genito-urinary tract, we find that inflammation is to be traced, as a rule, from the urethra to the prostate, bladder, ureters, and last of all, it may be, to the kidneys. Sometimes, however, the prostate is inflamed apparently as a purely idiopathic occurrence, and not from continuity of tissue with adjacent parts. This, excepting the cases produced by violence, as by instruments, &c., is probably extremely rare.

CAUSES.—Systematic writers on this subject enumerate many circumstances as giving rise to acute inflammation of the prostate. The relation between some of these and the supposed effect appears, however, to be less clear than the precise statements generally made might lead the inquirer to imagine.

The alleged causes may be arranged for consideration in three classes, as follows :—

(*a.*) Undoubted causes of acute prostatitis.

The pre-existence of acute inflammation of the urethra of any kind, but especially the gonorrheal, by continuity, as already alluded to. Urethral stricture, in an aggravated form, tending as it does to the production of inflammation and disorganisation of all the parts posterior to it, especially those more immediately adjacent, as the prostate and bladder. The direct application of irritating agents in the shape of strong injections, cauterization, and mechanical violence of various kinds. Inflammation of the bladder sometimes. Calculi of the bladder. The application of cold and damp to the perineum, as by sitting for a long period on moist ground, perhaps most frequently in gouty and rheumatic subjects. Urethritis has been referred to as a proximate cause, but it may also be the remote cause in the circumstances last enumerated, as well as in some of those which come under the next head.

(*b.*) Circumstances which cannot be stated with absolute certainty to be causes, but which may, with some amount of probability, be so regarded.

Horse-exercise is constantly said to be a cause of acute inflammation in the organ, by means of the concussion occasioned. Evidence is wanting, I think, to establish this. That it may aid in producing it when some inflammation of the urethra already exists, is quite possible. Certainly it cannot be said that hard riders, such as huntsmen, jockeys, but above all the cavalry soldier, from the nature of his seat, are in any notable degree more subject to it than are other men equally exposed to other and better recognised sources. Cantharides may, perhaps, occasionally act as a cause when taken internally, but probably never without primarily affecting the kidneys and also the bladder. Alcoholic drinks, especially when mixed with acids, as punch, may induce prostatic inflammation, gonorrhœa already existing, but only on this condition. Inordinate sexual intercourse, under the last-named circumstances, may probably be assigned also to the present category.

(c.) Circumstances stated to be causes by numerous authors, but respecting which there is either little or no evidence to render it probable that they are so.

Diuretic medicines, copaiba, cubebs, and turpentine, even coffee, and highly-seasoned dishes, are said to cause inflammation of the prostate. Drastic purgatives are similarly regarded by some. Irritations in the rectum, by ascarides, hæmorrhoids, &c., are enumerated as causes. All these may, and do, undoubtedly sometimes induce an irritable condition of the bladder, and some of them, perhaps, even some degree of inflammation of the viscus; but I am not aware of any authenticated case of acute prostatitis directly or indirectly occasioned by any of these agents. The morbid condition of the organ which exists in the presence of carcinomatous infiltration, or of tuberculous deposit, cannot be regarded as by any means identical with the affection now under consideration. Nevertheless, the diseases referred to are commonly enumerated as causes. Such a course appears to involve confusion, and to destroy the definite meaning of terms, which it is extremely desirable in all pathological studies to maintain distinct and inviolate as far as possible.

Sedentary habits are spoken of as a cause of the affection, but without the slightest shadow of evidence to support the assertion, as far as I have been able to discover. The same also may be said with regard to a constipated habit of body. It is probable that want of exercise and a torpid state of the bowels, both tend to induce a loaded condition of the veins of the abdomen, and among them those of the prostate; and in this manner a mechanical congestion of these vessels, which are prone to be large and dilated, as well as of the proper capillaries of the organ, is doubtless favoured, but how far this may be considered causal, even in a secondary or predisposing relation, of acute inflammation, it might not be very easy to state.

MORBID ANATOMY.—It is very rare to obtain an opportunity of studying the morbid anatomy of acute prostatitis, when it forms the primary, or the chief affection. When it occurs as a consequence of prior disease of the bladder or of the urethra, it is still not usual that the results can be examined anatomically, during or immediately after the acute stage. By far more common is it to meet with the organ when chronic inflammation or suppuration are present, and when it is presenting none of the phenomena indicating the presence of acute inflammatory action.

Nevertheless, these phenomena have been seen and noted. I have observed and recorded them myself in one instance.

They will be regarded under two heads.

a. The pathological changes which take place in the early and middle stages of acute prostatitis; conditions which do not necessarily tend either to extensive suppuration or to disorganization, but which may, and often do, end in resolution.

The organ is swollen to double or even to quadruple its natural size, and feels tense and firm to the hand. The external bloodvessels are loaded with dark blood. On making an incision from the front aspect into the urethra, the mucous membrane is seen to have a somewhat deeper tint than natural, although there is less change here than one might expect; the cut surface is redder also than in health. Pressure causes to exude a reddish, semiturbid fluid, a mixture of effused lymph and serum, of blood from loaded capillaries, of prostatic secretion, and a very small quantity of pus, that is to say, if the fluid is examined under a microscope some pus-globules will be observed. Making an incision through the lateral lobes, the same fluids exude, but in somewhat greater abundance than from the anterior part just examined. As the inflammation advances, the fluid contains a larger admixture of pus, and a section of the lateral lobes especially shows several minute spots of thickish pus, not abscesses apparently, strictly speaking, but the gland-crypts, whose lining cells are now secreting pus, and whose cavities are distended with it. Thus, as the morbid action continues, these phenomena pass insensibly into others, which indicate more permanent and organic changes in the organ, and which generally follow the preceding, if resolution does not take place. It is convenient to arrange these phenomena under the head of—

β. Pathological changes which occur in the later stages of acute inflammation of the prostate.

The continuance of inflammatory action, of the secretion of purulent matters, and of the organization of effused lymph, lead to the formation of isolated deposits of pus, or small abscesses in the substance of the prostate. These are found existing, either numerous and small, from the size of a grain of pearl barley to that of a pea, for example; or few, or even single, but larger than the preceding: in the latter case, sometimes, occurring from the coalescing of the smaller deposits, and the destruction of partitions of intervening tissue. It is to be observed that the pus of a prostatic abscess has generally, if not always, a peculiar character, namely, a glutinous or adhesive quality, and so differing from the diffluent creamy character of ordinary healthy pus. Sometimes clots of blood are found, the result of small hæmorrhages into the diseased gland-crypts, or other cavities. Portions of the organ may become softened, and be acquiring the condition of sphacelus; others have already become gangrenous. The mucous membrane of the prostatic urethra is reddened, sometimes thickened and velvety; or it has patches of whitish and membrane-like material closely adhering to its surface, exudations of organized lymph, the product of inflammatory action. Again, portions of the mucous membrane may be destroyed by ulceration or gangrene, and form the orifice of a cavity or cavities already described as occupying the substance of the prostate itself.

All these conditions I have repeatedly observed and noted in the examinations of the dead-house, after the existence of advanced acute prostatitis, usually, though not invariably, consequent on disease primarily occurring in allied and neighbouring organs, and implicating the prostate secondarily. They will be traced further in describing, hereafter, the pathological anatomy of prostatic abscess.

SYMPTOMS.—At the outset, a sensation of weight and fulness about the rectum and perineum is experienced, with some pain and uneasiness referred to the neck of the bladder. The patient requires to pass water more frequently than natural, and does so with an increase of the existing pain, especially at the close of the act. These symptoms increase; the pain becomes severe, then lancinating and pulsatile, and almost continuous; a sense of tension and swelling is experienced, and the anus and perineum are tender when pressed upon. Movements of the body become difficult on this account, as does also the sitting position. The act of relieving the bowels at stool produces considerable distress; still more so does the act of micturition; the stream of urine being generally small, and its passage necessarily prolonged, much straining accompanies it, and the pain is exquisitely acute. In these circumstances a finger introduced into the rectum encounters much opposition, however quietly it is carried through the sphincter; the anterior wall of the bowel is prominent, hard, and hot, and the outline of the prostate may be traced, not, however, without causing great suffering to the patient. An attack of piles may be induced, the close contiguity of the hæmorrhoidal and prostatic veins appearing to favour this result. At a later stage, if suppuration has taken place, the rectal swelling is softer, local throbbing is experienced, and should a catheter be passed, the patient will complain of excessive pain when the instrument reaches the prostatic part of the urethra. General fever, in a greater or less degree, manifests itself after the accession of the earlier local symptoms, rigors and exacerbation accompanying the onset of suppuration. Pains in the back and loins, as well as in the glans penis, and running down the thighs, are experienced, and not unfrequently a sensation of constant desire to go to stool. The mucous membrane of the bladder participates more or less in the inflammation; the urine is febrile in character, and generally contains mucus to some amount, occasionally in considerable quantity, the latter condition probably indicating some implication of the bladder. Besides this, there may be pus in the urine to a greater or less extent, from which it is deposited as a sediment on standing.

The chief signs upon which a diagnosis depends may be noted as follows: enlargement of the prostate, ascertained by rectal examination, the prosecution of which is extremely painful to the patient, with acute pain complained of when pressure of the finger is made upon any part of the swelling there. The act of defæcation is often productive of much distress; that of micturition still more so. If a catheter is introduced, exquisite suffering is caused when it arrives at the prostatic part of the urethra. Added to these there is constant and deep-seated, often throbbing, pain felt about the fundament. These symptoms alone, but especially when associated with a history of recent urethral discharge, which may have previously ceased or not, will suffice to determine the nature of the case.*

TREATMENT.—It is unnecessary to enter here into the full detail of that which constitutes the ordinary constitutional treatment of inflammation. In general cases the antiphlogistic regimen and diet, as commonly understood, should be observed. I have seen no reason to bleed from the arm, or to give calomel—except

---

* Some exceedingly well-detailed cases of acute prostatitis are given by Vidal de Cassis, who has paid special attention to this affection, in the Annale de Chirurgie. Paris, 1844.

as a purge—but prefer the administration of antimony and full doses of alkali, such as the bicarbonate or acetate of potash, to the extent of about two or three drachms during the twenty-four hours, if the former, and three or four, if the latter be employed. I have occasionally almost doubled these quantities in a strong subject, and I think with advantage. The dose of antimony may vary between the eighth and the sixth of a grain, to be repeated very frequently, say, not less than every two hours if the inflammation is severe. The bowels should be freely opened at the outset, and a gentle action upon them maintained afterwards. The hæmorrhoidal and prostatic veins very freely intercommunicate, and are so commonly found after death loaded with blood in most cases of inflammatory disorder affecting the urinary organs, that there can be no doubt as to the propriety of ensuring, as far as in this manner may be possible, a free circulation through the abdominal viscera.

The local treatment demands a more special notice. Bleeding from the neighbourhood of the affected part often affords greater relief than any other single agent, and the mode of abstracting blood must depend in some measure upon circumstances. In most cases the application of leeches is the most available means. They should be employed in large numbers, as from ten to fifteen, on the perineum and around the anus, the latter, perhaps, being the more important situation. An exceedingly expert cupper will, with no very great pain, and in a short time, obtain eight or ten ounces of blood from the perineum; but it is very rare to find any but the most dexterous and practised professors of cupping successful in accomplishing this, and in default the leeches will certainly be preferable. Leeches to the rectal surface of the organ, introduced by means of tubes devised for the purpose, have been recommended by some, but the effect to be produced in this way must be small, as one or, at most, two leeches can be applied at a time.* The bleeding over, a hot hip-bath

---

* The instruments which have been employed for this purpose are described and figured in the *Lancet*, Vol. xxxix. p. 645, and vol. xl. p. 299.

should be taken, for a few minutes only, then a large poultice or hot flannel placed on the perineum, and the patient be wrapped up warmly in bed. The hip-bath may be frequently repeated with advantage in the course of the treatment, but should never be applied at any one time for a lengthened period. From six to eight minutes is the longest time I think it right to permit a patient, who is suffering from inflammation of the prostate or bladder, to sit in the bath, which should commence at about 100°, and be raised to about 103° or 105°, during the period named. Such a method of using this most valuable agent appears to me much more advantageous than prolonging the sitting to fifteen or twenty minutes. The object of the bath is not to induce a flow of blood to the pelvic viscera, but, on the contrary, to expand and fill the vessels of the skin by a smart impression quickly made upon it, one which is also participated in to a certain extent by every part of the cutaneous surface. And it is by using the bath in the manner described that a general diaphoresis is effected, and a temporary congestion of the pelvic *surface* produced, with the result of relieving that of the deeper parts. During a prolonged hip-bath, it often happens that the perspiration at first induced becomes checked, congestion of the pelvic viscera is rather encouraged than reduced, and of course no relief or benefit, but, on the contrary, some uneasiness, or even injury, may be occasioned by this abuse, rather than use, of an admirable remedial agent.

In the course of a few days, the severe pain and the frequent micturition gradually subside, as a rule, often, however, not without the occurrence of an occasional relapse for a day or two, sometimes, indeed, even of a severe character. Such untoward circumstances are generally attributable to some slight indulgence on the part of the patient, especially to exercise too freely taken, or at a too early period. Hence it is necessary to restrain with firmness the activity which a young and generally hearty patient is ready to display as soon as his complaint begins to recede. Care and moderation in diet, with total abstinence from alcoholic stimulant must be enjoined for a time, and the season of convalescence should especially be watched and appropriately managed, as means must now be taken, not merely to restore the general health on the ordinary principles of nutritive and tonic regimen, but to reduce the bulk of the organ, enlarged as it is prone to remain from the effect of the inflammatory process. Speaking in general terms, the prostate will be found on examination after the lapse of about a month from the subsidence of the acute symptoms, as large as it was then, tender, but not by any means exquisitely so, firm and resisting to the finger, and, if necessity exists for passing a catheter, it will be found necessary to depress the handle of the instrument to a greater extent than in a healthy subject before the urine flows, while the operation is still attended with no little uneasiness, especially when the instrument traverses the prostatic portion of the urethra. There is more than the natural amount of exertion required in order to empty the bladder, and the stream is propelled with less than the usual degree of force. Such are the only effects remaining when the acute affection terminates in a healthy resolution. Unassisted by treatment, it is very long before the enlargement subsides; it is doubtful, indeed, whether in such circumstances it does ever do so completely.

CHRONIC INFLAMMATION OF THE PROSTATE.—The affection which I recognize under this title is by no means an uncommon one, even in its simple uncomplicated form; and when it appears as a secondary condition to existing disease of the bladder and urethra, it is unquestionably one of frequent occurrence. Nevertheless, its existence is barely recognized by some of the best known writers on prostatic disease, and by some it is not even named.

Chronic prostatitis is met with in three different phases; but the symptoms and pathological characters are much the same in each, and differ in degree rather than in kind. It may solely originate in an attack of acute prostatitis, and be due to a morbid persistence of unhealthy action, which shows no disposition to diminish after the acute symptoms have subsided; or, secondly, there may be a long and tedious resolution, naturally leading, but by slow steps, to the re-establishment of healthy action; and, finally, the condition may commence in the chronic form independently of any acute attack, in which case it may be the primary or sole existing complaint, or it may be dependent on disease of adjacent organs.

It is exceedingly common to meet with instances in which simple chronic inflammation of the prostate, producing enlargement, is regarded as an example of hypertrophy. Yet nothing can be more distinct than the two affections, if we compare their pathological history and characters. Chronic inflammation, however, by no means necessarily causes enlargement of the organ; indeed it is rather exceptional than otherwise to meet with it, all varieties considered. But when inflammatory enlargement does exist, it is almost invariably in the early or middle periods of life; while hypertrophy never occurs before the fiftieth, very rarely before the fifty-fifth year, and is not commonly manifested by symptoms before the sixtieth year. Inflammatory enlargement is almost invariably preceded by some urethral inflammation. Urethral discharge of a purulent nature, urine containing small flocculi, and pain during and after micturition, have been or still continue present. Associated with these, there is almost always an impaired condition of the general health. All these may be, and usually are, absent in hypertrophy during its earlier stages. Finally, inflammatory enlargement is due to the effusion of morbid products, lymph, pus, &c., into the substance of the organ; while a hypertrophic enlargement is due, as the term implies, to simple overproduction of the normal constructive elements of the prostate gland itself.

CAUSES.—The most fertile cause of chronic prostatitis is gonorrhœal inflammation, which has extended backwards and affected, more or less acutely, the prostate. Local cold and damp must be recognised as occasionally producing it; more rarely still it is due to mechanical injury inflicted on the urethra or on the perineum. Long continued indulgence in venereal excesses of any kind is undoubtedly a cause. That form which results from long standing and severe stricture of the urethra, of chronic cystitis, of calculus, either vesical or prostatic, is common enough, and needs no separate consideration; it is the mere result of existing adjacent disease, upon which it altogether depends.

MORBID ANATOMY.—I have examined several prostate glands which manifested a pathological condition resembling that which, when observed in other organs, obtains the name of chronic inflammation; such, for example, as occurs in lymphatic glands, superficial and deep, in the mesenteric glands, in the tonsils, in the uterus, and in the mucous membranes of the bronchi, stomach, intestine, and bladder. The prostate, so affected, may be found larger, or even smaller, than the natural size; for there is no necessarily constant deviation in this respect; the consistence, if any difference exists, is less firm, the texture is more open and spongy. The colour of the cut surfaces has a more dusky hue, sometimes with a redder tint. More fluid than natural is found in the gland-tissue, and freely issues on being pressed. This fluid is of a dirty hue, and if firm pressure is made, appears to be stained slightly red. In advanced cases, deposits of pus are found, varying from the size of a grain of pearl sago to that of a pea, but they are few in number, perhaps only one or two are seen, thus contrasting with the numerous disseminated small abscesses met with in the later stages of acute inflammation. The mucous membrane may be thinned and more vascular than natural, with the duct-orifices large; this being the case if there has been dilatation of the prostatic urethra as sometimes happens behind a stricture: on the other hand, it may be coated in places with organised lymph, giving it a roughened and opaque appearance. Or its own structures may be thickened, not reddened as in acute inflammation, but presenting after death a dull ashy gray, or even slaty hue, the last-named marking long persistence of the unhealthy action. Pus is, in such cases, often found filling the sinus pocularis, the ducts entering the urethra and sometimes a cavity filled with it, communicates also with the urethra, a chronic abscess in the prostate, or if not in it, it is not uncommon to find abscesses surrounding it; in other words, peri-prostatic abscesses depending on foregone disease of the prostate itself.

SYMPTOMS.—A patient who suffers from simple uncomplicated chronic inflammation of the prostate, complains of a little undue frequency in making water, of some muco-purulent discharge from the urethra, of dull pains in the perineum and about the anus, sometimes occasional, sometimes persistent, but almost always increased by exercise; often of pains in the thighs and legs, or in the sacral region; sometimes increased, but not invariably, by sexual intercourse. There is usually no pain in micturition until the end of the act, when it is occasionally, but by no means always, felt, and then it is never very acute like that of calculus. There is tenderness in the perineum, sometimes felt in the sitting posture; tenderness in the prostate itself to rectal examination; an irregularity in form is sometimes detected by the finger, but this is exceptional; and there is not necessarily any enlargement. The passing of a catheter gives more than usual pain when it traverses the prostatic urethra and neck of the bladder. The urine is a little cloudy, but on examination this condition is found to be mainly due to shreds of tenacious muco-purulent matter, and masses of epithelium, which have their origin in the prostatic urethra and not in the bladder, as may be ascertained by desiring the patient to pass water into two glass vessels, the first ounce or so into one, the remainder into the other, when all the purulent matter will be found in the former portion, while the latter is clear.

Further inquiry will frequently discover that the patient has little or no sexual desire, and he may or may not be the subject of frequent involuntary seminal emissions during sleep. The health is mostly impaired, and general debility is complained of.

TREATMENT OF SIMPLE CHRONIC PROSTATITIS.—When the dull pains about the perineum are pretty constant, and there is obvious tenderness of the prostate, both of these conditions being increased by exercise, I find nothing so serviceable as counter-irritation at the surface of the perinoum. But it is necessary to continue it for a period of four to six, or even sometimes of eight weeks, and if thus persevered in, benefit is almost certain to follow. It may be accomplished by rubbing a moistened stick of nitrate of silver on the skin in front of the anus, over the bulb of the urethra, so as to blister a portion of about 1½ inch long by 1 inch wide. The method, however, which I prefer, and now always employ, is an application to the part described of strong "acetum cantharidis," or of the ethereal solution of cantharides, the soreness to be kept up by the daily application of a piece of blistering-paper, or by occasional reapplications of the fluid originally used. It occasions some trouble to the patient at first, but by keeping applied to the part an abundant supply of simple ointment, little pain is experienced after the first blistering. In applying the blistering-fluid, care must be taken not to allow it to run downwards to the margin of the anus, nor should it come into contact with the scrotum.

At the same time the state of the digestion must be improved, the vigour of the system be promoted by tonics and generous diet, and exercise is to be permitted and increased by degrees as soon as the power of taking it without inducing pain is acquired.

The digestive organs being in tolerable condition the use of iron is almost always attended with benefit. The sulphate, combined with sulphate of quinine, and made into pills with some extract of rhubarb, and a little extract of nux vomica, is an advantageous form of tonic for these cases, in which it is desirable to maintain regular action of the bowels, impaired as it usually is by inability to take much exercise. A little of the watery extract of aloes can be combined, if it is necessary to employ more decided aperient action. The tincture of the sesquichloride of iron, is also a very efficient remedy, provided the bowels are not permitted to become unduly constipated. I prefer these forms of iron to any other, including the recently introduced hypophosphites, for the great majority of patients suffering from this affection. When the predominant symptom is frequent nocturnal emission, and there is much pain felt in passing an instrument through the prostatic urethra, nothing acts so beneficially, in many cases, as the application of a solution of nitrate of silver, commencing with about five grains to the ounce, and increasing it to twenty grains if necessary, by means of a perforated catheter containing a piston by which the fluid can be set free at the proper spot, and its action mainly limited to it; a matter which it is important to accomplish pretty accurately. In performing this little operation, the following course should be pursued. The bladder should first be emptied; and the opportunity of doing so should be employed to determine the exact length of the urethra, in the usual manner, while the urine is flowing. The instrument containing the caustic solution should then be passed immediately, the solution being discharged as soon as the perforated extremity has arrived within the prostatic urethra, the situation of which may be correctly inferred from the known length of the urethra, as well as by the undue sensibility which the part possesses. The immediate results usually are, repeated wants to pass water, which are painful and sometimes attended with slight bleeding; these subside in 24 hours, and possibly are succeeded by a little purulent discharge for a day or two. During the first few days it is not uncommon that the symptoms originally complained of increase, but afterwards they gradually diminish. Should this, however, not be the case, no fresh application of the caustic should be made until three or four weeks have elapsed since the previous one, as it is impossible to ascertain the effect in a shorter period of time. If necessary it may then be again applied as before, and with stronger solution. A third or a fourth application may be necessary. I have, however, never continued to employ it if not successful on the fourth occasion. Its success mainly depends on applying it freely and accurately to the prostatic urethra, and when this is ensured, it generally proves a very valuable remedy.

In these cases also, alterative and tonic treatment, such as that just described, and generous but non-stimulating diet are mostly required. It is scarcely necessary, perhaps, to add, that we are not to infer the existence of chronic prostatitis because the patient is the subject of seminal emissions to a morbid extent. This state may exist, and often does so, from altogether different causes, which, as they have no necessary connection with the prostate, or with any disease of it, will not be further considered here. I have referred only to the occurrence of this symptom when associated with others, which indicate the presence of a subacute inflammation affecting the organ.

When there appears to be little active disease of the prostate, and we have only to deal with the sequel of a severe acute attack, in which convalescence is progressing but very slowly, there is less indication for counter-irritation, and probably none for the employment of caustic in the urethra. One of the most prominent conditions in such a case, is the persistence of enlargement and induration of the organ, due to inflammatory deposit which took place during the acute stage. This is the enlargement which belongs to the period of youth, and differs totally from that which occurs in advancing years, and is known as senile hypertrophy.

This enlargement and induration may be reduced, and for this purpose it should be our aim to employ appropriate means without delay, inasmuch as the facilities for producing absorption of inflammatory exudations undoubtedly diminish in proportion to the length of time during which they have existed.

The general health being vigorous, those internal remedies which appear to possess a specific power to promote absorption of the effused matters should be employed. Such are the iodide and bromide of potassium. They may be administered conjointly or separately, the latter agreeing with the stomach sometimes when the former will not. Frequently either agent may be advantageously combined with fifteen or twenty grains of the acetate or bicarbonate of potash twice a day, or with two or three scruples of the tartrate of potash, if there is a constipated habit of body. These are the more indicated if the urine presents an

F

unduly acid, and therefore irritating character. If the enlargement is considerable, it is desirable to employ also local measures. No means should be left untried to restore the organ to its normal size, as it may deviate very considerably, and be productive of long-continued uneasiness in this condition. I have had recently under my care two patients, one at twenty-eight, the other at thirty-five years of age, with enlarged prostate; in the former it was very considerably so, and was caused by a severe attack of acute inflammation, which he experienced three years ago ; in the second, the enlargement, from the same cause, is so extreme as to render a rectal exploration even somewhat difficult from the great projection of the organ into the bowel. In both cases, the latter especially, the frequent occurrence of symptoms of irritability of bladder, and of pains in the perineum and loins, appeared to be due in great measure to this abnormal condition.

The local means which appear to be useful are, iodine suppositories, hip-baths, plain, and impregnated with iodine and bromine, in the same manner as advised in the eleventh Chapter. The treatment there recommended in order to promote absorption, may be followed, if the case is one for which the use of external as well as internal means is desirable. The local blistering of the perineum, as before described, is also sometimes useful. At the same time, everything must be avoided which might occasion derangement of the general health; there should be no unnecessary passing of catheters; walking exercise to be taken freely, but not riding on horseback, at all events for a considerable period after the acute attack. A considerable amount of perseverance in the use of these means is necessary, but very decided benefit may be anticipated from it.

Respecting chronic inflammation of the prostate, resulting from prior disease in associated organs, as stricture of the urethra, calculus, organic disease of the bladder or of the rectum, &c., it will be unnecessary to say more than, that treatment which is suited to the primary complaint will embrace all the necessities arising from the prostatic implication.

CHAPTER IV.

SUPPURATION OF THE PROSTATE.—ABSCESS, ACUTE AND CHRONIC.

Results of Inflammation.—Abscess.—Diffuse Suppuration.—ACUTE AND CHRONIC ABSCESS.—Morbid Anatomy of Acute.—Of Chronic.—Symptoms.—Treatment of Acute.—Chronic Abscess.—Cases.—ULCERATION.—Pathology of.

AN attack of acute inflammation may be followed not merely by chronic inflammation and induration, varying in degree and in tendency to persist, but, especially when neglected or occurring in a naturally weak or delicate constitution, in suppuration of the organ. And suppuration may take place either as an acute action, at a very early period in the case, or subsequently after a long persistence of chronic inflammation; constituting in the first, acute, and in the second, chronic abscess: it may also occur in a diffused manner among the tissues of the organ.

The existence of suppuration in a diffused form, and not in that of localized and limited deposits, is rare, and occurs only, as far as I have observed, in cases of acute inflammation affecting the bladder and prostate, and following long unrelieved retention of urine from organic disease, or surgical operations upon the urethra or bladder.

MORBID ANATOMY OF ACUTE AND CHRONIC ABSCESS.—It has been already mentioned that the prostate may be the seat of numerous small abscesses, apparently commencing in suppuration of the glandular crypts; that these may dilate, coalesce with neighbouring crypts, and so lead to the formation of cavities, varying from the tenth to the fifth of an inch in diameter. But abscess may form in the prostate by a different course of events.

Acute abscess of considerable size sometimes forms rapidly in acute prostatitis, and rapidly discharges itself, most commonly by the urethra. It is exceedingly rare to have an opportunity of dissecting such a case, because resolution takes place, or, at all events, the patient does not succumb, or if he ultimately does so, almost never during the acute stage of the complaint. Consequently, it is mainly from explorations made during life that we can ascertain anything of its morbid anatomy. The prostate examined by rectum feels full, tense, and much larger than natural. Suddenly, a discharge of matter by urethra taking place, it becomes somewhat smaller, and much less tense. Other signs, both objective and subjective, but not anatomical, existing to confirm the diagnosis of prostatic abscess, the fact of its existence in the substance of the gland is established; and its healing is equally so, since I have examined subsequently patients who have been so affected and have verified the fact of recovery, some undue hardness and increase of size usually existing for a very long period of time afterwards.

Recently, however, the rare opportunity of dissecting a case offered itself in one of the metropolitan hospitals. A man, aged 25, was admitted as a medical patient, with fever and pain in the loins, coming on within fourteen days from the commencement of a severe gonorrhœa. He was unable to pass urine, and his water was removed by the catheter. He gradually sank on the eighth day after his admission. At the post-mortem examination pus was found abundant in the urethra; the mucous membrane, however, was only moderately injected in places. A large abscess was found between the rectum and bladder, and continuity of passage existed between it and the floor of the prostatic urethra, where it opened by two large orifices close to the verumontanum. It had produced destruction of a considerable portion of the prostate gland itself, but parts not destroyed seemed tolerably healthy. (See Lancet, Oct. 27, pp. 408-9.) Whether, however, this abscess was truly prostatic at the outset, or periprostatic, it would still be difficult to decide. It largely involved the prostate, however, and is therefore illustrative of the present subject.

Prostatic abscess may also assume a chronic condition in a variety of ways.

1st. The acute abscess above described may not heal, but may assume, and continue in the chronic condition.

2nd. Abscess, never acute in its character, may slowly form in the prostate as the result of chronic inflammation in it, or in the neighbouring parts.

3rd. A tubercular deposit in the prostate may soften and be discharged, a chronic abscess resulting.

4th. Abscesses may occur from the irritation of a foreign body such as calculus, embedded in the prostate; or in the midst of malignant growths.

In examples of the first and second kinds I have seen, in a few instances, a cavity capable of holding from two to three drachms of fluid, situated in the prostate gland, and occupying either lobe, or, partially, both. The walls of the cavity are composed of a thick layer of organized material, with ragged flocculent surface, and pus adhering. The colour is grey or grey-slate, the same tint pervading a thin layer of the prostatic structure itself lying adjacent to the abscess, then shading off into the natural aspect. I have seen such a cavity, which was large and of long standing, possessing a dusky green hue. Sometimes a large proportion of the proper structure of the prostate has disappeared, as the effect of the abscess; indeed, occasionally, almost the entire organ has been destroyed under these circumstances. I possess two examples, in which nearly the whole of the prostatic substance being thus removed, the urethral canal traverses the hollowed gland, being continuous, nevertheless, with the membranous urethra on the one hand, and with the bladder on the other. This condition is shown in Plate XI., drawn from one of them when recent; the subject of it being that of a man under my care at the Marylebone Infirmary. The case is recorded in the Pathological Trans. Vol. V. p. 208, and also in this chapter, page 74, together with the other and almost precisely similar example of the same disease.

An excellent example of this somewhat rare condition exists also in the Museum of the College of Surgeons, Edinburgh, No. 2090, XXXII. c.

In an example of the third kind, the walls of the cavity present similar appearances, and it occupies usually portions of one or both lateral lobes; but portions of pale yellowish, or greyish-yellow, curd-like matter, mixed with a little pus, may adhere to the walls, or may, indeed, fill the cavity if not discharged during life. Both these conditions I have seen, but they are nevertheless of rare occurrence; see Chapter XV., on Tubercle of the Prostate.

Fourthly, abscess may be associated with prostatic calculi from the irritation produced by the foreign body. The membrane-like envelope which lines the cavity in which the calculus is situated, may become with inflammation a pus-secreting structure. Small collections of pus are commonly met with in making sections of encephaloid tumours of the prostate: but these need no special description.

SYMPTOMS.—The occurrence of suppuration, as has been before stated, may be suspected when after the first six or seven days the acute symptoms do not subside, when the pain and difficulty of micturition and defæcation increase, if rigors occur, and the patient is very restless and feverish, complaining of great tension, and of a pulsating sensation in the perineum and at the neck of the bladder. The fact is determined, if by rectal examination, the swelling there increases, and communicates to the tip of the finger a sensation of softness and elasticity, in place of the firmness and resistance which were noted before. The act of examining also, although necessary to be performed, is excessively painful; yet, by devoting time and care, it may be made very much less so than would otherwise be the case. Pressure in the perineum may also reveal tenderness and fulness in that situation. The natural course of abscess in the substance of the prostate is generally spontaneous evacuation by the urethra. I have seen it occur immediately after the passing of a catheter, when the patient's condition, in consequence of the tumefaction caused by its presence, has rendered instrumental assistance necessary in order to empty the bladder. Occasionally the matter is evacuated by the rectum. In the case of a gentleman of middle age, recently under my care, two ounces of healthy pus were suddenly evacuated from the enlarged prostate when the patient was at stool, a smaller quantity being evacuated every day for some time after, from the opened cavity of the abscess. In this case the enlargement, which was very considerable, was due to a previous acute attack of prostatitis, complicated with obstinate and narrow stricture of the urethra. Some years ago, I met with an example in the case of a gentleman 45 years of age, who came from New Orleans, United States, to place himself under my care with a most obstinate and irritable stricture of the urethra, of which he was finally relieved by the external division, which I performed as soon as he was in a suitable condition for it. He had had chronic inflammation of the prostate for some time before, and the organ was much enlarged. Subsequently it rapidly suppurated, and he also at stool evacuated not less than eight ounces of pus from it with great relief. He completely recovered from it. I do not deny the possibility that the abscess in the last-named case may have been situated between the prostate and rectum, as it is by no means always easy to pronounce positively upon this point. In the former case it was undoubtedly prostatic, and the urine passed for some time subsequently by the rectum through the resulting sinus. The spontaneous evacuation of matter by the rectum is perhaps as favourable, generally speaking, as through the urethra. It may be followed by a troublesome urethro-rectal fistula, but not necessarily, or even usually so. On the other hand, although the opening of the prostatic abscess into the urethra may soon close, the walls of the cavity having granulated and united by adhesion, yet if this does not take place, the sac will probably long remain open, and become a receptacle for urine, giving rise to fresh collections of matter around, from the inflammation so produced. Extravasation rarely, if ever, has to be feared, the parts exposed to urine being defined and thickened by exudation matter. Nevertheless, in many cases I believe the last-named course, that is, by urethra, is that on which we must rely. The only treatment we can adopt in order to prevent it, matter being already formed, is to make an artificial opening in the perineum as early as possible.

TREATMENT.—This mainly consists in providing for the evacuation of the pus as soon as abscess has formed, since no special treatment is indicated beyond that which the acute prostatitis demands, already described.

The incision requisite must be made with some boldness, in the median line, in the known direction of the prostate, inclining a little below its situation in the healthy state, since its bulk is chiefly increased in the direction downwards, a fact, however, which is supposed to be perfectly ascertained by the rectal examination. The forefinger of the left hand having been introduced into the bowel, a long, straight, and narrow bistoury, the cutting-edge of which is upwards, should be thrust into the raphé, about three-quarters of an inch anterior to the anus in the known direction of the swelling, and the incision enlarged in a straight line upwards, to a slight extent, so as to give a fair patulous opening for the discharge of matter. The depth to which such an incision must be carried cannot be less than an inch and a-half, it may be two inches; less than the former will be probably useless, and if so, unnecessary and injurious.

But it must be sufficiently obvious that the surgeon should be well satisfied of the nature of the case, that is to say, of the existence of a collection of matter in the situation referred to, before he decides on making this attempt to evacuate it by artificial opening. In cases of doubt, we must await the result, confining ourselves to palliating the symptoms which arise, and not unfrequently nature will clear up the case by discharging the pus through the urethra. Under the various circumstances which may be met with, the surgeon must employ his own judgment in deciding upon the course to be pursued, and it must be confessed that occasionally a case will present in which the indications of that course are not very marked or decided.

Occasionally, but rarely, a prostatic abscess has been observed to come forwards spontaneously by the perineum. Such, at least, is stated to be the case. It is contrary to the principle understood generally to hold good with respect to the course of matter confined in all parts of the body, viz., to seek exit in the direction of least resistance; and at all events it usually fails to perforate strong fascial partitions, other shorter and easier routes existing. In the case supposed, the matter must find its way through the deep perineal fascia. It is not improbable that in some of the alleged cases, this barrier was behind the abscess, and not between it and the surface at all, a distinction not always easy to establish. Very rarely prostatic abscess may burst into the peritoneal cavity. Mr. Adams gives an example of this in a case of tubercular prostate.* It is probably a very rare occurrence under any other circumstances.

CHRONIC ABSCESS.—The discharge of matter from an acute abscess of the prostate is sometimes followed by long-continued suppuration. Chronic abscess may arise also by itself, though not very frequently; and it may occur as the result of confirmed or neglected stricture of the urethra. In the former case either with or without any existence of tubercular deposit. The latter-named affection will be considered hereafter.

It is by no means always easy or even possible to diagnose the existence of a chronic abscess of the prostate during life; in some instances, however, it may be recognized by rectal examination. A part of the prostate may yield sensation of fluctuation to the finger in the rectum; and if this is obvious in a case where the history and symptoms have been such as to render the occurrence of chronic abscess there probable, it is advisable to take means for evacuating it. This may be done by passing a grooved sound into the bladder, and making an incision into the spot detected, a proceeding which is rendered easy by the support which the sound affords to the parts; a rectal speculum may be used or not as the surgeon prefers. It by no means necessarily follows that

* Anat. and Dis. of the Prostate. 2nd ed. p. 128.

we are to open the urethra in these cases, although it has frequently been done without producing any mischief in the shape of fistula afterwards. Some interesting illustrations of this fact have been given by Mr. Meade, of Bradford.*

When we are in doubt as to the physical signs of abscess, and only suspect its presence, the indications are almost solely those which relate to the patient's general health, and this, as in almost all cases of chronic abscess in other parts of the body, requires the greatest care. The digestion must be strengthened, and then generous diet administered—cod-liver oil, tonics, and pure air, especially at the sea-side, will generally aid materially the progress of the case. Chronic abscess very rarely calls for an incision from the perineum, and the less instrumental interference by urethra also the better. It must generally heal very slowly, since, after it has broken, urine is apt to pass into the cavity, maintaining irritation and pain, and hindering the reparative process. When it is due to the presence of calculus, the removal of the foreign body must precede a cure.

Case No. I.

Prostatitis and Cystitis after Intemperance and Exposure to Cold, resulting in Abscess of the Prostate

J. P. Aged 54. Oct. 1854. A man of intemperate habits since 20 years of age. Two or three attacks of gonorrhœa when young. For some years past has had occasional pain when passing water, but nothing which he considered serious.

Six months ago he spent a Sunday at Greenwich, drinking, and rode to London in the evening on the top of an omnibus, the weather being very wet and cold.

Next day he felt very ill, and on Tuesday morning, having passed no water since Sunday, he sent for a medical man, who relieved him with the catheter without difficulty. He was unable to pass water without an instrument after this, and subsequently became an in-patient of St. Mary's Hospital. Some time after he was dis-

* *Med. Times*, Oct. 20, 1860. A Paper on Inflammation and Abscess of the Prostate Gland.

charged, passing his urine by the catheter, which he had been taught to pass for himself.

Dec. 23, 1853.—He was admitted under my care at the Marylebone Infirmary.

Present state: he requires to pass water every hour, which he can do only by means of the catheter; and he suffers severe pains about the perineum, pubes, and loins, if the relief is not afforded. No tenderness in hypogastric region; nothing particular noted after rectal examination; no tenderness within reach of finger there. General condition weak. Complexion sallow, expression of much suffering.

Urine: slightly acid, deposits some mucus, a good deal of pus, and is slightly albuminous. Has had no hæmaturia.

No. 8 catheter passes easily into his bladder. After several observations it is noted that *some urine flows before the instrument passes quite six inches;* after passing it further, more can be drawn off: no calculus, but a tender and rugose condition of the bladder is detected.

During the following three months, he notably improved after several times washing the bladder with warm water, particularly after injecting solutions of nitrate of silver, increased from half-a-grain to one grain to the ounce. He retained his urine three hours, the mucus disappeared, and the pus greatly diminished.

In April, 1854, he succumbed to an attack of pneumonia. Postmortem fourteen hours afterwards.

On removing from the body the penis, bladder, ureter, and kidneys entire, and laying open the urethra from above, it was found healthy as far as to the prostatic portion. Here a large cavity presented itself, capable of containing 10 or 12 drachms of fluid. It undermined the mucous membrane of the urethra, opening into the canal by an aperture the size of a florin, situated on the upper part: thus the floor of the urethra alone remained, forming a kind of bridge through the cavity, which extended below, above, and on either side of it. This cavity is bounded by the capsule of the prostate, the substance of the organ having disappeared. Passing through the cavity is the right ejaculatory duct found to be dissected out entire. It is dilated, admitting a No. 9 catheter until it leaves the prostate, where it opens into the sac of an abscess. The left duct has disappeared, but the opening by which it entered the urethra remains. On examining the base of the bladder, a sac is seen occupying the entire interval between the two vasa deferentia and the vesiculæ seminales, but apparently not communicating with either. There is, nevertheless, a free communication between this cavity and the urethra by means of the right ejaculatory duct, for the catheter above-mentioned passed directly into it. This cavity is capable of containing about six drachms of fluid, and may be either the sac of an abscess in the cellular tissue, or one originating in the right vas deferens itself. The latter supposition seems to be more probable from the appearance of the parts.

The walls of the bladder are much thickened, and the mucous membrane is much injected, exhibiting reddish, brownish, and greenish tints, and bright crimson arborescent injection in patches. The ureters are a little above the natural size. The kidneys are above the usual size, and present the appearance of interstitial deposit, with much fat, under the microscope.

Case No. II.

LONG-STANDING URETHRAL OBSTRUCTION, INCONTINENCE. ABSCESS OF PROSTATE AND RIGHT VESICULA SEMINALIS.

J. T. Aged 73. Admitted to Marylebone Infirmary, under my care, May 23, 1854. For many years has had much difficulty in micturition, and incontinence; beyond this no information can be obtained, as he is evidently fast sinking, apparently exhausted by disease.

Present state: urine passes from him constantly, and he keeps a vessel in bed according to his habit. There is a large opening in the scrotum, which gives exit to much pus. The resident officer had attempted to pass an instrument, but without success; it was stopped at five inches from the orifice. The man is almost pulseless, and evidently dying.

Post-mortem. Bladder much thickened; very rugose; mucous membrane exhibits fine reddish brown and few slatish tints. On laying open the urethra a part in the bulb was found with thickened walls, narrowed, and much lymph deposited on the surface, partially fixing two calculi, each about the size of a small pea. In the prostatic part was seen an opening into the sac of an abscess, formed by the capsule of the prostate, precisely like that just described in the preceding case; with the floor of the urethra bridging over it. Behind it was another abscess, involving the right vesicula seminalis. Both kidneys much diseased, and containing large cysts. As far as the abscess of the prostate is concerned, the engraving, Plate XI., representing Case No. I. would suffice to describe the condition very correctly.

ULCERATION OF THE PROSTATE.—The urethral mucous membrane of the prostate suffers ulceration as a consequence of previously-existing lesions. Pure idiopathic ulceration is probably extremely rare. Thus, when an abscess opens into the urethra, the tissues around become ulcerated and the opening remains during a certain space of time; such cases have chiefly furnished the examples of ulceration described by authors. A like condition exists in the discharge of tubercle, or in the sprouting of malignant growth. A slight tubercular deposit into the mucous membrane of the prostatic urethra only has been noted as leading to an ulceration there. It is commoner to find it so affecting the lining of the bladder and the vesical orifices of the ureters. So also, after the impaction of a calculus, either entire or in the form of a fragment, ulceration may take place. After long-continued and aggravated stricture of the urethra, followed by dilatation of the canal, passing through the prostate, and thinning of its mucous membrane, I have not unfrequently observed chronic ulceration affecting it and the subjacent tissues. It may be occasioned during urinary retention and extravasation. I have seen cases in which ulceration of the prostatic urethra and neck of the bladder has occurred from maintaining a catheter in the urethra and bladder during too long a period, for the sake of treatment. An example of this may be seen in the Museum of the Royal College of Surgeons, London; No. 2551; it is described by the catalogue as a "patch of lymph where a catheter rested." Ulceration may occasionally be found existing after any long-continued vesical disease in the elderly and enfeebled, and especially in paralysed patients, from cerebro-spinal disease or injury. It may go on to sloughing, with considerable destruction of prostatic tissue in any of these cases.

After all, there is nothing special in these forms of ulceration as affecting the prostate; it is a morbid action affecting alike its tissues and those of all other soft parts in the organism when their vitality is impaired by certain morbid conditions, local or general.

There is no specific form of ulcer affecting the prostate; nothing peculiar to it as an organ, as there is in the stomach, for example. There is no ground either for believing it to be the subject of any specific chancrous ulceration, except in those very rarely observed instances in which such an affection has made its way along the urethra, and even into the bladder.

Further, as ulceration almost never exists apart from some other morbid condition of the prostate, bladder, or urethra, and never, except as a consequence of some prior lesions there, it is manifestly impossible to offer any description of the symptoms characteristic of it. Neither in relation to symptoms nor to treatment of ulceration in the prostate, is there any specific indication beyond that of treating the general and local symptoms as they arise, in accordance with principles already laid down under the head of acute and chronic inflammation, in the preceding chapter; and by removing the cause, if it be a removable one, such as stricture of the urethra, or calculus of the bladder or of the prostate, which has given rise to the ulceration.

CHAPTER V.

HYPERTROPHY OF THE PROSTATE—ITS ANATOMICAL CHARACTERS.

External Characters of Hypertrophied Prostate. — Parts chiefly affected. — Amount of Enlargement produced.—Changes in Urethra and Neck of the Bladder which result.—Internal or Structural Changes produced.—Varieties of Hypertrophy.

HYPERTROPHY OF THE PROSTATE. — Among men who have passed the prime of life, or, to speak more definitely, those who have passed the fifty-fifth year, or thereabout, as far as the most carefully-prosecuted researches tend to show, a peculiar affection of the prostate is commonly, but by no means generally, met with. The organ enlarges by a slow and gradual process, and almost always produces more or less obstruction to the discharge of urine from the bladder. This enlargement, as will be seen when we consider the details of the subject, is not a product of inflammatory action, nor is it caused by any results of inflammation corresponding with those which constitute inflammatory enlargements in other organs; that is, the albuminous, fibrinous, or other exudations, such as occur in lymphatic glands, for example; but it is due to an increased formation of the normal tissues of the prostate, and may therefore be fairly classed as an hypertrophy.

MORBID ANATOMY.—It will be convenient to study the morbid anatomy of this affection under the following heads:—

I. The external physical characters belonging to the Hypertrophied Prostate.

II. The Parts of the Prostate chiefly affected by Hypertrophy.

III. The amount of Enlargement produced by Hypertrophy.

IV. The Anatomical changes in the Prostatic Urethra and neck of the Bladder induced by the Hypertrophied Prostate.

V. The internal or structural changes observed in the Hypertrophied Prostate.

I. EXTERNAL PHYSICAL CHARACTERS.—It has been already seen that it is scarcely possible to describe, in exact terms, the limits, in respect of either weight or size, of a healthy prostate, any addition to which must necessarily and consequently be regarded as indicating hypertrophy. By some observers a weight exceeding six drachms has been regarded as always abnormal. Absolute proof of this appears to be wanting, but I think that seven drachms may fairly be considered as indicative of hypertrophy when found in a subject of about 60 years of age. I have never seen such a prostate which did not present other indications either of external physical, or of internal structural, changes, corroborating the suspicion aroused by the fact of weight. On the other hand, I have seen a prostate weighing less than six drachms, which afforded unmistakable evidence of the existence of hypertrophy in the conformation, &c. The possibility of this can be easily understood when we recollect that in some individuals a prostate of less than four drachms is of normal size.

The first external character generally observable in a hypertrophied prostate is undue fulness, and an unnatural tendency to rotundity. Almost always the gland is *thicker* than natural. Neither the length nor the transverse measurement may be necessarily increased, while the measurement from before backwards (erect position) is almost invariably increased. Hence the fulness remarked. The surface-markings described in the section relating to healthy anatomy disappear; the external indications of bilobular formation diminish. If a section now be made through the anterior part (anterior commissure) so as to lay open the urethral canal longitudinally, the lateral lobes appear fuller than natural, and to protrude a little into the passage, so that their opposed sides press lightly upon each other.

In a more advanced stage, the enlargement is more considerable; either lateral lobe may predominate, or the portion behind the verumontanum, "median portion," may be largely developed. The form or outline of the gland may be very irregular and unsymmetrical from the presence of projecting portions, which may result from two conditions—first, either lobe may protrude greatly beyond the others, as just alluded to; or, independent tumours may be found embedded in it, or only partially so, springing from the surface, and forming very salient projections in any direction, most commonly, however, towards the cavity of the bladder. When these are present and largely developed, the original form of the prostate is often altogether obscured, and a mass of rounded prominences, sometimes numerous, and generally irregularly placed, surrounds the neck of the bladder. (See Plate VIII.) In any of these conditions the urethra, as it passes through the prostate, may be diverted a little right or left of the middle line, by pressure from the most enlarged part; the prostatic portion may be increased in length (necessarily with the increased length of the prostate itself), and the antero-posterior diameter of the canal is often much increased by reason of the enlargement of the lateral lobes. To all these changes the capsule accommodates itself, and increases commensurately with the increasing gland.

In external colour there is no difference ordinarily observed; unlike to malignant enlargement, in which heterologous formation and increased vascularity are accompanied by marked yellow, red, and violet tints in some variety of shade.

In consistence, there is usually some change; the hypertrophied prostate is most commonly more firm and dense to the touch than the healthy: one experiences a sensation suggestive of structures firmly bound within a tight or stretched envelope, the truthfulness of which is apparent if a section is made, when the contained parts protrude, or the containing parts recede, so that the cut surfaces become more or less convex. In a few exceptional examples this does not occur, but it is undoubtedly the rule.

II. The parts of the prostate which are chiefly affected by hypertrophy.

There is no part of the prostate which may not be affected by hypertrophy. The lateral lobes, the median portion, or middle lobe, the anterior and posterior commissures are all capable of exhibiting this change, although in different degrees of extent, as regards both their several liability to be effected, and the extent which development may attain when the affection has taken place.

The data for forming a conclusion on this subject, and those are now numerous, show that no one portion of the organ exhibits any marked predominating tendency to enlarge above the remaining portions, either in point of early manifestation, rapidity of growth, or extent of development.

Writers, and surgeons generally, seem to attribute the first disposition to enlarge to the median portion ("middle lobe"), as well as to regard this as the source of the most considerable development in point of size. As regards the locality of commencing enlargement, I do not think the median portion exhibits the change at an earlier period than the lateral lobes; but, as regards the rate of development, the former, perhaps, progresses more rapidly than the latter. At the same time, most of the preparations contained in four of the principal museums of London, including that of the Royal College of Surgeons (each one of which I have carefully examined, and possess written notes respecting), amounting to a total of 123 specimens, exhibit about an equal development, in size of at least three portions of the organ; that is, of the lateral lobes and median portion, while in many the anterior commissure is correspondingly enlarged. I have classified the preparations referred to in distinct groups, marked by the direction or situation in which the enlargement is chiefly manifested, a method which will afford the means of obtaining a comprehensive view of the question under consideration. The first group consists of the four varieties of form which most commonly occur; the second group of three varieties of form which are decidedly rare.

Common forms of the affection.
{
I. General enlargement of prostate, that is, both lateral lobes and median portion pretty equally enlarged, is present in 74 preparations of the 123.

II. General enlargement of prostate, but the median portion enlarged in greater proportion; in 19 preparations.

III. General enlargement, but the right lobe predominating, and very decidedly larger than the left; in 8 preparations.

IV. General enlargement, but the left lobe predominating, and decidedly larger than the right; in 11 preparations.
}

Uncommon forms of the affection.
{
V. The lateral lobes only enlarged; in 5 preparations.

VI. The anterior commissure only, or chiefly enlarged; in 3 preparations.

VII. The lateral lobes and anterior commissure enlarged, not the median portion; in 3 preparations.
}

The several series of preparations tabulated in Chapter II., furnish results very closely corresponding with the foregoing, including two series of 64 preparations by myself, and another, by a wholly independent observer, Dr. Messer, formerly of the Royal Naval Hospital, Greenwich. All these present results more valuable than those obtained from museums, because they consist of average and not of selected specimens. Among 100 prostates from subjects of 60 years old and upwards, Dr. Messer found 35 which were enlarged. He reports, that in 17, or about one-half of the number, "all these lobes were pretty equally enlarged;" that in 14 both lateral lobes were chiefly affected; that in one case the left lateral and middle lobes were mainly affected; that in one the right lateral lobe was mostly, that in one the left lateral lobe was alone, and that in one the middle lobe ("median portion") was chiefly affected: in all 35.*

* See a valuable report on this and other questions connected with the Anatomy and Diseases of the Prostate, by Dr. Messer. Trans. Med. Chir. Soc. vol. xliii. p. 153.

The general inferences which must be drawn from all the facts adduced are—

(α) That the lateral lobes and the median portion, or middle lobe, are equally liable, or nearly so, to be affected with hypertrophy.

(β) That the posterior commissure is generally involved with the preceding enlargements, and in proportion to the extent which they manifest.

(γ) That the anterior commissure is not frequently affected, but nevertheless is so in rare instances.

(δ) Lastly, that development takes place at about an equal rate in each of the three principal divisions; in some cases the lateral lobes appearing to enlarge more rapidly than the median portion; in others the contrary effect is found, and perhaps in a rather larger number of cases than in the preceding.

III. The amount of enlargement produced by hypertrophy.

It has been shown that a weight of about 400 grains, or nearly 7 drachms, must be regarded as an example of hypertrophy, the average weight of a healthy prostate being about 270 to 300 grains, or $4\frac{3}{4}$ drachms. Now the weight may be considered a very fair index to size, and is sufficiently so for all conceivable practical purposes, since the amount of increase in one corresponds (on trial) very closely with the amount of increase in the other. The proof of this can be made by calculation from the amount of water displaced by immersion, an experiment very easily applied to a few cases. Where, as often happens, the structure is denser than in the healthy prostate, a slight increase may be allowed in calculating the true amount of hypertrophy. However, it is scarcely conceivable that such exactness can be required in any given case.

Taking weight then to correspond with the amount of hypertrophy, and as very fairly indicating the increase in size in all cases, it is unnecessary to record rectilinear measurements of so irregularly-formed a body as a hypertrophied prostate most commonly is, but to furnish, under the topic here considered, the amount of augmentation in weight which the hypertrophied prostate does ordinarily attain, and that which it may occasionally and extraordinarily exhibit.

My own dissections (see table) afforded 20 examples of hypertrophied prostates, from persons at and over 60 years of age. Those only are reckoned to be so which presented physical evidence of the change, without regard to weight. These 20 ranged from 6 drs. 20 grs. to 18 drs. The average weight is about 9 drs. 15 grs. The most common weight is from 7 drs. to 10 drs.

Dr. Messer's dissections afforded 35 examples of hypertrophy, from persons at and over 60 years. The 35 ranged between 6 drs. 15 grs. and 48 drs. The average weight was 15 drs. 2 grs. The most common weights ranged between 7 drs. and 14 drs.

It is clear, in considering all these cases, that an *average* product furnishes no reliable figure, because the accidental presence of one or two unusual examples so greatly affects an average, unless the numbers dealt with are extremely large. The *most commonly occurring* weight, when the complaint may be presumed to have existed some ten or twelve years, appears to be between 9 and 12 drachms, or more than double the natural size. Nevertheless, many cases exceed this limit considerably. The largest example of hypertrophy known is 9 or 10 ounces, or about 75 drachms. The size of such a mass, forming a preparation now in the Museum of University College, is nearly that of a cocoa-nut; but such an example is extremely rare. It is well represented by Plate VII. In malignant disease even these limits are exceeded. Other observers, including John Hunter, Sir E. Home, Mr. Howship, Sir Chas. Bell, Sir B. Brodie, Mr. Stafford, Dr. Gross, Mr. Adams, Dr. Hodgson, MM. le Drs. Civiale, Mercier, Leroy d'Etiolles, and Rokitansky, have spoken in their writings of the amount of hypertrophy in general terms, but not one of them in precise terms; and as it does not appear that they have pursued any exact researches in relation to this question, it is not thought necessary to allude further here to their statements respecting it. A large orange is the simile furnished to describe the largest example observed by Mr. Howship.* Sir Charles Bell speaks of one "as a monstrous enlargement of the prostate gland, probably the largest in this country."† This is now in the museum of the College of Surgeons, Edinburgh (No. 2071, xxxii. B), and may be described as having the volume of a medium-sized cocoa-nut. Dr. Gross describes a specimen as weighing 9 ounces.‡

IV. The anatomical changes in the prostatic urethra and neck of the bladder, induced by the hypertrophied prostate.

The most important result of enlargement, at all events in any of its four common forms, is obstruction to the flow of urine. Very rarely is it otherwise, although there appear to be some cases, few and exceptional, in which the condition of the vesico-urethral orifice is so altered, that the bladder is unable to retain the urine, which consequently runs off as fast, or nearly as fast, as it enters the viscus from the ureters; a subject which will be more fully discussed hereafter. Admitting, then, that the result almost uniformly met with is *obstruction*, producing more or less retention in many instances, it is obviously one of the most important parts of the present enquiry to ascertain the influence of enlargement upon the form, size, and direction of the urethra; since success in affording relief to retention by the passing of a catheter must depend upon adaptation in the instrument itself, or in the method of applying it, to the mechanical obstacles which a distorted canal presents.

The first effect to be noted is one which is common to all the first four forms of enlargement: viz., increase, sometimes considerable, of the antero-posterior diameter of the prostatic urethra. Associated with this is diminution of its lateral or transverse diameter; so that the canal becomes a narrow or chink-like passage, instead of one which, when distended, is of about equal

* Diseases of the Urinary Organs. Howship. London, 1823, p. 299.
† Treatise on the Diseases of the Urethra, &c. London, 1822, p. 423.
‡ Urinary Organs, 2nd ed., p. 688.

diameter in every direction. The lateral lobes, increasing, not only encroach laterally upon the canal itself, but gradually carry upwards that portion of the urethral wall which is constituted by themselves, and that to such an extent that I have seen the slit-like opening produced by transverse section in such a case, measuring three-fourths of an inch from the pubic to the rectal limit of the urethra.

The length of the prostatic urethra is also materially increased by the same forms of enlargement. The increased magnitude of the encompassing body in every direction, involves an addition to the length of the passage which passes through it. This would of course be the case were that passage continued in its ordinary straight direction. But it is often rendered tortuous also, which further contributes to the augmentation of its length. In some preparations which I have examined, the urethra has measured three inches from the orifice of the bladder to the membranous portion, instead of one inch and a quarter, which is the normal length.

The next effect is a deviation from the natural direction; and this varies with each form of enlargement. Thus, where there is enlargement of, or outgrowth from, the median portion, the form of all kinds most generally present, a change in the

FIG. 3. FIG. 4.

direction usually commences about the middle of the prostatic urethra, its posterior wall being carried upwards, or upwards and forwards, in the erect position of the body, producing a more or less angular curvature in the place of a nearly straight line.

Examples of this deviation are shown in the adjoining figures, which, although diagrams, represent in profile the form of the actual specimens from which they were taken (figs. 3, 4, 5, and 6). The direction, in early stages of enlargement, is usually

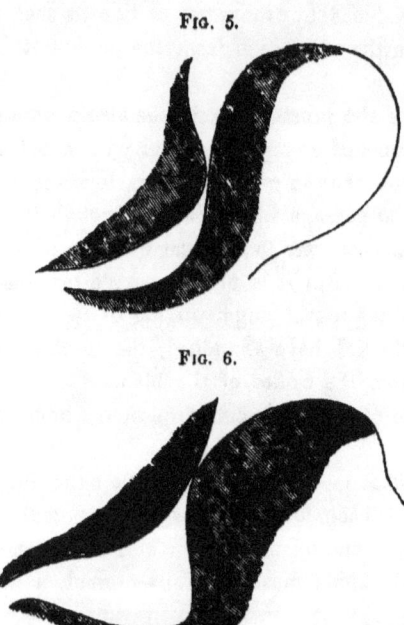

Fig. 5.

Fig. 6.

more or less that of a simple curve, but in more-advanced stages it may be almost angular, so that in some instances a complete step has to be surmounted at the neck of the bladder, before an instrument will enter the cavity (figs. 4 and 6). Now, when associated with this undue development of the median portion, there is a predominating enlargement of either lateral lobe, it is obvious that the lateral direction of the canal will be also changed. Thus, if the right lobe predominates, there will be a lateral curve of the urethra, the convexity of which is presented to the left, and *vice versâ*. And as the predominating lateral lobe is almost always found in connection with a large median portion, and is usually more or less blended with it, the direction of the urethra will be upwards and to the right, or to the left, as the case may be (fig. 7). Sir E. Home, up to the date of publication of his first volume on the prostate gland, had never seen predominating enlargement of the right lateral lobe, and he inferred, as a rule of some importance in relation to the introduction of catheters in enlarged prostate, that such an enlargement, and by consequence, that a deviation of the canal to the left either did not occur, or was extremely rare. He met with an enlarged right lobe, however, before the publication of a second volume on the same subject, but still regarded it as uncommon, and in this light it has been viewed, I observe, in the latest works on this subject. There is, however, no ground for supposing that there is any difference in the liability of either lobe to the affection, since among the existing specimens the predominating lobe is to be found in nearly equal numbers to the right and left respectively. But in the absence of any predominance of a lateral lobe, where the median portion is largely developed, a similar kind of deviation is often met with, only it is not necessarily confined to one side, but may exist equally on both. The vesical end of the urethra being divided by a large median outgrowth of pyriform shape, a passage is left on either side of it, giving to the canal there the form of the letter Y (figs. 8 and 9). In these diagrams, drawn, like the foregoing, from actual specimens, the line which the catheter must take is indicated by dotted lines. And the degree of vertical direction associated with it frequently depends upon the amount of mucous

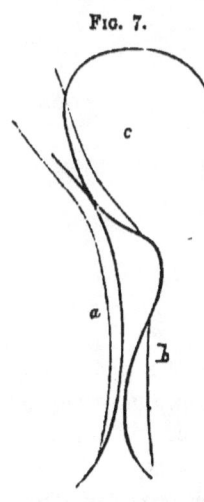

Fig. 7.

a. Right lobe of prostate, considerably enlarged.
b. Left lobe, less so.
c. Tumour from median portion blended with left lobe, and consequently deflecting the urethra to the right side. The course which an instrument must take in such a case is indicated by dotted lines.

and submucous tissue drawn up by the growth, on each side of which, at the vesical orifice, it forms a kind of semicircular bar. It is hardly necessary to allude to the importance of calling to mind, when instrumental aid is required, the frequency with which lateral deviation is found to exist in largely-developed forms of prostatic obstruction.

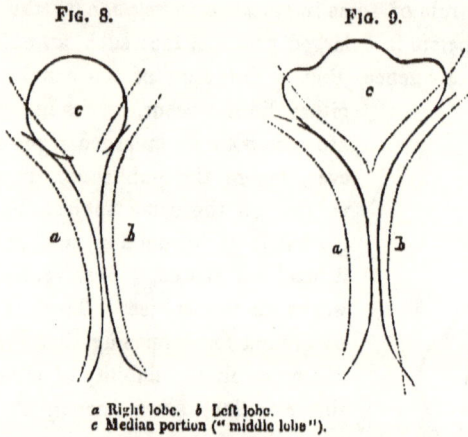

FIG. 8.   FIG. 9.

a Right lobe. *b* Left lobe.
c Median portion ("middle lobe").

The next deviations are those to be observed in the form of the internal meatus, or vesico-urethral orifice, which in its healthy condition is too well known to require description. When, however, the posterior median portion is predominant, the vesico-urethral orifice acquires a crescentic form, the convexity of which is directed upwards. When the right lobe considerably exceeds in size the left, the crescent has its convexity to the left side, and so on. In some preparations, where two or more irregularly-enlarged lobes are combined, the orifice is very much distorted, presenting an elongated and tortuous outline. Sometimes it appears to be overlapped altogether, when an outgrowth from the posterior median portion affects a valvular form, or has a narrow peduncle. In these circumstances, which, however, are not very commonly met with, the valvular portion appears to be forced against the neck of the bladder by the effort of micturition, and the obstruction rendered still more complete. Plates III. and IV. show this condition on a small scale, not unfrequently met with, but in both cases the effect was almost complete obstruction. Occasionally the tumour may be very small, yet if it is engaged in the internal meatus the same result takes place; thus in the instance represented at fig. 10, where it was extremely so, the patient was almost incapable of passing water by his own efforts.

It occasionally happens, but very rarely, that one result of prostatic enlargement is undue patency, and not obstruction, of the urethro-vesical orifice, and this circumstance, regarded, as it has been, as not infrequent, is supposed to explain the occasional occurrence of genuine incontinence, the bladder being empty, in the place of retention of the urine. Submitted to the test of extended anatomical research, the following facts appear. First, it is very rare to find expansion of the internal meatus. Secondly, when it does occur, it is almost invariably associated with distended and hypertrophied bladder, proving incontestably that obstruction was present during life, and that retention, not incontinence, was the result.

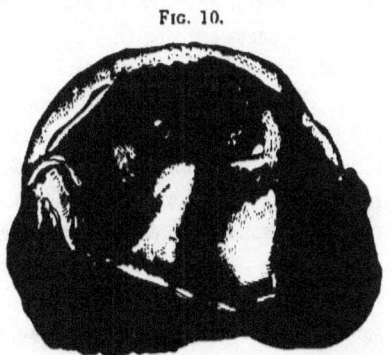

Fig. 10.

Internal meatus of urethra: two or three small tumours producing nearly total obstruction.

The expansion of the vesico-urethral orifice, which prevents the bladder from retaining urine, is accounted for in the following manner:—The lateral lobes being considerably enlarged, the tumour of the median portion, instead of projecting backwards into the bladder in the usual manner, enlarges between the hinder parts of the two lateral lobes themselves, and opens them out as by the action of a wedge, giving to the meatus an expanded and triangular appearance. I have carefully examined the histories attached to preparations exhibiting this peculiarity, and where these have been wanting have verified the existence of a hypertrophied and distended bladder, and in no instance have I been satisfied that real incontinence, that is, unassociated with retention, existed during life. Anterior to the vesical orifice, it may further be added, there is usually a sufficient degree of encroachment of the lateral lobes upon the canal to produce considerable obstruction and habitual retention of urine. I am quite aware that the condition and its alleged result sometimes coexist; but that such cases are frequent I have no hesitation in denying.

The following is an example of it. A man who died at the Royal Naval Hospital, Greenwich, at the age of 84 years, did not suffer from any retention of urine during life, yet, at his post-mortem examination, Dr. Messer found a prostate weighing no less than 26½ drachms; it contained many tumours. Dr. Messer makes the following remark respecting this question:—

"It may be remarked that, in cases where the hypertrophy is advanced, and the tumours tend inwards, the urethra becomes greatly expanded, and the interspaces between the rounded sides of the projections into it, serve as channels for the urine to pass away by. This condition will frequently explain, I have no doubt, the occasional absence of symptoms of obstruction in cases where the prostate is known to be considerably enlarged." *

But there is another point of view from which the development of the enlarging prostate may be regarded. The tendency is, in some instances, strongly manifested in a direction towards the centre of the organ, or the neck of the bladder. In others, it appears to affect an opposite direction, to become developed very largely at its periphery. In the former, which may for brevity's sake be denominated *centric* hypertrophy, the outflow of urine may be very materially obstructed before the prostate has increased much in the matters of weight and size. In the latter, which may be described as *eccentric* or peripheral, a very large development may take place, and an enormous prostate may be encountered in the rectum, and yet little obstruction to the course

* Op. cit. p. 153.

of the urine will be manifested. This, it is almost unnecessary to remark, is a form far more favourable for the patient than the other. I have had several opportunities of observing the contrast which cases belonging to these two classes present to each other. It lately occurred to me to have the opportunity of examining a very striking example of the latter kind. The case was that of an elderly gentleman, for whom on one occasion in consultation I passed a catheter. He expelled his urine very frequently, and with difficulty, but emptied the bladder completely. There was no residual urine on introducing the catheter after the act of micturition; a fact verified, not only by myself at the time referred to, but by his ordinary attendant on numerous occasions. Nevertheless, the prostate formed an enormous tumour in the rectum, presenting an unusually marked example of enlargement there when digital examination was made.*

V. The internal or structural changes observed in the hypertrophied prostate.

This portion of the subject has been entirely re-written. The essential facts are nearly the same as those given in the first edition; but they are arranged in a different manner—it is believed in a better one. This change is due to the complete re-examination of the question, which I thought it right to institute in preparing an essay for the Jacksonian prize.

A. The naked-eye characters.

When we make a section of one of the thicker portions of the mass presented by a considerably hypertrophied prostate, such being usually one of the lateral lobes, certain peculiarities of appearance are observable. These may be noticed under two heads: viz., first, those common to all hypertrophied prostates; and, secondly, those which are met with in only some of them.

First.—Characters which are common, although in greater or less degrees of prominence, to all hypertrophied prostates.

When a section of the gland is made,

* See also the enormous prostate, Plate VII., and referred to, p. 85, in which obstruction to the flow of urine was only occasional.

(*a.*) The surface bulges irregularly;
(*b.*) It is, also, more or less parti-coloured; and
(*c.*) It is striated in places with small tortuous blood-vessels.

(*a.*) IRREGULAR BULGING.—There is a general protruding of the cut surface of the prostate, above the level of the border of the divided capsule, a condition not seen in section of the normal organ. It is obvious, also, that portions of different size, and sometimes of spheroidal form, are more tumid and prominent than the rest; these are by no means always independent tumours, as we shall presently see.

(*b.*) PARTI-COLOURING.—It is equally obvious that the shades of colour affecting the cut surface, have a greater range in the hypertrophied than in the healthy prostate. Always parti-coloured to some extent as the latter is, the former is more so; varying from pale dingy yellowish-grey and buff to yellow, the last sometimes deep in tint. Red patches sometimes intervene between paler portions; and small black spots, collections of dark concretions, are seen interspersed in follicles or in ducts.

(*c.*) The redder portions just referred to are generally easily resolved by the naked eye into fine arborescent vascular injection; deeper in tint than that seen in the healthy organ.

Secondly.—Characters which are met with in some hypertrophied prostates, but not in all.

In some glands, the spheroidal prominences displayed on section are easily enucleated: they are then firm, pale, and destitute of much moisture. These are simple tumours, contained in the structure of the prostate; formations so commonly met with, and playing so important a part in connection with the pathology of the prostate, as to demand a separate consideration. (See Chapter VI. relating to Tumours.)

Other prominences are loose in texture, partially separable from the surrounding structure, but by no means to be enucleated. These have often quite as much of the yellow tint as the surrounding parts; and they have a moist succulent character. They are single gland-lobules, which seem to have become larger than the neighbouring gland-lobules; and they are much larger than those found in the normal prostate.

This condition is by no means uncommonly met with. It seems to indicate that a true hypertrophy may affect a portion of the organ, while adjacent portions are normal or but slightly affected. Thus one lateral lobe may be considerably hypertrophied, while the other is scarcely altered in size. When single lobules are affected in this manner, it suggests the existence of a condition intermediate between general hypertrophy and isolated tumour, and furnishing an important link between them.

There is another very marked difference observable in examining hypertrophied prostates, which separates them into two somewhat distinct categories.

A typical example of the one kind exhibits, on section, an abundance of fluid exuding from the cut surface, and slight pressure greatly increases the quantity issuing. Similar fluid issues freely from the orifices of the prostatic ducts in the urethra.

A typical example of the other kind, shows no fluid on section, neither on pressure is there any, or scarcely any, from the orifices just named.

These conditions are easily explained by the varying amounts of active gland tissue present in the two cases, as will be seen under the next division of the subject.

Again, in some instances small cavities are disclosed by section; these are sometimes empty, and sometimes give issue to a drop of thick yellow fluid, exactly like pus to the naked eye. These are by no means necessarily abscesses, and are not so unless some inflammatory action has been present, a condition studied under its appropriate heading. We may so far anticipate the next section of this chapter as to say that this yellow matter is, when not pus, the prostatic concretion in a more concrete state than natural, and containing the cell and allied elements, with a very small proportion of "liquor" or fluid medium.

B. The structures of hypertrophied prostate, as examined by the microscope.

Regarding the elementary fibres of different kinds, of which the normal prostate has been shown to be composed (see Healthy Anatomy), we may commence by affirming that the hypertrophied prostate, when submitted to the closest microscopical scrutiny, presents no new, or other elementary, structure whatever than those there described. The existing normal structures are simply augmented in quantity. They may, however, although not necessarily, exist in different relative proportions to each other, or they may be arranged in a form differing from that which obtains in the normal state.

These structures are, as we have already seen,

(a.) *Fibrous structures;* viz., pale muscular fibre, connective tissue, and elastic tissue; and—

(b.) *Glandular structures;* viz., basement membrane forming recesses and excretory ducts; both lined with epithelium, polygonal and prismatic, and containing secreted matters.

The former (a), have been termed the basic or stromal part of the organ: the latter (b), the glandular part.

Structurally regarded, the prostate exhibits abnormal development in four varieties or types, all of which are fairly included by this term hypertrophy.

Such, at least, is the result of an examination of very numerous specimens, which being necessarily destroyed by the process, were not included in the tables.

The four varieties of hypertrophy, structurally considered, may be defined as follows:—

1. Simple increased development of all the component tissues of the organ in about equal ratio.
2. Excess of development in the stromal portion over the glandular portion.
3. Excess of development in the glandular portion over the stromal.
4. Re-arrangement of the structures, stromal and glandular, in the form of tumour.

The first form of hypertrophy is that which is generally observed in those examples in which the size of the organ is not very greatly increased, although the morbid action may have continued for a long period of time. The normal relative proportions of different parts are not greatly disturbed; there is general fulness; increase of weight, to perhaps about double that of a healthy organ; and, generally speaking, there is no single portion of it which much predominates in size over the other portions. Neither on section, nor by pressure, does any undue amount of secretion appear, neither does there seem to be any deficiency.

Under microscopic examination, such a specimen shows few characters differing from those observed in the normal organ. There is some dilatation of the gland-follicles, a condition present in all hypertrophied prostates that I have examined. Many of these are blocked up with yellowish semi-fluid contents, which appear to be the prostatic secretion in an inspissated form, and the well-known "prostatic concretions." This semi-fluid matter, although often resembling exactly pus to the naked eye, is resolved by the microscope into the following constituents:— a clear liquor, loaded with epithelium, both prismatic and ovoid or polygonal (the former from the ducts, the latter from the vesicles) globules of highly-refracting matter resembling fat, but probably not so, much amorphous granular matter, and small amber-coloured, semi-transparent concretions. These concretions are also found lodged in the ducts, and very often at or near their orifices in the urethra.

2. Excess of development in the stromal portion over the glandular portion.

This is no doubt the form in which hypertrophy most commonly exists. It has been said to be the invariable condition which constitutes hypertrophy of the prostate; this, however, is certainly not the case. In the larger, and in the very largest examples, we may say those weighing from 2 ounces upwards, this form of hypertrophy is almost the only one met with; that depending upon a considerable formation of tumours (class 4) being sometimes excepted.

Under the microscope, we meet, in well-marked specimens of the class now under consideration, with large portions of stromal structure, unaccompanied, or not penetrated, by any glandular structure. Portions of the greyish material which pervades the mass may be unravelled under water on the glass slide, and no gland-structures be found in it: in other portions these latter are found in very small quantity. The larger portions of stromal tissue, unmixed with glandular, are situated usually near the periphery of the organ, in the lateral lobes, lying beyond the glandular lobules, and forming sometimes a thick stratum between them and the capsule.

It is hardly necessary to observe that this form is to be distinguished from that enlargement which results from inflammatory action, and which is considered by itself in the Third Chapter. That action occasions a deposit of its own proper products; but there is no reason to believe that it has any power to generate any of the natural structures of the prostate. Their overgrowth is a true hypertrophic formation; while the lymph of inflammation is the result of a morbid effusion, and as such becomes in course of time in part removed by the unassisted efforts of nature.*

It is common to find, especially in the larger specimens, cavities measuring one or two lines in diameter, lined with smooth membrane, often empty, sometimes containing several concretions. Abundance of these latter are commonly found in and among the glandular parts of the organ, where also may be found some collections of the yellowish semi-fluid matter described above.

---

* Dr. C. H. Jones appears to have been the first to point out the fact that senile hypertrophy was frequently due to increase of the fibrous element rather than of the glandular.—*Medical Gazette*, August 20, 1847.

It is this form of enlargement to which the term parenchymatous hypertrophy has been applied, and particularly by Dr. Hodgson, of Glasgow, in his excellent monograph on this subject lately published.

Dr. Messer regards the enlargement as "produced principally by hypertrophy of the fibrous tissue which naturally exists in the organ. The gland-follicles also become enlarged, and more numerous, but do not affect the size of the organ to the same extent as the fibrous development."—*Op. cit.* p. 150.

3. Excess of development in the glandular structures over the stromal.

This condition is exceedingly rare. I have examined certainly one, if not two, such specimens, and record them here in consequence. The gland, in the first instance, weighed about 14 drachms, and hypertrophy appeared to effect each part in a pretty equal degree. When cut, much fluid exuded, the structure was extremely succulent, and the yellowish glandular parts obviously predominated over the greyish stromal part.

Under the microscope, gland-structure was seen to abound throughout the organ, and gland-products pervaded almost every part. The gland-follicles were a little enlarged, that is, to about double their average size; besides these, several cavities, containing the yellow semi-fluid matter, were present. There can be no doubt that in this case the glandular structures existed in a larger proportion to the stroma, than that which is met with in the normal prostate. A second specimen was examined, the condition of which approximated to the preceding, but it was not quite so well-marked.

4. Re-arrangement of the normal structures of the prostate, fibrous and glandular, in the form of tumours.

As before stated, we have here to deal only with the elementary constituent tissues of the organ; but they no longer exhibit the primordial arrangement, as they do in those forms of hypertrophy which have been already described. In these cases, the tissues assume new modes of arrangement; rounded masses, often quite isolated from the adjacent prostatic structure, by which, nevertheless, they are surrounded, are found in all parts of the organ. Some of them are adherent to the adjacent parts, continuity of structure being evident at limited portions of their periphery.

If we examine microscopically one of these bodies, the stromal tissue of the prostate is at once identified, and is found forming almost the whole of the mass; although some small proportion of glandular structure is generally discoverable also; but this, unlike the fibrous stroma, differs usually from the corresponding elements in the healthy organ, by being less completely developed, exhibiting often an imperfect or aborted condition of the gland-vesicles and follicles.

These tumours are so frequently associated with hypertrophy of the prostate, and are so peculiar in character, that their full consideration will be deferred to a separate chapter, which immediately follows this. Enough has been said here to indicate their general characters. The fact that they are composed solely of elements entirely identical with those normally belonging to the prostate itself, has alone entitled them to be named in connection with hypertrophy, while certain independent characteristics which they possess, demand that they should not be regarded as merely one form of that action. We now proceed to their anatomical analysis in detail.

CHAPTER VI.

TUMOURS AND OUTGROWTHS OF THE PROSTATE.

SIMPLE TUMOURS.—Frequency of their occurrence.—Described by several Anatomists.—Examples cited.—Their Physical Characters and Intimate Structure.—OUTGROWTHS.—Nature of.—Analogy between these Affections of the Prostate and Tumours of the Uterus.—Rare Form of Polypus of the Prostate.—Cases.

SIMPLE TUMOURS AND OUTGROWTHS FROM THE PROSTATE.—In considering both the external characters and the intimate structures of the hypertrophied prostate, it has already been necessary to allude to, and even in some degree to describe, certain tumours often met with in the organ, as well as the not infrequent occurrence of distinct outgrowths, usually springing from that part known as its median portion.

It may be premised that these formations are always of slow growth, and generally of moderate size; they exhibit benign or non-malignant characters, and are thus wholly unlike any of the cancerous formations which will be hereafter discussed as occasionally affecting the prostate. They are found almost always in the hypertrophied prostate, but are, nevertheless, occasionally met with in the organ of normal size. Their presence is much more common, I believe, than is generally supposed; indeed it cannot be doubted that they are present in a large majority of the cases of hypertrophied prostate.

Thus, in three-fourths of the hypertrophied specimens in my own series, tumours were present, as well as in two or three which were not hypertrophied: of the latter, No. 181 is an example; here they are extremely minute, although quite distinct. Dr. Messer reports the occurrence of isolated tumours in no less than 27 out of 35 hypertrophied prostates.

Of 70 specimens of hypertrophied prostate in the Museum of the Royal College of Surgeons, in seventeen the isolated tumours are so clearly discernible, that the careful observer cannot fail to see them in the preparations as they stand in the containing vessels. There can be no doubt that many more would be found similarly affected if the test of dissection were applied. In a large proportion of the remainder there is an outgrowth affecting more or less a pyriform shape, projecting from the median portion.

It is manifest from the examination of all these morbid specimens, that we meet with two distinct classes of new formations in connection with the enlarging prostate.

They may therefore be considered as follows:—

A. Tumours which are generally imbedded in the substance of the prostate, but the structures of which are isolated from those which surround it.

B. Outgrowths which are continuous in structure with the parts of the prostate from whence they spring, but which manifest a tendency to become partially isolated, by assuming a more or less polypoid form and maintaining attachment to the parent organ through the medium of a pedicle only.

Both these conditions differ materially from that already considered as simple hypertrophy affecting the organ either generally or partially. The class of outgrowths appears to offer examples of a morbid formation midway between isolated tumour and general hypertrophy.

We will study, first,

A. The isolated tumours of the prostate.

The fact of their occasional appearance has long been recognized. Sir E. Home describes and very clearly depicts them in his work.* He believed them to be the remains of extravasated blood-clots in the substance of the gland, and he conceived that

---

* Practical Observations on the Treatment of Diseases of the Prostate Gland, vol. ii. London, 1818. Vide pp. 17; 21–25; 273; 277; 285. Plates i. ii. iii. iv. v. vi. and vii.

they were produced by the rupture of vessels, and generally occasioned by violent exercise, especially by hard riding.

Within the last few years the nature of these tumours, and their relation to the prostate, has been the subject of more extended inquiry.

Cruveilhier has carefully examined them. This author has a remarkable passage, in which he foreshadows some of the views which some later writers have more completely developed. The original, which appears in his "Anatomie Pathologique," Livraison xvii., descriptive letter-press belonging to Plate II., p. 3, published at Paris about 1833-4, is as follows. It may first be premised, that the subject of the description was a prostate gland considerably hypertrophied, taken from a patient who had been cut for stone by the high operation in one of the hospitals, who died soon afterwards, and had been submitted to a minute postmortem examination.

"The tissue of the gland was very easily torn, and in so doing, portions of an irregular spheroidal form were separated; the largest of these had about the magnitude of a middle-sized nut. A section of the gland exhibited surfaces of circular outline, of which each belonged to one of the spheroids.

"Each of these spheroids was evidently a glandular vesicle or lobule (*grain*) hypertrophied. The tissue of each lobule presented a cellular texture" (to the naked eye understood); "the cells, of unequal magnitude, were filled with the prostatic liquor. Several of the larger (glandular) vesicles were converted into cells (or crypts) and contained a matter, yellowish and thick, like purulent matter.

"These large vesicles or lobules, although perfectly independent or isolated, were held together by a framework or bed of intervening tissue, evidently muscular in its character, and which I can compare to nothing so well as to the tissue of the gravid uterus. The prostatic envelope, very easy to isolate from the muscular fibres of the bladder, was constituted by a rather thick and whitish muscular layer." . . . "Thus we see that the prostate was made up of large vesicles or lobules scattered throughout a bed of muscular tissue, which furnished a thick envelope for them. Each glandular lobule was cellular in its character."

Velpeau called special attention to these tumours of the prostate; and although he appears not to have minutely examined their histological resemblances to other tumours, he pointed out what he believed to be an analogy between them and the fibrous tumours of the uterus.

He says, speaking, in his lectures, of certain cases of death which had occurred, during the session in his *clinique*, from enlargement of the prostate and its consequences:*—"When these patients die, they always present tumours in the prostate. Let us dwell for an instant on this subject, since it merits particular attention. I have attempted to show the relations which these tumours, affecting certain individuals, hold to those fibrous bodies or tumours which are developed in the uterus. You know that the fibrous tumours of that organ develope themselves sometimes very near the uterine mucous membrane, from which they soon project; that they often become engaged in the uterine neck; and from thence protrude into the vagina, where they constitute what is commonly called 'fibrous polypus' of the uterus; that at other times they are situated very near the serous covering of the uterus, become prominent into the abdominal cavity, and give rise to so-called peritoneal fibrous tumours. If, on the contrary, they are developed in the substance of the uterine walls, and remain so, they form there the 'fibrous tumours' properly so-called, which sometimes present masses of extraordinary size and weight.

"I perceive a great analogy between these prostatic tumours and the 'fibrous tumours' of the uterus.

"1. There are some of these fibrous prostatic tumours which are developed in the direction of the vesical cavity, which become absolutely pedunculated precisely as do the fibrous polypi of the

* Leçons orales de Clinique Chirurgicale. Par M. le Prof. Velpeau. Tome 3^{ieme}, pp. 478-9. Paris, 1841.

uterus, and become enveloped by the vesical mucous membrane. These, with a peduncle more or less extended, may acquire the volume of a nut or even of half of an egg.

" 2. These fibrous tumours may be developed in the very substance of the prostate, and may acquire a volume similar to that of the preceding.

" 3. Finally, they may be developed at the perineal or at the rectal surface of the prostate, and may project towards the perineum, towards the side of the pelvis, or into the rectum.

" Observe then, a great analogy in the matter of situation. But there is a great analogy in the matter of structure also. These are veritable 'fibrous tumours,' and not mere 'hypertrophy of the prostate' as has been said. I do not regard these tumours as degenerations of the organ, but rather as new productions. It is not surprising that this (hypertrophy) should be the common idea respecting the prostatic tumours, since for a long time it has been customary to believe that the 'fibrous tumours of the uterus' were nothing else than hypertrophy of the normal tissue of the womb."

Rokitansky originally took these tumours to be simple fibrous formations analogous to the "fibrous tumours," loosely so-called, appearing in other parts of the body. Subsequently, when studying the subject of bronchocele, and examining the isolated masses embedded in, and situated near to, the hypertrophied thyroid gland, which were evidently masses of gland-tissue like that of the parent gland, he remarked a similarity of relationship between the prostate gland and its tumours, that is to say, those now under consideration. He found, on examination, that these formations were composed of basic structures, identical with those forming the prostate itself, but that the glandular elements were less perfectly formed, less completely developed, as a rule, than those of the normal and healthy gland.

He observed, also, that the rounded masses were not always confined within the limits of the prostate gland itself, but that they were sometimes found beyond, as outlying formations, analogous in his opinion to those occupying a like relation to the thyroid gland.*

The outlying masses referred to are occasionally met with, although far less commonly than those which are embedded in the substance of the gland. Mr. Paget relates a striking example.— He writes: " Near the enlarged prostate, similar (that is, to those of the thyroid) detached outlying masses of new substance, like tumours in their shape and relations, and like prostate gland in tissue, may be sometimes found." A case follows of " a man 64 years old, who for the last four years of his life was unable to pass his urine without the help of the catheter. He died with bronchitis; and a tumour, measuring $2\frac{1}{2}$ inches by $1\frac{1}{2}$, was found lying loose in the bladder, only connected to it by a pedicle, moving on this like a hinge, and when pressed forwards, obstructing the orifice of the urethra. Now, both in general aspect and in microscopic structure, this tumour is so like a portion of enlarged prostate gland, that I know no character by which to distinguish them."† Speaking of the embedded tumours, Mr. Paget says:—

" In enlarged prostates they are not unfrequently found. In cutting through the gland, one may see, amidst its generally lobed structure, portions which are invested and isolated by fibro-cellular tissue, and may be enucleated. . . . They lie embedded in the enlarged prostate, as, sometimes, mammary glandular tumours lie isolated in a generally enlarged breast. They look like the less fasciculate of the fibrous tumours of the uterus; but to microscopic examination, they present such an imitation of the proper structure of the prostate itself, that we cannot distinguish the gland-cells, or the smooth muscular fibres of the tumours, from those of the adjacent portions of the gland. Only their several modes of arrangement may be distinctive."‡

* Zur Anatomie des Kropfes. Wien, 1849, p. 10. Von Prof. Carl Rokitansky.
† Lectures on Surgical Pathology, vol. ii. pp. 8-9. By Jas. Paget, F.R.S. London, 1853. The preparation is now in the Museum of St. Bartholomew's.
‡ Idem, p. 264.

Several examples of these tumours have been examined and described with care by others within the last few years. The following examples will aid us in considering the subject, and are therefore quoted.

At the Pathological Society of London, Mr. Fergusson exhibited two tumours, one "the size of a filbert, the other that of a horse-bean," removed from the prostate gland in the operation of lithotomy;* and Mr. Shaw showed a similar one, "the size of a moderately large hazel-nut," developed in the centre of the left lateral lobe. "The surface of the tumour was very smooth, and it was embedded in a cavity, the sides of which were also smooth, and the connection between them was so slight that the tumour could easily be enucleated from the prostate, which body it resembled in structure; the only difference perceived by the microscope being that the gland was traversed by small wavy fibres which were not visible in the tumour." †

Mr. Henry Gray exhibited a similar tumour, occupying the centre of an enlarged lobe of the prostate from a man aged 62. It was "circular in form, about the size of a hazel-nut, contained in a thick capsule of fibrous tissue, from which the tumour was easily turned out; . . . . . . it was very firm in texture, its structure consisting of cœcal pouches filled with epithelium, connected to each other, and surrounded by a fine filamentous tissue." ‡ Mr. Gray remarks the analogy which exists between these tumours and some of the forms of mammary glandular tumour, and he adduces two cases in which a glandular tumour existed within the limits of the external capsule of the prostate gland, which, in its growth, had projected into the cavity of the bladder, and had been erroneously supposed to be an enlarged lateral lobe.

The following were exhibited by myself: one, a preparation in which several of these tumours, each about the size of a large

* Path. Soc. of London, Feb. 19, 1849. Vide Report in the Proceedings for the Third Session, p. 83.
† Path. Soc. of London, May 7, 1849. Proceedings, pp. 83-4.
‡ Trans. of Path. Soc., vol. vii. p. 252. 1855, Nov. 20.

pea, were embedded in various parts of the prostate. These were "lighter in colour, and denser in structure than the adjacent parts," and appeared to be composed of fibrous structure, and not to contain glandular elements.* The second was a specimen of partially-outlying tumour, which contained glandular tissue, and also the "concretions" so often found in prostatic structure.† Also a specimen of tumours embedded in the anterior commissure of the prostate, which were "made up mainly of the organic muscular fibres characteristic of prostatic structure, together with a small proportion of imperfectly-formed glandular elements intermixed. Each was isolated by a capsule of fibrous tissue."‡

Since that time I have examined numerous other specimens of isolated prostatic tumour. They corroborate the views which I expressed in a paper on the subject at the Medical and Chirurgical Society in the session of 1856–7, viz., that some are almost, if not entirely, made up of the constituent fibrous structure of the prostate, without gland-elements, and that these are by no means common. That, generally, there are present, in addition to the fibrous structures, some gland-elements more or less imperfectly developed.

Among specimens sent to the College of Surgeons, I submitted some to section and microscopical examination. The latter showed the basic structure to be identical with that of the healthy prostate, viz., bundles of pale muscular fibre, with connective tissue and a little elastic tissue; very little glandular structure seen; a few depressions or pouches, and some ovoid epithelium (same as gland-epithelium in the normal prostate), and a few prismatic epithelial cells (same as exists in the ducts of the prostate). To sum up all that is known on the subject of the tumours in question:—

In a considerable number of hypertrophied prostates, rounded tumours are found, which are more or less isolated from the surrounding tissues of the gland. These tumours do not appear to

* Trans. Path. Soc., vol. vii. p. 254, 1855, Dec. 4, illustrated by a plate.
† Idem, p. 256, 1856, May 6.
‡ Idem, vol. ix. p. 293, 1857, Dec. 1.

affect any particular part of the prostate more than another, and may be found in any part of it. Perhaps they are more numerous in the lateral lobes, especially at their posterior extremities, than elsewhere. Occasionally they are embedded in an enlarged median portion. It often happens that the small multiple eminences so frequently seen at the neck of the bladder in the site of the uvula, are due to these small tumours there, situated under the mucous membrane and few submucous fibres, there being no enlargement of the median portion. When the swelling is large and single, it is more commonly hypertrophy or outgrowth of the part. Rarely they are found in the "anterior commissure" of the prostate. Sometimes they appear just under the capsule, and so spring from the surface, carrying, in an outward direction, the capsule as a covering, but nevertheless escaping altogether the contour of the gland, and looking almost like an independent or outlying formation. Occasionally they are really outlying, *i. e.* separated by an interval from the prostate itself. A space of half an inch has been seen to intervene between such a tumour and the adjacent gland; a narrow line of duct, with other vessels, and a little tissue alone connecting them.

When embedded in the substance of the prostate, as most commonly happens, they occasionally appear to be almost unconnected by continuity of tissue with the adjacent structures, and they may then be easily enucleated. Most commonly some fibres exist, uniting them to these, and in a few instances the uniting tissue is considerable in quantity.

The size of the embedded tumours ranges between a tenth of an inch and about five-eighths of an inch in diameter. Their density is rather greater than that of the prostatic tissue proper; they are firmer to the touch, and more compact. If divided while retained in their original position, as when a section made through a prostate gland passes through one of the tumours, its cut surfaces protrude above the plane of the surrounding surface, and present a slight convexity. They are for the most part paler in tint than the prostatic structure proper. Commonly a thin layer of the tissue which immediately environs them is a little compressed, so that in some cases the tumour appears to lie in a kind of cyst, which is occasionally very well marked. When not isolated, but "outlying" in situation, the tumour may attain a larger size; for example, the case quoted, page 106. Their vascularity is evidently less than that of the surrounding prostatic structure.

Examined by the naked eye, or under the microscope with powers of different degrees, we can discover no structural peculiarities as compared with the prostate itself; they possess all the elementary tissues common to the normal prostate, and they possess no tissue not belonging to it.

Respecting the arrangement of these tissues, the basis of the tumours appears to be the fibrous basis or stroma of the prostate itself, a structure already described under the head of "Normal Anatomy;" an admixture of unstriped, soft, pale muscular fibre and connective with a little elastic tissue, closely interwoven. Interspersed with this, there are present in most cases, small cavities containing flattened polygonal or spheroidal epithelium, like that seen in a pouch at the extremity of a prostatic gland-duct, and sometimes, also, some prismatic epithelium. These cavities are sometimes solitary, sometimes slightly branched, and sometimes of an elongated or tubular form. In a few instances there is very little, or, perhaps, no such glandular tissue to be found; generally, however, a careful search will discover it. In some of the outlying tumours, the glandular structure is more perfectly developed—in some it is quite so—and a duct is furnished which evidently carries secretion to the appointed destination.

With such a mode of formation, all that is known respecting their slow mode of development, and their indisposition to manifest any morbid or other changes whatever, except that of simple increase in size, it is quite certain that they possess none of the characters of malignant growths, but are simple or benign in their tendencies.

There can be no doubt that the embedded tumours occur both in the hypertrophied and non-hypertrophied prostate, though much more commonly in connection with the former condition. There appears to be no ground whatever for the theory suggested by Velpeau, that a hypertrophied condition of the prostate is altogether dependent on tumour-formation. There may be, as already shown under the subject of "Hypertrophy," simple hypertrophy of the prostatic tissues without any tumour-formation at all. But there is, at the same time, no doubt that their presence is more common than their absence in the hypertrophied prostates of elderly people.

It may be that there is a tendency to the production of these tumours in the prostates of all, or nearly all, elderly subjects; and that when the disposition to hypertrophy exists also, the tumours participate in the disposition equally with the rest of the tissues, and thus become objects of attention to the pathologist.

B. Outgrowths which are continuous in structure with the parts of the prostate from whence they spring, but which manifest a tendency to become partially isolated, by assuming a more or less polypoid form, and maintaining attachment to the parent organ through the medium of a pedicle only.

We have already seen that any one portion of the prostate may exhibit undue development, while surrounding parts are either but slightly or not at all affected by any such action. The outgrowth from the median portion is the most familiarly known example of this. It is then generally composed for the most part of the ordinary structures of the prostate, although sometimes containing a smaller proportion of the secreting elements than belongs to the organ in the normal state, and it appears to enjoy activity of function in common with the rest of the prostate. It assumes a pyriform shape even in its earliest stage, and is always continuous in structure with the adjacent prostatic tissues from which it springs. It has its own special ducts, which traverse the pedicle to open in the urethra, and in its substance may be almost invariably detected those concretions which are found in the adult prostate. It may vary in size from that of a pea to that of a middle-sized pear, and at the outset exerts a perceptible influence on the neck of the bladder, the lower or posterior border of which is gradually elevated as it increases. Ultimately it finds its way into the cavity of the viscus, where it is truly polypoid in shape.

Occasionally this pyriform mass is connected to the main body by so long and slender a pedicle, that it appears at first sight to be a separate or outlying portion. Such was the condition in a case I recently examined.

The glandular structure of these outgrowths is generally, as might be expected, more perfectly developed than in the isolated tumours. Concretions, as just stated, are commonly found in various stages of progress embedded in the former, but I never found this to be the case in any of the latter; and the reason appears obvious. In the one case there is an actually-secreting structure with ducts of exit in a state of activity; in the other the structure is rather an imitation, or imperfect development, of a secreting apparatus, and consequently it cannot be supposed that any functional office is performed by it.

The occurrence of outgrowth, although most common in the part described, is not invariably confined to it. A projecting growth may occasionally spring from the posterior part of either lateral lobe, and has been even observed to arise from that part of the prostate which lies above or anterior to the internal meatus of the urethra.

It must not be overlooked that general hypertrophy of the tissues of the prostate may, and commonly does, co-exist with tumour or outgrowth—almost invariably with the latter. And, doubtless, the outgrowth is only a more marked expression of the same disposition which pervades the organ, in some measure, perhaps, determined by the form and nature of the cavity towards which the protrusion is directed; the existence of the cavity of the bladder probably permitting a development which would not be possible in other directions, where masses of solid structure oppose such extension. With circumscribed tumour, also, there is often general hypertrophy, but not invariably.

A consideration of the facts exhibited under the subject of tumours and outgrowths, serves as an appropriate introduction to the remark that analogies of a very remarkable kind exist between the characters and relations of these two forms of tumour and those which affect the uterus. We have already seen that Velpeau suggested the idea some years ago, and that he rested the analogy upon the correspondence which he assumed to exist in the two sexes between the uterus and the prostate, from a belief that the two organs originated from the same centres of development in the early condition of the ovum, coupled with the fact that both are liable, in after life, to exhibit tumours presenting similar external characters.

The ground of analogy, which is derived from regarding the uterus of the female and the prostate of the male as the morphological equivalents in the two sexes, is not, perhaps, the strongest that might be adduced. I shall indicate one which I think is still more conclusive, as well as other points of analogy, the combined result of which will render the correspondence more obvious.

Firstly.—In studying the typical plan on which the entire genito-urinary apparatus of the two sexes is constructed, the most recent labours of modern philosophical anatomists confirm the view that the analogue of the uterus, or rather of the uterus and vagina combined, in the male, is the prostatic vesicle or utricle. This is the view taken by Leuckart, in a recent article written for the "Cyclopædia of Anatomy and Physiology." It has been also maintained by Dr. Simpson, in a very elaborate "Memoir on Hermaphroditism and Sexual Malformations generally," which first appeared in the same work, but which is now republished, with considerable additions. An extract from this is

I

appended in a note below.* The prostate, then, although not of itself the absolute equivalent of the uterus, contains it in the utricle, situated as this cavity is in the very centre of the organ.

Secondly.—The point, however, on which I would lay greater stress is, that the prostate and uterus are organs whose bulk is constituted by the same tissue, namely, the organic muscular fibre. No other organ in the body besides these two is similarly constructed by thick masses of this structure; elsewhere it is distributed in membranous layers. This analogy of structure is, perhaps, in relation to the pathological question before us, stronger than that of identity of origin in early fœtal life, since it has more influence, doubtless, in determining the appearance of tumours and outgrowths of similar character, than any other circumstance.

* "Few, or indeed none, of the eminent anatomists who have in later years studied the subject of the prostatic vesicle or utricle, as Huschke, Leydig, Rathke, Leuckart, Bischoff, Arnold, Wahlgrew, Külliker, Duvernoy, Goodsir, and Allen Thomson, have at all doubted that this organ is a representative or analogue, in the male organization, of the genital canals of the female.* But different opinions have been expressed as to whether it morphologically represents the vagina, or the uterus, or both. H. Meckel at one time, and in opposition to almost all other authorities, suggested and maintained that it was the analogue of the vagina, rather than of the uterus. Weber considered it as the male prototype of the female uterus; and still more lately Bimbaum and Louckart have shown that this organ may be more truly held as the morphological equivalent of the whole sinus genitalis, both the uterus and the vagina—an opinion now generally shared in by those who formerly took a different view of the subject. Huschke has sometimes found the lower or vaginal portion of the male utriculus separated from the upper and dilated end by a constricted point, as if indicating its division normally into uterus and vagina. Indeed, it is only in accordance with this last doctrine that we can understand the relative positions and modes of junction of the genital and urinary canals in some monstrosities, and the fact of the great variety of forms and shapes which the male uterus or prostatic vesicle assumes when it is found—as so often happens—preternaturally enlarged and disproportionately developed in different kinds of hypospadiae and hermaphroditic malformation."

* "Some of the various diseased states attributed to enlargement, &c., of the third lobe of the prostate gland will be yet found, I believe, to be morbid states of this prostatic vesicle. To the minds of some 'the Investigation of the Diseases of the Male Uterus' would appear to be almost a paradox in thought and words."—*Obstetric Memoirs and Contributions.* Edited by Drs. Priestly, of London, and Storrer, of Boston. Vol. ii. pp. 318, 319. London, 1856.

Thirdly.—The two organs thus similarly constructed, are very frequently the subjects of tumours, identical both in external and histological characters. Thus, in the uterus we find these formations nearly or completely isolated, made up of organic muscular fibres, with connective tissue, embedded in the substance of the organ, or standing out in relief from either surface.* In the prostate we meet with precisely the same tumours, and they are similarly disposed. Although, on the high authority of Rokitansky before referred to, an analogy has been pointed out as existing between these embedded tumours of the prostate and those of the mammary gland, I confess that the grounds of that analogy appear to me less complete than those which indicate their relation to the fibrous tumours of the uterus as just suggested. The prostate differs very materially from the mamma (and, in a corresponding degree, resembles the uterus) in being mainly constituted by tissue, designed to exert a mechanical power; while the mamma is simply a secreting organ, or gland. The prostate is a muscular organ, but permeated by glandular tubes and follicles. Were the small glandular tubes found in the inner wall of the uterus, prolonged more deeply into its substance than they are, the analogy between the uterus and prostate would be complete. The organic muscular tissue appears to have a tendency to become the nidus of isolated masses of like tissue, in structures formed by it; the type of these being found in the uterus. In the prostate we have the same phenomenon, plus certain imperfectly-formed gland-tissues, but the addition may be fairly regarded as an accident, depending on the presence of glandular elements in the muscular organ in its normal state. Hence the amount of gland-tissue so intermixed with the tumour is extremely variable in different specimens. The fibrous tumour, we know, in whatever part of the body it occurs, is very prone to imitate in some measure the tissue in which it is placed. Thus,

* Vogel established this fact in relation to the structure of the so-called fibrous tumours of the uterus in 1843, and it has been confirmed by Dr. Oldham, Dr. Robt. Barnes, Dr. Bristowe, and others.

as Mr. Paget remarks, spiculæ of bone may be frequently observed, when it is situated in bony structures;* a disposition which, I believe, accounts for its acquisition of some gland-elements when it appears in the prostate.

Fourthly.—In the uterus we are familiar with another form of tumour, which, springing from the interior, and forming a polypoid growth there, is much more intimately connected with the uterine structure than the variety just described, perfect continuity of tissue existing between it and the polypus. So from the median portion of the prostate we meet with an outgrowth, tending in form to become truly polypoid, which continues its development in the direction of least resistance, and exhibiting complete continuity of structure with the prostate itself. It contains also the glandular elements proper of the organ in varying proportions.

It may be further observed that all these outgrowths and tumours, among the latter especially those of the fibrous kind, may remain of so small a size, both in the uterus and in the prostate, that the bulk of the organ is not sensibly increased, and no signs indicating their existence during life are produced; while on the other hand they may increase to an enormous extent, so as to exceed by very many times the natural size and weight of the organ in which they originated, and give rise to the most alarming derangements of function.

Fifthly.—The two organs are subject to considerable hypertrophic enlargement, mainly consisting of their constituent fibrous elements. And in both, this condition may be associated with some tumour-formation, or it may exist independently of it. In the latter case, the hypertrophy may be general or local, affecting the whole or certain parts of the organ; and, when thus local, affecting particular spots more commonly than others. These remarks apply equally to the prostate and to the uterus.

Sixthly.—The two organs are liable to these changes after the prime of life has passed. Bayle, whose observation is quoted by

* Lectures on Surgical Pathology, vol. ii. p. 136.

Rokitansky, and verified by Dr. Robert Lee, says that 20 per cent. of women after 35 years of age, have fibrous tumours of some size in the uterus.* I have found prostatic tumours in 30 per cent. of males after 50. In women, however, the tendency to this formation declines after 50, although it cannot be said to cease. Nevertheless it is exceptional after that period. It is generally regarded as most active during the term of uterine functional activity, or rather during the latter moiety of the time. The age at which the reproductive function of man is most vigorous is certainly not that at which a like tendency in the prostate is evinced; but on the other hand it may not be forgotten that the term of productiveness is not limited in the case of the male, as in the opposite sex. And still further, it may, I think, be fairly admitted that our acquaintance with the prostatic function is not at present sufficient to forbid, but on the other hand rather to encourage, the supposition that it is possible that its activity may not in any way diminish, if it be not augmented, during the middle and later years of life, when the hypertrophic disposition is manifested. One thing is certain—the prostatic secretion, whatever its purpose, does not appear to be at that time less plentiful, judging from the state of the organ after death, than at any previous age.

There is one form of tumour, or rather of outgrowth from the prostate, the occurrence of which is extremely rare, but which may be mentioned before closing the chapter. It has no relation to any of the preceding varieties. It is a polypus springing from the verumontanum. I know of three instances only. The first is in the museum of St. Thomas' Hospital, No. BB. 8, a small polypus about half an inch long, and two lines in breadth, springing from the verumontanum in a child, and directed backwards

* Rokitansky makes the age still later. Fibroid tumours of the uterus, he says, "are unusual up to the thirtieth year, and present themselves most frequently shortly after the fortieth year."—*Manual of Patholog. Anat.*, Sydenham Soc., vol. ii. p. 298.

towards the neck of the bladder. The second is mentioned by Rokitansky, but not described, as a solitary case which he had seen.*

The third occurred in my own practice, and was exhibited by me at the Pathological Society of London in 1856.†

It was about five-eighths of an inch in length, soft in consistence, and at its base was continuous with the apex of the verumontanum, lying in the urethra, which it appeared to fill, and reaching to the neck of the bladder. It was composed of the elements of fibro-cellular tissue, with a few organic muscular fibres intermingling at its base. In some parts were seen, near the centre of the growth, some minute crystal-like bodies, having very much the appearance of uric acid, yellowish in tint, and rhomboidal in form, with a few octahedra. They proved to be crystals of some earthy carbonate. The tumour was covered with mucous membrane, and columnar and spheroidal epithelium (see fig. 11).

Fig. 11.

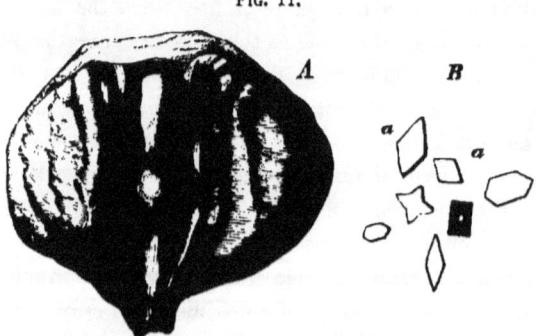

The only point in the history of the patient which related to the presence of the polypus, was, that he had for some time been in the habit of passing water with greater frequency than usual.

* Path. Anat., Syd. Soc., vol. ii. p. 235.
† Trans. Path. Soc., vol. vii. p. 250, with engraving.

ILLUSTRATIVE CASES.

Case No. III.
Enlargement of the Prostate due to Isolated Fibrous Tumours; Diseased and Sacculated Bladder.

J. P., aged 74, admitted to the Marylebone Infirmary, October 27, 1855, under the care of Mr. Henry Thompson. Two or three years past he has suffered with the ordinary symptoms of chronic cystitis. Now, there is much irritability of bladder; no retention or incontinence. No. 10 catheter passes with perfect ease. The bladder slowly empties itself of pus and mucus after urine is drawn off. No calculus. Urine thick, opaque, mixed with much pus, slightly acid on several occasions, albumen considerable, no renal casts or crystals.

Only a slight degree of enlargement of the prostate is recognizable from the rectum. Patient suffers very little pain, but is extremely weak, and appears to be gradually sinking. Death occurred on the 7th of November.

Post-mortem, six hours after death.—Bladder and part of urethra removed entire. The former is elongated, and projects upwards behind the pubic symphysis; the walls are thicker than natural. Projecting into the cavity, at its neck, is a lobulated tumour about the size of a small walnut, of yellowish colour and smooth aspect, contrasting strongly with the crimson hue and corrugated surface of the lining membrane of the bladder (Plate VIII., fig. 1). The cavity was contracted, and corresponded with about two-thirds only of the organ removed. The upper third proves to be a sac, opening by a very small orifice into the bladder proper, and an opening made into it gave exit to about an ounce of pus and mucus. The tumour springs from the prostate, which, though not enlarged towards the rectum, bulges into the urethra considerably, and also into the vesical cavity.

Both kidneys are diseased, and contain several small collections of pus, besides some ordinary cysts.

On cutting into the left side of the prostate (marked "left" in the drawing) the section of a rounded body embedded in the gland is seen. This body is four-tenths of an inch in diameter, is isolated from the surrounding tissues, separable from them in places with a blunt probe, but in others closely adherent, or united by prolongations of tissue common to both. It is lighter in colour, and apparently denser in structure than the adjacent parts.

Several other bodies of a similar character are found corresponding with the protuberances seen in the prostate before making sections of it. Three of the largest of them are dissected out (Plate VIII., fig. 2). The mucous membrane was first divided over them and turned aside; then a few longitudinally-disposed fibres, which in places peeled off from the rounded bodies like a capsule, but which at other times appeared to run into their substance, and to be continuous with it. One of these bodies, corresponding with a protuberance on the right side of the gland marked (a) in both drawings, is isolated in the preparation, and bands of fibres appear to unite it with the fibrous constituents of the gland behind. Small portions of tissue detached from any part of the bodies of which section has been made, exhibit under the microscope the elements of the organic muscular and connective tissues, closely packed in bands. Acetic acid developes a number of the rod-shaped nuclei lying longitudinally among the fibres.

An isolated mass of the true glandular elements of the prostate may be seen in one of the sections—that on the right side. It is obvious to the naked eye by exhibiting a yellower tint than the surrounding parts. Under the microscope it presents the ordinary glandular elements, as well as some of the minute "concretions" so called, often seen in the prostates of elderly people. None of these structures are found in the rounded bodies before described.

A large proportion of the bulk of the organ is made up of a whitish, fibrous-looking tissue, which intervenes between masses of the gland-structure, almost surrounding and isolating them. This tissue, under the microscope, is seen to consist chiefly of the connective filaments, intermixed with organic muscular fibre. The rounded bodies are made up of the same elements, and are more distinctly connected with this tissue than with the glandular parts, and may be, therefore, considered good examples of the ordinary fibrous tumours of the prostate. In this specimen some are embedded in the mass of the organ; others protrude beyond its periphery into or towards the cavity of the bladder. They do not partake of the characters of the proper glandular structure, nor do they appear to have any close relations with it.

Case No. IV.

Enlargement of the Prostate: Numerous Tumours: Several Calculi in the Bladder.

A gentleman, aged 65, who had experienced some difficulty in micturition about twenty years. Marked symptoms, amounting sometimes to retention, had existed about six years.

In the commencement of the year 1857 he came under the care of my friend Mr. Sampson, of Ipswich, who suspected the existence of calculus in addition to organic obstruction of some kind. In May I had an opportunity of examining the patient in consultation with that gentleman, and we were satisfied of the existence of very considerable prostatic enlargement; but the condition of the bladder was at that time too irritable to admit of any sounding or operative proceedings for stone. There was more or less retention of urine requiring daily instrumental relief, and there was a depressed condition of the vital powers. It is unnecessary for the present purpose to enter upon the details of the case.

Death from exhaustion took place at the end of June. At the post-mortem examination the prostate was found enormously enlarged, the anterior or pubic portion being that which exhibited by far the greatest increase in size; this portion formed an irregularly-shaped nodular mass, about the size of a hen's egg, and contained numerous embedded tumours, each about the size of a kidney-bean. On examination I found these to be made up of fibrous material containing a considerable proportion of the glandular elements of the prostate. Nine small calculi were also found in the bladder, lying in a deep depression behind the enlarged prostate.

CHAPTER VII.

THE CAUSES OF HYPERTROPHY OF THE PROSTATE.

The Subject of Causes obscure.—Its Investigation extremely important.—Most Circumstances alleged to be Causes must be rejected as such.—Present Views Stated.—Inflammation not a Cause—nor Stricture nor Calculus—nor Venous Stasis.—Gout, Rheumatism, and Syphilis not Causes.—Sexual Excess.—Prostatic Enlargement not Analogous to Glandular Hypertrophy, nor to Hypertrophy of other Muscular Organs depending on Increased Function.—Enlargement of Prostate and Uterus shown to be Identical in Nature and probably in Causation.—Perhaps a Necessity of their Common Structure.—Ascertainable Conditions under which Enlargement occurs.—Analysis of Results arrived at in relation to Age.

NEXT in importance to the discovery of some unquestionably successful means for the cure of enlarged prostate, perhaps, indeed, equal to it, would be a recognition of those circumstances which stand in the relation of causes, remote and proximate, to this remarkable affection. While some authors acknowledge that a considerable degree of obscurity attaches to the subject, others do not hesitate to express decided opinions in reference to it, confidently enumerating many things which they conceive to be undoubted causes; and all appear to agree in reference to some few of these, or, at all events, as to the existence of a strong probability in favour of so regarding them. It appears to me, however, that in order to accept the etiological views of this affection which are at present current, a good deal must be taken for granted; and that if we require a fair amount of evidence before we admit into the category of causes the circumstances and conditions usually recognized, we shall probably reject them all, or very nearly so. However discouraging to the practitioner such a result may at first sight appear—however unsatisfactory such a confession of ignorance may be deemed by the student—we may rely upon it, that, if it be a true one, it is the necessary and important preliminary step to a better state of knowledge on this subject. The admission of a single circumstance into the list of causes which cannot be sustained there by something better than a fanciful belief, conventional custom, or by, it may be, the "impression" or the "conviction" of an author, unsupported by testimony, must assuredly become a stumbling-block in our progress towards truth. Better were it to sacrifice the apparent completeness which often seems to be thought essential to a pathological treatise, if it can only be obtained by collecting all the suggestions and speculations that have ever been associated with the subject in the literature of the past; and rather to exercise—however much the literary character of the work may appear to suffer—a vigilance, lest we admit too much, than an anxiety to press into our service every line resembling a contribution, under the semblance of information on the subject. Nothing would be easier, on such a principle, than to swell this chapter into a volume, and for the simple reason that so little is known of its subject, so much conjectured. But that the reader may possess a sketch of the opinions of some of the most experienced authors in reference to it up to the present time, I will state them as briefly as possible by way of quotation, and then attempt to enter upon an independent examination for ourselves.

John Hunter says nothing directly in relation to causes, but states that he has "seen hemlock of service in several cases. It was given upon a supposition of a scrofulous habit. On the same principle I have recommended sea-bathing," &c.* It should be remarked, however, that the distinction between the enlargement of the prostate in earlier years, usually consequent upon inflammation, especially gonorrhœa, and the senile affection at present under consideration was not then made. In the chapter quoted, these two widely-differing conditions are spoken of indif-

* A Treatise on the Venereal Disease. 2nd edit. London, 1788. p. 174.

ferently; yet, on the authority of this passage, hemlock has been largely administered in the last-named complaint.

Sir E. Home, who enjoyed large opportunities for the observation of these cases, was of opinion that the chief predisposing cause consisted in "the slow return of the blood from the neck of the bladder, arising from the disadvantageous situation of the veins respecting the heart," inducing habitual congestion of those vessels; and that this was rendered more powerful by the undue indulgence in the pleasures of the table, or in any habits which "increased the circulation of the blood in those parts." The most common and influential proximate cause he considered to be the effects of horse-exercise, producing "rupture of vessels in the internal parts of glands," establishing thus "a great analogy between this complaint and apoplexy." * At the same time he believed prostatic enlargement to be one of the changes natural to old age.

Mr. Wilson, in his lectures at the College of Surgeons, in 1821, having stated that he has "met with several cases which confirm the justness of the observation" of Sir E. Home, respecting the liability of individuals of full habit to the disease, observes, "that it appears to occur most frequently in those persons, who, either from living a life of strict celibacy, have not used the genital organs so much as nature seems to have intended, or who have injured both the genital and the urinary organs by a life of excess." Finally, he adds that, "many persons have suffered much from the enlargement of the prostate gland who have lived a moderate and quiet life, without approaching to either of the above-mentioned extremes." †

Sir Charles Bell gives no opinion as to the remote or predisposing causes, but believed that a predisposition to prostatic

---

* Practical Observations on the Treatment of the Diseases of the Prostate Gland, vol. ii. London, 1818. pp. 9, 10.

† Lectures on the Structure and Physiology of the Male Urinary and Genital Organs. London, 1821. pp. 331, 332.

enlargement existing, one of the most frequent and important exciting conditions would be found in any source of irritation to the bladder, inducing repeated contractions of the organ. Whatever the occasion of these, which was indifferent as regards the ultimate result, the "muscles of the urethra" were set in constant action, the effect of which, according to his view, was to draw backwards the middle lobe or median portion, to which he stated them to be attached, and thus to produce the elevation so frequently observed to form an obstruction to the omission of urine by the vesical neck.*

Sir A. Cooper says, "the enlarged prostate is the consequence of age, and not of disease." †

Sir Benjamin Brodie regards enlarged prostate as an almost invariable accompaniment of advanced age, assigning it a place in that category of phenomena which marks the decline of life. Thus he says, "when the hair becomes grey and scanty, when specks of earthy matter begin to be deposited in the tunics of the arteries, and when a white zone is formed at the margin of the cornea, at this same period the prostate gland usually—I might, perhaps, say invariably—becomes increased in size." Hence no other circumstances or conditions than the general one of declining life are mentioned by him in the light of causes.‡

Mr. Samuel Cooper, in his Dictionary, after reviewing various statements, sums up with the following opinion:—"It seems to me better to confess that the etiology of this complaint is unknown * * *;" but he adds, "I have known several persons afflicted who had led very sedentary lives." §

Mr. Coulson states, on the authority of others, the usually recognized causes, but expresses no decided opinion in favour of any one of them.∥

---

* Medico-Chirurgical Transactions, vol. iii. 1812, pp. 171 to 189. Illustrated by three plates, showing dissections of these muscles.
† Lectures in Lancet, vol. iii. 1824, p. 239.
‡ Lectures on the Urinary Organs, 4th ed., pp. 163-166, and 186, 187.
§ 7th Edition, p. 1122.
∥ Diseases of the Bladder and Prostate Gland, 5th ed., p. 589.

Dr. Gross, of Louisville, doubts the influence of some of the agencies usually assigned as causes of hypertrophy of the prostate, but thinks there is no doubt that it may be induced by the following:—"Habitual engorgement, protracted and frequently-repeated sexual intercourse, irritation resulting from the presence of a vesical calculus. Finally," he says, "the protracted or frequent use of stimulating diuretics, of wine and alcoholic drinks, exposure to cold, the repulsion of cutaneous diseases, gout and rheumatism, external violence, the frequent introduction of the catheter, and habitual straining at stool, as in chronic diarrhœa and other affections of the bowels, may all be enumerated as so many exciting or predisposing causes of this affection."*

Desault speaks of enlarged prostate as being "very common in elderly people, and in those who have had many attacks of gonorrhœa; nevertheless, it is not always a result of venereal taint." He believed that it might "sometimes arise in the scrofulous and other cachectic habits." †

Amussat adopts the older views, which had long been current among continental surgeons; and these his opinion may be regarded as well and briefly expressing. "Syphilis, the presence of a foreign body in the bladder, the existence of strictures in the urethra, are its most ordinary causes. It is observed especially in elderly persons who have long used sounds or bougies, which they introduce themselves. In this case, the swelling of the prostate is occasioned by chronic inflammation produced by the contact of instruments." ‡

Civiale devotes a section to the special consideration of causes, in which he declines to consider speculative questions relating to the supposed analogy between the prostate and the uterus, or the resemblance which has been suggested to exist between its enlargement and that of the thyroid gland or liver, &c.; and enu-

* A Practical Treatise on the Diseases, &c., of the Urinary Bladder. By S. D. Gross, M.D. 2nd ed. Philadel. 1855. pp. 688–691.
† Œuvres Chirurg., de P. J. Desault, t. iii. p. 238. Edit. 3º. Paris, 1813.
‡ Leçons sur les Retentions d'Urine. Par Dr. Amussat. Paris, 1832. pp. 199, 200.

merates those which he believes to be proximate or exciting causes; placing first in order of influence the presence of calculus in the bladder. Next come organic strictures of the urethra and the difficult micturition which results. Much stress is laid upon this, and Cruveilhier's remark that stricture and prostatic enlargement rarely coincide, is quoted for the purpose of refuting it; Civiale stating that numerous facts exist attesting the accuracy of his view of the question. On the other hand, he admits that the influence is not constant, since urine, arrested by the stricture, may hinder, by means of the pressure reflected backwards, the prostate from becoming enlarged. The improper use of instruments in the urethra is placed next on the list. He combats the notion that venereal excesses have any intimate relation with the prostatic affection; and believes that authors have been far too ready to admit their influence without examining the question.*

Mercier, who discusses the subject at length, regards as predisposing circumstances, " all those which most favour stagnation of the blood. Persons of soft and lymphatic habit, with the cellular and adipose systems largely developed, possess, generally, very lax and unresisting venous tissues; and observation shows that such are most frequently the subjects of prostatic engorgement." * * * "I believe there is a certain relation between weakness of the inferior veins and hypertrophy of the prostate; this explains why the affection appears sometimes to be hereditary." He considers sedentary habits to favour greatly prostatic enlargement, stating that shoemakers have formed one-third of his cases in hospital practice; after these come house-porters, weavers, and tailors. He adds that it is no less true that active men are also victims to it, and questions whether a vertical position of the body, much prolonged, may not produce the same effects as a sitting one. Finally, admitting the effect of blood stasis, he inquires, does this act " by rendering nutrition more active, just as a limb which is much exercised acquires great development, or does it rather

* Traité pratique sur les Maladies des Organes Génito-Urinaires. Par le Dr. Civiale. Ed. 3°. Paris, 1858. pp. 244 et seq.

retard the process of decomposition, rendering less easy the separation of elements which otherwise would be eliminated?" He confesses his inability to answer this question.*

In considering this subject for ourselves, it will be desirable first to examine the alleged causes of senile enlargement of the prostate, and show why many of them have no title to be so regarded. And first, inflammation must be eliminated from the category. Let us contrast distinctly the enlargement of youth or middle age with that of advancing years. Nothing can be more calculated to originate erroneous views than the habit common to most authors of disregarding this important distinction. Thus, "the complaint" is said to be "very common among elderly persons, but occasionally met with also in early manhood." But no two affections can be more different than those which are thus confounded. One category cannot be made to contain them both. In youth the organ becomes enlarged by interstitial plastic effusion, the result of inflammatory action. In age, there is an unnatural development of the prostatic tissue itself. Histological examination of its elements, already sufficiently considered, shows that the redundant parts are in no way due to the inflammatory process, in any of its modifications. There is no proof that the enlargement which is constituted by undue production of the fibrous and muscular elements, whether in the diffused form or in that of tumour or outgrowth, is a product, direct or indirect, of inflammation in any part of the canal. The newly-formed tissues are not the result of morbid deposit dissimilar to the organization to which they are added, but of an immoderate development of the elements proper of the part. The action of inflammation, and the deposit of its products in a tissue, so far from favouring growth, is directly antagonistic to such a process. A prostate, therefore, which has been enlarged by inflammatory effusion, is, *cæteris paribus*, most probably less

---

* Récherches Anatomiques, Pathologiques, et Thérapeutiques, sur les Maladies des Organes Urinaires et Génitaux, considérées spécialment chez les hommes agés. Par L. Auguste Mercier. Paris, 1841. Chap. iv. pp. 218-233.

likely subsequently to exhibit a hypertrophic tendency. Nutrition is thus impeded, not encouraged. Inflammation must, therefore, be excluded from the list of causes.

Stricture of the urethra, and calculus of the bladder, are frequently stated to give rise, probably by irritation, to hypertrophy of the prostate. Respecting the first, the fact, as determined by numerous observations of the dead body, and careful examinations of the living, is, that a co-existence of stricture and senile enlargement of the prostate, is certainly not common. Obstruction to a catheter encountered beyond the stricture, and produced either by enlarged lacunæ, dilated urethra, or undue development of the muscular structures at the neck of the bladder, constituting more or less of a barrier there, is common enough, and has often been attributed to the complaint in question—but erroneously. Calculus of the bladder may, in a similar manner, induce the last-named state, which arises under circumstances of prolonged irritation whatever its cause (see Chapter XIX.), but not the hypertrophied prostate. Were the latter a result, we should surely sometimes meet with it in childhood and youth, periods of life at which stone is prevalent.

Habitual engorgement of hæmorrhoidal and prostatic veins is very confidently held by many as among the best-established causes of enlargement. In this manner, sedentary occupations are considered as predisposing causes. Anything which tends to obstruct the return of venous blood from the pelvis, such as mesenteric or hepatic obstruction, or the like, is ranged under this head. Great stress is laid by some writers upon the venous enlargement and stasis which elderly persons in particular are not uncommonly the subjects of. Undoubtedly, hæmorrhoidal swellings are thus produced, and most frequently. But is there any analogy between this effect and that observed in enlarged prostate? between congestion and thickening of tissues, by exudation from overcharged blood-vessels, and the new formation of normal structures? Do varicose veins lead to the latter result in any other part of the body? Assuredly not. The effect of

K

venous stasis in the leg may often be seen in thickening of the integuments, and distension of the capillaries from which the veins arise, in the occasional occurrence of inflammatory action and consequent exudations or deposit into interstices of structure, but never in the increased production of pre-existing normal and healthy tissue. True hypertrophy, outgrowth, or tumour—not one of them has venous congestion for a cause. Venous congestion impairs structure, diminishes its vitality, and, often enough, predisposes to ulceration the tissues affected by it, so that the slightest injury produces the destructive process, but it never augments the vital force, or stimulates growth. On these grounds, then, it must be dismissed from the list of causes of hypertrophy of the prostate.

Gout and rheumatism have been made to perform a part in the category of causes in this, as in that of almost every other obscure affection; but without the smallest foundation in observed facts. Elderly people often have rheumatism, and are subject to enlarged prostate. I confess, after a careful investigation of the subject, that I know of no closer relation between the two affections. Nor can I say more of gout. That either diathesis has any causal relation to the prostatic complaint I do not believe.

There is not the slightest foundation for regarding syphilis as a cause. Perhaps it is not possible to speak with the same degree of confidence in regard of sexual excess. Much influence has been attributed to the effect of habitual indulgence of this kind; but, from the fact that the affection has been observed to occur in individuals known to have been remarkable for chastity, the opposite extreme of continence has been regarded also as exercising a similar influence. In regard to the first, it appears reasonable to believe that repeated use may induce hypertrophy here as elsewhere; while, without entering upon the question of the prostatic function, it is impossible not to associate the organ with the sexual act; and, admitting these, it appears not to be easy to escape the inference that hypertrophy is likely to result from sexual excess. Yet facts do not favour this view; hypertrophy does not exist when the function is in greatest vigour, and is not called into immediate existence by the most licentious excesses indulged during the prime of life. And it must be admitted that, when in any part of the body a hypertrophy is developed, it is coincident with, or, at all events, immediately follows, the increased action which induced it. Such is the universal law, and illustrations of its action must be familiar to all.

Supposing, for argument's sake, that we regard the prostate as a secreting gland in the fullest sense of the term, its enlargement cannot be considered as affording a parallel, or even a very similar pathological result to that which occurs in hypertrophy of other such glands. All its component tissues are not augmented in their relative proportions. There is no analogy between its enlargement and that of a hypertrophied kidney, for example. Considerable augmentation in bulk may take place in the prostate, when the glandular elements appear not to be increased at all, or very little, and that in either case, whether defined tumour be present or not. But that the glandular element may also be largely increased is no less a matter of fact.

On the other hand, suppose that it be regarded as a muscular organ, which is permeated by a glandular apparatus, and its hypertrophy may be compared to that of the uterus or bladder, both of which, when in that condition, maintain the glandular tubes and follicles which belong to their lining membrane unaltered. A consideration of the structure and position of the prostate has suggested that its function is partly a mechanical one, and, so far, analogous to the two organs just named. It has been regarded as an important portion of a muscular apparatus bearing pretty nearly the same relation to the seminal fluids, as regards the act of propulsion, as the bladder does to the urine.*

But that the enlargement of the prostate is not a mere muscular hypertrophy, induced by increased action, and corresponding with

* It is no part of the design of this work to enter upon the difficult subject of the function of the prostate. In my work on Stricture of the Urethra, which gained the Jacksonian prize of the Royal College of Surgeons for the year 1852,

the degree of augmented function excited, is proved by the facts just adduced, of its non-appearance during the terms of youth and prime manhood.

Nevertheless, viewing the phenomenon as involving a hypertrophy of the involuntary muscle, a condition the existence of which cannot be disputed, whatever be the opinion held as to the function of the organ, we may inquire whether the causes of such hypertrophy in other parts of the body, similarly constituted, have been ascertained; and if so, whether anything may be gained by analogical reasoning, in elucidation of the subject before us.

There is but one other organ in the body which is similarly constituted as regards the nature of the constituent tissue, and in the manner of its aggregation, a fact enlarged upon in the preceding chapter. The uterus, like the prostate, is composed of the inorganic muscular tissue distributed in thick strata, so as in either case to form a thick mass, not in thin planes, as found in all the other organs in which this tissue appears. The tendency to become the seat of local and general hypertrophy, of isolated tumours and outgrowths of a special character, which both organs equally manifest, has also been demonstrated. Starting from this remarkable fact, it is difficult to resist the inference that this tendency to overgrowth, this disposition to generate fresh elements identical in character with those proper to the structure of the organs, has a source common to both, and perhaps inherent as a kind of structural, or perhaps functional necessity. The capability of this structure for exhibiting rapid and enormous increase under certain circumstances, is admirably exemplified by what

---

I stated my belief that its function was partly that of a muscle, that it was an important element in the apparatus designed to expel the seminal fluids—*vide* pp. 31 and 47—and further investigations have but confirmed that view.

Professor Ellis says, in a paper already referred to,—"It (the prostate) may be considered as only an advanced portion of the circular layer of the bladder, though it must have the power of acting independently of the vesical fibres, as, for instance, in the propulsion of the seminal fluid. Its chief office will probably be to hurry on the semen, and deliver this into the grasp of the voluntary muscular fibres of the constrictor urethræ."—*Med. Chir. Trans.*, vol. xxxix. p. 332. 1856.

happens to the gravid uterus. A dormant force is awakened through the presence of the impregnated ovum, and the weight and bulk of the organ is in a few months increased tenfold. Active determination of blood is coincident, and doubtless supplying the materials of nutrition, but not venous congestion, nor any one of the numerous alleged causes of prostatic hypertrophy already referred to. But the uterine function having ceased temporarily or permanently, the organ diminishes and returns, sooner or later, nearly to its original size. During the latter moiety of the term of reproductive activity, the uterus is exceedingly prone to develope formations, identical in structure with its own, but more or less isolated from the parent tissues, either in the form of tumours or outgrowths, and these are associated with general development of the normal parts of the organ. These phenomena are observed, perhaps, with greater frequency in the virgin than in the impregnated female, showing that they do not depend upon any force called into play by pregnancy, but on one irrespective of it, and possibly inherent in the structure of the organ, or associated intimately with some function peculiar to it.

It is an interesting circumstance that the prostate, male homologue of the uterus, should exhibit analogies in many points of view with the latter organ in regard of its tendency to overgrowth. The most obvious explanation, and the conclusion which, after a careful examination of the subject, is that which appears to me better supported than any other, seems to be offered in the simple fact now completely established, that the structure in both is exceedingly prone to develope (as already shown, Chap. VI.) among its component elements, minute independent or isolated formations, possessing an organization identical with itself; which formations, in the majority of cases, do not increase beyond a certain very limited size, and do not interfere with the performance of any known function in either sex, but which in exceptional instances, continue to be developed, for the most part only, during a certain limited period of life, say, in general terms, between thirty-five and fifty in the female, and between fifty and seventy in the male; in the one case appearing in the form of uterine hypertrophy or tumour, in the other in that of prostatic hypertrophy or tumour. Whether the formation of these products is anything more than a contingency of structure, that is, whether it be connected with any functional action common to the structure in both cases, is more doubtful.

Anatomical examination of the enlarged organs, prostate and uterus, demonstrates the arteries and veins to be both enlarged, the latter, probably, as a result of the former. An increased supply of arterial blood is coincident with the increasing size of the organ; but whether the vascular determination precedes or closely follows the commencing development it would not be easy to affirm.

Are there any circumstances in the mode of life, or of pre-existing disease, which we are warranted by reasonable evidence in considering causes of prostatic hypertrophy? I know of none. The fact that almost all known causes of diseases in general have been alleged to be so of this one in particular—what is it, in reality, but a tacit expression of the same opinion? Every diathesis—gouty, rheumatic, tubercular, syphilitic, has been arraigned as the offending cause. Every form of local excitement possible to the pelvic viscera has been similarly held accountable. Thus it follows that the bearing of any single circumstance becomes neutralized in the concourse of numbers. Every proposition finds its refutation in the presence of some other one among the multitude.

The origin of hypertrophy being thus attributed to a necessity of structure, no doubt all circumstances which tend to induce active determination of blood to the locality may aid in its development. Thus we find emotional excitement of a sexual kind, and actual excesses, over-stimulating food, sedentary habits, horse exercise, and many other conditions having a like tendency, enumerated among the causes of this affection. But the *initial* step in the causation of hypertrophy is, I believe, independent of, and probably uninfluenced by, any of these circumstances, although they doubtless tend to increase already-existing disease. And thus it is that much may be done by judicious treatment, by well-directed management, to retard the progress of enlargement, that disposition being exceptional to the majority of cases, even where the seeds of overgrowth exist, as in the form of minute tumours or commencing outgrowth. All that tends to diminish the local supply of arterial blood to the organ may be held to favour the condition of *statu quo*, or slow increase. This, however, is not the place to enter further on the subject of treatment; the allusion made is sufficient to illustrate the question under consideration.

We have now to inquire what are the peculiar conditions, actually ascertainable by inquiry, under which hypertrophic enlargement of the prostate, or tendency thereto, is developed.

It never appears but in advanced years. But it is not, therefore, a natural or necessary concomitant of age. It is, on the contrary, a condition which a very large majority of elderly men escape. Contrary to the generally-received opinion, its occurrence is not normal but exceptional. An analysis of various particulars given of the dissection of 164 prostates from individuals at and over 60 years of age (see Chap. II.), presents an opportunity of determining with tolerable accuracy the average proportion of cases in which after that year has been reached, the organ may be expected to become the subject of hypertrophic and atrophic changes.

Of 164 individuals, ranging between 60 and 94 years of age, in 97 the prostate was unaffected by either change.

Of the remaining 67, eleven were affected by atrophy, and 56 by hypertrophy.

Of the 56 cases of hypertrophy 26 were marked specimens, *i. e.* examples which weighed 10 drs. and upwards (10 drs. to 48 drs.). The other 30 were slighter but undoubted examples (5¾ drs. to 9¾ drs.).

Of the 164, 57 were the subjects of isolated tumour in the prostate. In about one-fourth of these cases the organ did not exceed the natural weight, they are therefore not ranked with hypertrophied prostates; in the remaining three-fourths, tumour and hypertrophy coexisted.

Of the 56 cases of hypertrophy, nearly one-half made no complaint respecting symptoms during life. It is to be recollected, however, that it is exceedingly common for slight deviations from the normal function to take place in elderly persons without complaint being made; it is not, therefore, to be inferred that no symptoms of the existing enlargement were present. Most probably, an inquiry would have elicited the fact, that it was the habit of many of these patients to rise two or three times during the night to relieve the bladder, and that during the day, their frequency of micturition was greater than that of the healthy subject.

With the present data, therefore, it appears that actual hypertrophy of the prostate exists in about 34 per cent. of men at and above 60 years of age.

That it produces marked symptoms in about 15 or 16 per cent.

The following results of this investigation are extremely interesting.

The average age of the 108 individuals unaffected with hypertrophy, was 75·02 years.

The average age of the 56 individuals affected with hypertrophy, was 75·03 years. For all purposes these two amounts must be regarded as identical.

The average age of those in whom the enlargement was considerable, 26 in number, was 74·46 years.

The average age of those in whom the enlargement was less considerable, 30 in number, was 75·53 years.

It must be borne in mind that all these cases were placed in favourable hygienic conditions, and had the benefit of constant and careful medical supervision. Had they wanted both or either of these advantages, the result must have been greatly different. Prolongation of life in the later stages of the affection depends mainly on the concurring existence of the two conditions named.

This fact is one of considerable importance, and ought to be known to those who are the subjects of this complaint, since they are often prone to regard it as necessarily tending to shorten the natural term of life. They are entitled to the benefit which a more cheering view of their condition and prospects cannot fail, in most instances, to impart.

Among the 108 unaffected by hypertrophy, were individuals of greater age than any among the affected portion; for example, three above 90 and many above 80 years of age. But among the 56 hypertrophied, there were no less than 14 of 80 years and upwards, one reaching 87.

Besides the 164 cases thus analysed, were 20 specimens below 60 years of age, which formed part of my first series of 50. Among them were 13 cases between 50 and 60 years; only one of these exhibited any sign of hypertrophy: the age was 56 years; the enlargement was slight, evidently in the early stage, and affecting pretty equally the whole organ.

The period of life between 55 and 65 is that during which the affection most commonly begins to be developed. I have never been able to meet with an instance of its occurrence at so early a period as 50 years of age. On the other hand, it appears rarely to commence after 70. Where it exists, the disease has generally made considerable progress before 70 or 75. Consequently, it is met with but unfrequently in later years, and is exceptional after 80 or 85. It is not altogether unknown, however, rare examples having been met with at a much more advanced age. Dr. Beith has placed on record a case in which prostatic enlargement, and sacculated bladder as its consequence, formed the only abnormal conditions observable in the body of a man who died at the age of 103.*

It may then be regarded as established by these facts, that hypertrophy of the prostate, so far from being a change invariably, or even usually present in old age, is an exceptional condition. And it may be further regarded as highly probable, that

* Trans. Path. Soc. 1850-51, p. 124.

a slight tendency thereto, almost, if not quite, unrecognizable during life, may occur in about one out of three individuals after 60 years, and that a marked enlargement may be met with in one out of seven or eight, at that age and upwards.

It is germane to this subject to remark here, that, on the other hand, atrophy does not appear to be an effect of old age when hypertrophy is not present, as has been alleged. This subject will be further examined in a subsequent chapter, devoted to the consideration of Atrophic changes in the Prostate.

CHAPTER VIII.

THE SYMPTOMS OF HYPERTROPHIED PROSTATE.

Onset of Symptoms sometimes very gradual.—Sometimes sudden.—Phenomena first noticed.—Those which subsequently occur in their order.—Incontinence or OVERFLOW.—Characters of Urine.—Nature of "Ropy Mucus."—Complication with Calculus.—Bladder often much distended before existence of the Complaint is discovered.—Last Stage.

IN the earliest stage of hypertrophy of the prostate, there are no symptoms sufficiently marked to attract the attention of the patient. It is probable, indeed, that a very considerable period, varying in different cases from a few months to some years, is passed between the actual commencement of enlargement, and the occurrence of anything which is observed to be unusual in the act of micturition, or of any kind of derangement arising from the organic changes which are taking place. The length of time which elapses between the onset of disease and the manifestation of symptoms, depends mainly upon two conditions; first, upon the character of the enlargement itself; and secondly, upon the constitution of the patient. In respect of the first condition, we shall see hereafter, in what degree the nature of the enlargement influences not only the severity, but the kind of symptoms produced; and learn also how it is that even a considerable enlargement of one portion of the organ may produce little or no inconvenience for a long period, while a much slighter increase otherwise situated, may be the cause of great disturbance, both locally and generally. Secondly, there is in these cases, as in all others, that natural idiosyncrasy of the individual patient, which in one instance enables him to withstand the inroads of disease, and to a great extent adapt his constitution to altered circumstances, and in the other disposes it to yield, without much attempt to rally, to the morbid influences to which it is subjected.

Taking into account the variations which are thus encountered in the observation of numerous cases, the symptoms exhibited may be described as taking generally the following form and order of appearance.

It occasionally happens, as will be hereafter shown, that the occurrence of complete retention after some obvious exciting cause, as exposure to cold, &c., is the first announcement of the existence of prostatic enlargement. When, however, this is not the case, one of the earliest signs generally observable is a manifest diminution in the force with which the urine is ejected. The urine also makes its appearance less quickly than natural after the effort to expel it has been made, and a certain hesitation or uncertainty is experienced, before a stream is fairly established. The size of this is not necessarily much smaller than it was in health, but it cannot be projected so far by the ordinary amount of effort, neither can its force be much augmented by additional effort, a circumstance, occurring as it does in numerous but by no means in all cases, which is unlike to that which is observed in stricture of the urethra, where, however feeble the stream, increased action of the bladder almost invariably tells to some extent upon it. Indeed, there is no doubt that, in some forms of prostatic enlargement, increased efforts to make water do but augment the difficulty in micturition. A powerful contraction on the part of the bladder appears in such cases to add to the obstruction at its neck, a result which it is not difficult to understand when we observe the form which examination of the parts sometimes reveals. The desire to pass water becomes more frequent than natural, not greatly so at first, and the relief afforded by the act of micturition is less complete; there is not the satisfaction following it which is ordinarily experienced in complying with the dictate of nature in a state of health. Frequently the want occurs again in a few minutes, especially after the first effort on rising in the morning, when the bladder has become distended during sleep. In course of time, however, the act must be more repeatedly performed, and the period of night is no longer exempt from the calls to pass water. Pains in the groins, testicles, and thighs are sometimes, but by no means commonly, complained of at an early period. A sense of weight, fulness, or indefinable uneasiness about the perineum, rectum, and hypogastrium is felt, which the patient soon, almost instinctively as it were, learns to refer to the neck of the bladder. He is often subjected to annoyance by an unpleasant odour in the urine, which is new to him, and describes it as being strong and disagreeable—a symptom which to the feelings of some men is particularly repulsive. As the expulsive efforts to pass water become of necessity more vigorous and frequent, irritation of the rectum is experienced in a greater or less degree; the contents of the bowel are more frequently passed from inability on the part of the patient to prevent the act of defæcation accompanying that of micturition; and tenesmus, protrusion of the mucous membrane or prolapsus, and hæmorrhoidal swellings, are apt to result. The associated action of the bowel is the more likely to take place, if the enlargement of the prostate is developed in a direction backwards into the bowel, rather than forwards into the bladder. In this manner, an ever-recurring sensation of the existence of some matter in the bowel requiring removal is occasioned, and unavailing efforts are made to obtain relief at stool. Much stress has been laid by some writers, following J. L. Petit, who seems to have originated the idea,* on the appearance of flattened stools, as an indication of this form of enlargement, but as far as I have been able to judge, without sufficient grounds. I do not doubt that in some cases the motions may be passed in the form described, but should rather believe this to be an effect not of the protruding prostate, but of some action of the sphincter ani, which exercises a much greater influence upon the form than the organ in question can do. Of this I am certain, that in patients who have

* Traité des Maladies Chirurg. Tome iii. p. 24. Paris, 1790. Ouvrage l'osthume. Par J. L. Petit.

continued long under my observation, whose prostates have been found to protrude backwards very considerably, from rectal examination made by myself, this appearance has not been presented. Again, I have not less certainly observed it in others who have had neither stricture of the bowel nor enlarged prostate to occasion it. As the disease advances, pains directly associated with the condition of the bladder begin to form a more serious part of the sufferer's complaints. The constant aching and gnawing sensation behind or about the pubes which almost invariably attends on distended bladder or severe chronic inflammation of that organ, becomes one of the most trying consequences of its impeded function. Soreness and smarting in the direction of the urethra are frequently felt; and an aching or shooting pain extending to the glans penis, in which it is most acutely perceived. At the same time it is not uncommon that some muco-purulent discharge appears by the urethra; this varies with circumstances, sometimes appearing after exposure to cold or damp; or with an attack of retention; or with exacerbation of all the symptoms, depending, perhaps on a constipated state of the bowels; and then subsiding rapidly and altogether under appropriate treatment. At such times it occasionally happens, as I have myself observed, that the irritation involves the testicles, which become unusually tender, and slightly swollen; while the urine is expelled with greater difficulty, or requires to be drawn off for a few days by the catheter. Vascular excitement of the penis, producing frequent erections, is also at times a concomitant symptom. But as the complaint has thus been manifesting gradually but most unmistakably the signs of progress, it is certain that the relief to the bladder in the act of micturition has been slowly becoming less complete, although the efforts to obtain it have become most painfully frequent, and wearing to the constitution. While in some rare cases, exceptional to the general rule, the effect of the growing enlargement of the organ may be to open or loosen the neck of the bladder, and thus permit a real incontinence, a condition in which the viscus

SYMPTOMS OF ADVANCING HYPERTROPHY. 143

has been rendered unable to contain more than a small portion of the urine; the result is by far the larger portion is, that the neck, the urethro-vesical outlet, is abnormally closed, and requires a preternatural amount of contractile effort in order to drive the fluid through it. Hence the bladder is never emptied, its contents not being expelled below a certain level: the act taking place only when to the natural power of the organ is added the weight of a quantity of fluid adequate to distend it, conjoined with the extra pressure which is derived from the mechanical elasticity of its walls under these circumstances, and the action of the abdominal parietes, telling of course with greater vigour in proportion as the size of the body to be acted upon is increased. It is not difficult to see that these evils, if unrelieved by art, must inevitably increase: the capacity of the bladder yielding to the constantly-augmenting demand upon it. Hence at length the organ becomes habitually filled, and the surplus only flows off at each act of micturition, and that often not by a stream, but rather by a succession of drops. At night, when voluntary control is suspended by sleep, urine drains away, to the great discomfort of the patient. At length the same thing happens by day also, and the condition of the sufferer becomes painful in the extreme, his person and clothes exhaling a most offensive odour of the stale and diseased secretion, his participation in society of any kind being wholly impossible, while the keenly-felt consciousness of having become a source of annoyance to those about him is a most painful addition to his own personal sufferings. This condition is generally described as incontinence, a misapplication of the term, as we shall hereafter see, which has been productive of fatal errors in practice. A much better term is that employed by the French surgeons, namely, "regorgement" (overflow); and this I shall for the future employ, as aptly indicating a condition which, so far from being one in which the bladder *cannot retain*, is one in which it *retains too much*. For this reason, in my work on "Stricture of the Urethra," published in 1853, I employed the phrase "retention with incontinence" to designate the condition described, from the inapplicability of the word referred to. But the shorter term of Overflow is certainly preferable, and I shall make no apology, therefore, for seeking to naturalize the term in this its English form.

A sign which should be looked for in such cases is the existence of dulness on percussion above the pubes, and the degree, if present, to which it extends. Frequently in chronic retention the bladder may be felt, and its limits defined by the dull note elicited as high as the umbilicus, although more frequently to the extent of three or four fingers' breadth above the pubes.

As the complaint advances, it becomes a matter of difficulty to commence the act of micturition, much straining taking place before the obstruction at the neck of the bladder is overcome, and the urine begins to issue. The patient is sometimes compelled to take a position more favourable than that of standing erect, and he leans forward in order to make the urethro-vesical orifice of the bladder the most depending portion, this having been tilted up, and raised considerably above the level of the floor of the viscus, when much enlargement of the median portion of the prostate exists. His legs also are extended in order to obtain a firmer base of support during the powerful efforts he is obliged to make, although often fruitlessly. Indeed, so forcible are the exertions which patients undergo in these circumstances, that a hernial protrusion has not unfrequently been thus caused. Meantime the constitution exhibits the effects of the advancing local disease, of pain and broken rest. The patient loses flesh, becomes pale or sallow, his appetite fails, and already advanced in years, he "ages" rapidly. There are frequent febrile disturbances, and the strength rapidly decreases. Very slight irregularities, or exposure to adverse circumstances, disregarded with impunity in health, produce extreme distress from the severity of the symptoms occasioned. Attacks of complete retention are impending on these occasions, and thus also are embarrassed the vital functions of the kidneys, which have by this time become impaired through the long-continued impediments to the discharge of their secretion. Hence uræmic poisoning, inducing coma and death, is one of the modes by which the fatal event sometimes takes place, more especially when the sufferer has been a victim of the disease unchecked in its course, and unrelieved by art.

The occurrence of hæmorrhage to a trifling extent is a frequent accompaniment of the complaint. It thus sometimes relieves congestion, and may be to a certain degree salutary: often it occurs after exposure to cold, sexual excitement, or other circumstances which tend to produce a vascular determination to the pelvic viscera. Sometimes it results from the imprudent use of the catheter; and it may then happen to an alarming extent.

The characters presented by the urine are important, and should be carefully noted. They are such as mainly depend on decomposition of some of its constituents from abnormal retention, mixed with the products of chronic inflammation of the bladder. Accordingly the first deviation from health noted by the naked eye is that it is no longer transparent, but a little cloudy, often pale, of a greenish tint, with a few shreds or flocculi suspended in it. More or less of mucous deposit slowly falls and floats rather than settles at the lower part of the vessel containing the urine, when it is set by. A thin pellicle forms on the surface, more or less whitish and opaque, sometimes iridescent. In later stages the mucus increases in quantity, and appears as the glairy, tenacious, slimy, adhesive matter so well known to be associated with chronic inflammation of the bladder; not miscible with the urine, it adheres to the side of the vessel, and follows the urine as it is poured out from one to another, in a mass which it is difficult to separate. In advanced cases this mucus sometimes exhibits traces of calculous, generally phosphatic, matter, in the form of small amorphous masses of soft consistence and whitish colour. These may be drawn out into long linear forms, and give a streaked appearance to the deposit. When the mucus has subsided there is sometimes deposited upon it an opaque creamy-looking layer, which is unaltered pus, mixed with crystals of the triple phosphate in varying quantity.* In any stage of the complaint, before such matters are observed in the urine, or long after their appearance, the urine may be darkened in colour from admixture with blood. The tint is not red, or rarely so, except from recently-effused blood, and then depending, perhaps, upon the employment of instruments. It is much more commonly of a reddish brown, or dirty hue, which the colouring matter of the blood assumes after mixture with urine, especially that which has become somewhat decomposed.

The chemical reaction of the secretion is at first neutral, then alkaline in various degrees of intensity. The odour is pungent, ammoniacal, often fœtid, sometimes extremely so. These characters depend, to some extent, upon the quantity passed, that is to say, upon the degree of dilution with water, in which the solid constituents proper to the excretion are passed. This often varies considerably in the same patient from day to day; the measure being sometimes below, but more generally much above, according to my experience, the natural or healthy standard.

Under the microscope may be observed tesselated and spheroidal epithelium, blood corpuscles, and pus corpuscles; globules of a granular appearance, resembling the latter, but three or four times as large, are also frequently seen; these exhibit a tripartite nucleus with acetic acid, and are commonly found in cystitis, from whatever cause. Besides these organic elements, there are usually

* The so-called mucus, which is frequently passed in very large quantities from the bladder, is, according to my experience, usually a compound of pus and mucus, in variable proportions. It has by some been considered as merely pus rendered viscid by the addition of alkali. After repeated careful observations, chemical and microscopical, I find that its composition may be very different in different cases. In some it appears, after the addition of very dilute acetic acid, sufficient only to neutralize the alkalinity of the fluid, to be made up almost entirely of corpuscles, with multiple or irregular nuclei: in such cases, I presume it is altered pus, and little else. In other instances the proportion of the corpuscles is very small in quantity when compared with that of the amorphous or very faintly striated viscid fluid in which they are suspended, which appearance leads me to conclude that the secretion proper of the mucous membrane predominates in the mixture over the purulent formation.

many of the prismatic crystals of the triple phosphate of ammonia and magnesia in great variety of form and size, and the amorphous granular matter, free, or adhering in thin flakes, of phosphate of lime. Other crystalline forms may be present also, such as uric acid and oxalates, but the first-named constitute the typical forms which are present in the urine of these cases. Albumen is not present until introduced either by the admixture of blood or pus; or, in later stages, by invasion of the kidneys by disease, either of an inflammatory character, through extension; or as a result of the disorganizing processes occasioned by regurgitation of urine from the bladder, and the pressure and interference with its excretory forces so occasioned; or by organic changes acting through the agency of a vitiated circulating fluid and impaired constitution. Under these circumstances, the casts of the uriniferous tubes will most probably be detected at times by the microscopic observer, and albumen will be thrown down in larger quantities on application of the usual tests. But it is by no means to be regarded as a common occurrence that degeneration of the kidney is associated with enlargement of the prostate, nor, indeed, that the local condition has more than a small influence in producing the renal disease.

As a result of long-continued disorder in the urinary apparatus, and of the changes in the urine itself, which have been thus described, it is not surprising that the formation of calculus not uncommonly takes place. Its presence will be suspected if sudden impediment is frequently experienced to the stream of urine; if there is much pain about the neck of the bladder, at or following the act of passing it; if the pain at the end of the penis is unusually severe, and if the blood and pus are very frequently observed, or are disproportionate in quantity to the degree of urinary retention or obstruction manifested, and especially if fragments of calculous matter have from time to time been passed. But it is true that the existence of calculus is sometimes masked by the prostatic disease; in the first place because many of the symptoms are common to the two disorders; and secondly, because the conformation which the neck of the bladder assumes in the latter affection, tends to prevent the occurrence, in some measure, of the most distinctive symptoms of stone; inasmuch as the foreign body is less liable to be engaged in the vesical neck, but lies back deeply behind the enlarged prostate. Hence, neither the stopping of the stream of urine, nor the pain immediately following micturition, may be very obviously present under the circumstances referred to; and yet a calculus may exist in the organ, and be the source of irritation both to it and to the system at large, giving rise to the secretion of large quantities of pus and mucus, of which it, and not the enlarged prostate, may be in greater part the cause.

The same absence of symptoms is favoured by atony of the bladder, the coats of the viscus not contracting upon the calculus, and so not producing the exquisite pain occasioned by the grasp which follows the act of micturition, in the bladder not so affected.

But, although such is the ordinary course of unrelieved disease of the urinary organs, resulting from enlargement of the prostate, the earliest signs of this affection are not always to be observed, as has been already remarked, in the gradual order of progression here presented. On the contrary, it is by no means an unprecedented occurrence for an attack of complete retention to reveal its existence for the first time. It is not that the progress of the disease should necessarily have been rapid under these circumstances; but that the impeded urinating function, although, probably, of long duration, has not excited any very obvious symptoms. It is still commoner to discover that the impediment has long existed when its presence has been least suspected. A patient, for example, who passes a sufficient quantity of urine daily, without particular effort, and more frequently than natural, who is conscious, also, it may be, of some little escape involuntarily during sleep, or even in the day on making any effort which requires straining, or, in other words, a strong contraction of the abdominal muscles, is very apt to think that his water passes with unusual—indeed, perhaps, with too much freedom; and the last thing in the world he dreams of is the existence, in his own person, of any obstruction in the urinary outlet. The condition of such a patient, too, has sometimes been overlooked, even by his medical attendant, and no specific investigation of the bladder is made. The apparent freedom of micturition has masked the real malady, and the treatment is directed only to those symptoms which have been productive of most discomfort or anxiety to the patient; it may be, only to the general *malaise*, or some febrile condition not uncommonly resulting from the hidden cause. The march of events, however, must ultimately arouse suspicion as to the state of the bladder; a catheter is passed, and, greatly to the astonishment of the patient, and sometimes scarcely less so in the view of those who have long watched him closely, some thirty or forty ounces of urine, or even a very much larger quantity, may be drawn off, notwithstanding that the act of micturition has been just performed. Now, it is during the prevalence of such a state of things that unaccustomed exposure to cold and damp, or undue indulgence in alcoholic drink, or in sexual excitement, may suddenly produce congestion of the already enlarged prostate; and a condition of complete retention, thus induced, may be the means of discovering the existence of the affection for the first time. After this the habitual distension may be greatly lessened in degree by the daily use of the catheter, but it rarely or never happens that the bladder is able to regain the power of evacuating its contents completely, as it does after simple over-distension of its coats, resulting from unnatural retention, when there is no organic obstruction at the neck.

The last stage may sometimes be indicated more by the signs of a gradual decline of the powers of life, than by those of advancing obstruction; on the other hand, the final symptoms are sometimes those of rapid depression, consequent on sloughing of a portion of the organ, and repeated hæmorrhages, with or without the infliction of instrumental injury; or of gradual exhaustion, from constant discharge of pus and mucus from the bladder; or, lastly, of urœmic poisoning of the system, from the failure of the eliminating function of the kidney.

CHAPTER IX.

THE EFFECTS OF ENLARGED PROSTATE IN RELATION TO THE FUNCTION OF MICTURITION.—RETENTION.—INCONTINENCE.—ENGORGEMENT AND OVERFLOW.

Retention of Urine, more or less considerable, the general result of Enlarged Prostate.—Contrast between Retention and Incontinence.—Retention due to Obstruction, not to Paralysis.—True Paralysis of the Bladder extremely rare, except from Lesion of a Nervous Centre.—Overdistention and Atony of the Bladder.—Tabular View, showing various degrees of Obstruction and corresponding results.—Engorgement and Overflow.—Importance of last-named Symptom.—Commonly confounded with Incontinence.—When does real Incontinence exist?—The effects on the act of Micturition produced by the various organic changes in the Bladder, Ureters, &c., which occur as the result of Enlarged Prostate.

MECHANICAL obstruction, which may be situated either at the neck of the bladder or in the urethra, is the chief cause of chronic retention of urine; that is, a state in which the patient, being unable by his unaided efforts to empty the bladder, retains there a certain portion of urine, varying considerably in quantity, just as the amount of obstruction itself varies, unless the fluid be withdrawn by artificial means. By far the most common cause of this condition is enlarged prostate; and when the obstruction is considerable, it may be that no urine at all is passed by voluntary effort. The bladder then becomes permanently distended, unless the catheter be employed; and the fluid gradually increasing in quantity at length opens out the orifice, and flows off spontaneously. To designate this phenomenon the term Incontinence was originally applied, and is still employed by many, although it has long been well known that the condition so described is in reality the very reverse of *incontinence*, since the bladder already contains too much, and the surplus only overflows, the viscus often retaining much more than its capacity, in a state of health, would admit of. The bladder is in fact *engorged*, and the urine *overflows*.

The phenomenon of involuntary micturition in elderly persons is very frequently accounted for, not on the ground of existing obstruction, but on that of paralysis affecting the bladder. It is supposed that either the neck or the body of the bladder may be separately paralyzed, the remaining portion retaining its normal supply of nervous influence—a pathological state the existence of which it would be very difficult if not impossible to prove, and which, if it be an actual occurrence, is certainly extremely rare. On this theory, however, it is said that when the neck is paralyzed and the body unaffected, the vesical outlet becomes patent and incapable of contracting, and that the urine flows off as fast as it escapes from the ureters, while the bladder itself remains empty. The term Incontinence has been employed to designate this condition also, although it presents a state which is the exact reverse to that already so described. But in this case, whether a nervous lesion be the cause of the phenomenon or not, the term is appropriate, because the bladder is unable to retain; the condition may, therefore, be very accurately described as one of incontinence of urine.

On the same theory, also, retention of urine is supposed frequently to be caused by the converse form of paralysis of the bladder, that is, when the neck retains its nervous supply, and the body losing it, becomes unable to expel its contents. Thus it will be seen that the term Incontinence comes to be frequently applied to precisely opposite states of the containing function of the bladder. Hence the misunderstanding, the difficulties, and the grave errors in practice which sometimes occur, especially to the student, in connection with this subject.

Now, without discussing here at length the question already raised, as to whether these local deprivations of nervous influence do, or do not take place—in the one case affecting only the neck of the bladder, at other times the body alone—I have no hesitation in affirming that in a great majority of the cases in which habitual retention of urine, with overflow of a surplus portion, exists, the cause is palpable and physical, and not impalpable or dynamical; a fact which in each individual case may be ascertained by examination; in other words, there is *an organic obstruction in some part of the urethra*, situated either at its commencement in the neck of the bladder, where it is usually constituted by enlarged prostate; or in a portion of the canal anterior thereto, where it usually takes the form of permanent or organic stricture.

We are, I believe, indebted to Mercier of Paris for first calling attention forcibly to the important fact that organic obstruction, not local paralysis, or impaired nervous supply, is the great, and almost universal, cause of the various states which are described as retention and incontinence, when existing in elderly individuals who present no sign of impaired nervous power in other parts of the body. To his able discussion of the subject I would refer my readers for the arguments in favour of this view.* It is not entered upon here, because I have preferred regarding it as a question of fact, rather than as a theme for abstract reasoning. The obstruction is, or is not present; if the former, it may generally be verified. Experience alone has led me to reject the impalpable cause, and to appreciate the material one, and to an extent sufficient to warrant me in referring to the fact alone for corroboration of the assertion made above.

In 19 cases out of 20, excluding two classes of cases which will be immediately named, the symptoms described are invariably associated with permanent obstruction of some kind, as

* See Recherches Anat. Path. et Thérap. sur les Maladies des Organes Urinaires et Génitaux, considérées spécialement chez les hommes agés. Part II. chap. i. Par L. Aug. Mercier, Paris, 1841. Also, Recherches sur les Valvules du Col de la Vessie, Paris, 1848. By the same. Chapter iv. And more recently by the same author, a paper, " Sur l'Inertie, ou Atonie de la Vessie," &c. Gazette Médicale, 1854.—The local paralysis theory was defended by Civiale, in a reply to the above, in the Moniteur des Hôpitaux, Feb. 8, 1855. Mercier's rejoinder appeared in the same journal, April 10 and 12. See, also, these two memoirs, with some additions, in his latest work, Recherches sur le Traitement des Maladies des Organes Urinaires, &c. Paris, 1856.

may generally be verified during life, or, more perfectly, after death. The exceptional cases referred to are, first, those in which there is a cerebral or spinal lesion of an organic kind, which paralyzes more or less completely the nerve-functions of motion, voluntary and involuntary, of sensation, or of sensation and motion combined, of the whole body below the situation of the injury. In such cases the body and neck of the bladder are alike affected, still the result is retention and overflow of surplus urine,—the very condition, it may be remarked, which is affirmed on the local paralysis theory (as above shown) to be the result of paralysis of the body and non-paralysis of the neck. Under certain circumstances, paralysis of the bladder may be produced by reflex action, no organic lesion of nervous centres being present; as in severe shock following injury, and the like.

The second exceptional case is that in which, under the influence of certain circumstances, a healthy individual voluntarily retains his urine for a considerable period, in spite of urgent desire to pass it. The not unfrequent result is that the muscular expelling apparatus of the bladder is overstretched, loses its tone, and is more or less unequal, for a certain period of time, to perform contraction in a normal manner. More or less chronic retention results, and may continue unless relieved in the ordinary way. To this state also the term paralysis has been employed; but as there is no evidence whatever that the lesion consisted in any loss or impairment of the *nervous* force transmitted to the viscus, it is indicative of a better pathology, and conduces to a better apprehension of the case in hand, to call it simply as it is, Overdistention, or, if preferred, Atony of the bladder. It is perfectly gratuitous in such a case to imagine a lesion in any part of the nervous system to account for this phenomenon, and it is, therefore, undesirable to speak of it as a paralytic condition of the bladder.

The phenomena produced by obstruction at the neck of the bladder caused by enlarging prostate, in connection with the function of micturition, may be briefly recapitulated as follows, and connected step by step with the organic changes which arise in the progress of the affection.

OBSTRUCTION AT THE NECK OF THE BLADDER FROM ENLARGED PROSTATE, PRODUCES		ORGANIC RESULTS.
a. Increased efforts to expel urine through obstructed orifice;	giving rise to	Corresponding Hypertrophy of the muscular parietes of the Bladder.
b. Inability to effect a complete contraction of the Bladder; and consequent *Retention* of a certain portion;	,,	Dilatation of the Bladder corresponding with the increasing amount of residual urine, the result of augmenting obstruction at the neck.
c. Increased inability to effect contraction of the Bladder from the increasing dilatation and consequent overstraining of the muscular coats;	,,	Atony more or less complete of the muscular parietes.
d. The Bladder becomes *engorged* with urine, the neck is dilated, and the surplus *overflows*, irrespective of the will of the patient;	,,	Organic changes, with a destructive tendency, in the mucous membrane of the Bladder.

Thus we arrive at a true appreciation of terms, which require no explanation, or limitation of their signification, and which, therefore, when applied to the morbid states described, simply and clearly speak for themselves ;—Retention, Engorgement, Overflow. The term Incontinence should not be employed to denote alike one condition in which the bladder is full to overflowing, and another in which it is organically incompetent to contain at all, but should be reserved to designate the latter state alone.

I say "organically" unable, because, when from any cause of inflammation, such as calculus, tumour, &c., there is frequently-repeated expulsive effort on the part of the bladder, and, consequently, but little retentive power, the term Irritability of the bladder is generally employed, as indicating a condition of activity, and therefore one very distinct from Incontinence, which is, on the other hand, universally understood to imply the result of a passive or quiescent condition, resulting from organic change or malformation, in which the urine dribbles off, being neither retained nor expelled.

Real Incontinence, then, is a rare occurrence in the adult male. That it is so is one of the most salutary and important lessons which the student can learn. It should be held as an axiom, the importance of which it is impossible to overrate, that AN INVOLUNTARY FLOW OF URINE INDICATES RETENTION, NOT INCONTINENCE. How often has the overflow of surplus urine from an engorged bladder concealed the real evil from an inexperienced practitioner, and induced the patient to believe that his "water was too abundant, or passed too freely," and wanted repressing rather than withdrawing! And what has been his astonishment, when, the true state of matters being recognized by his attendant, the introduction of a catheter has given exit to some pints, it may be, of the retained urine!

But does Incontinence, that is, organic inability on the part of the bladder to retain urine, ever occur as the result of enlarged prostate? M. Mercier states that he has not infrequently found it so occasioned, in this manner. He believes that when the enlargement of the organ is uniform, so that each lateral lobe and the median portion (middle lobe) are pretty equally augmented in size, the last-named part acts as a wedge, that it separates the two lateral lobes, and opens out the neck of the bladder, so as often to prevent the bladder from retaining the urine. He gives particulars of four cases under his own care, in each of which, at an examination of the body after death, the bladder was found empty, and contracted in size, while the prostate was enlarged considerably, but equally, so that the internal meatus was patent and of a triangular form. During the latter part of life, incontinence had been present in each case, and there had been no retention of urine.*

As has been observed in a preceding chapter devoted to the morbid anatomy of the affection, I have not seen a marked instance in the extensive collections which our museums afford, presenting a parallel to these cases of M. Mercier, and must, in consequence, believe it to be a rare result. On the contrary, the bladder is almost always dilated considerably; or, at all events, is not less than the natural size. Most rarely is it much contracted; nor on examining the numerous recorded histories attached to these preparations, does it appear that real incontinence had been verified during life. It may be remarked further, also, that, generally speaking, the equal development of the three portions does not necessitate the opening out of the internal meatus, since the outgrowth from the posterior median portion is almost always directed backwards towards the cavity of the bladder, and presents no appearance of acting as a wedge *between* the lateral lobes, from which, indeed, on the contrary, it seems rather to diverge, as if forced out by their lateral pressure. Since the publication of the first edition I have had in two or three instances an opportunity of verifying the existence of the condition described. And, although it has been carefully looked for, the infrequency of its occurrence corroborates the views I had previously expressed respecting its rarity. I am truly glad, however, to be able to confirm M. Mercier's views as to its existence.

A result which almost uniformly occurs from considerable hypertrophy of the prostate, is elevation of the urethro-vesical

* Recherches Anat. Path. et Thér. sur les Maladies des Org. Urin. et. Gen. Par L. Aug. Mercier. Paris, 1841, pp. 261-273.

orifice above the floor of the bladder. The entire neck of the organ is pushed up behind the pubes by the enlarging mass. The most dependant portion of the cavity is no longer on a level, or nearly so, in the standing position of the individual, with the outlet by which the urine has to pass; for the base of the bladder not being involved in the change, a depression, more or less deep, according to the degree of prostatic enlargement, exists behind the neck, in which urine may remain, after it has ceased to flow through a catheter lying just within the bladder. Hence the posture of the patient will sometimes affect the flow of urine; and accordingly, he finds that when kneeling, or in the prone position of the body, he is able to pass water after ceasing to do so in the upright posture. In some few instances, the outgrowth from the median portion having become pyriform, and possessing but a narrow pedicle, the mass is so movable as to fall forwards upon the vesico-urethral orifice in the manner of a valve, and close it very effectually when the individual is in the upright or in the prone positions. Thus a patient finds by experience that urine passes more freely when he is lying on his back, than by any other method. Such a circumstance may lead to the belief that this form of enlargement exists, a condition not difficult to verify, by means of a proper exploring sound (see the following chapter on Diagnosis).

Sometimes, however, when there is not much hypertrophy, the same obstruction occurs from the presence of simple tumours in the prostate. While, as we have seen, a prostate may become largely hypertrophied without causing obstruction to the outflow of urine, one of these small tumours developed at the neck of the bladder may produce complete inability to micturate. The small, firm, rounded projection which it constitutes, fills up the internal meatus, closes it almost entirely against the natural efforts to relieve the bladder, and renders necessary the artificial aid of the catheter. At the same time, these circumstances existing, little or no enlargement of the prostate may be present or ascertainable by rectal exploration, and little or no difference in the length or the curve of the catheter is required for the case. These conditions, which are undoubtedly exceptional, I have nevertheless not unfrequently observed during life and verified after death.— (See fig. 10, p. 91 : also Plates III. and IV.)

We have thus a condition quite distinct from hypertrophy, but liable, like it, to produce obstruction to the outflow of urine, and chronic retention and engorgement of the bladder, in elderly subjects—a condition which requires similar management also, in all respects. The practical lesson is one which relates to diagnosis; and it is simply this, that we are not to affirm the absence of obstruction of the neck of the bladder, and infer the existence of paralysis affecting that viscus, when it is unable to empty itself by voluntary efforts, because we discover no enlargement of the organ by rectal or by urethral examination. The obstruction may be, and in these circumstances it not unfrequently is, due to the existence of a small rounded tumour developed in the substance of the prostate, but projecting considerably into the urethral canal, and usually, although not invariably, in or near to the situation of the uvula vesicæ.

Other circumstances may be briefly named as results of enlarged prostate, and affecting the function of micturition; viz. those arising from the hydraulic pressure exerted upon the entire urinary track, behind the point of obstruction. I shall merely name them here, as the whole subject is treated at length in my work on "Stricture of the Urethra," chapter the second. The changes thus produced in the containing and excreting organs are, in the main, almost the same, whether the obstruction to outflow is situated in the course of the urethra (stricture), or at the neck of the bladder (prostatic).

Hypertrophy and dilatation of the bladder are the commonest, indeed almost the invariable, results of prostatic enlargement. Hypertrophy and contraction, on the other hand, are not uncommon in stricture of the urethra. Sacculation of the bladder often follows, being produced by protrusion of the mucous membrane outwards, through small interstices between the enlarged fasciculi of the muscular coat. The sac commencing as a mere indentation, becomes pouch-like, and, in the course of long-standing disease, is distended into a spheroidal cavity, capable sometimes of containing several ounces of urine. (See Pl. IV.) The presence of these sacs may be suspected during life, when, after the bladder has been completely emptied by a catheter, certain changes of position in the patient's body, in the course of a very few minutes, set free some three or four ounces of very turbid urine. Mr. Guthrie believed that the presence of a sac was proved by the occurrence of what he denominated "the fluttering blow of the bladder," communicated through the catheter to the hand of the surgeon. The phenomenon thus described by him I believe I have repeatedly experienced in circumstances in which there was no evidence of the existence of sacculi, and, indeed, every reason to conclude they were not present; I cannot, therefore, admit it as a diagnostic sign. The ureters may become dilated to a very considerable extent, and their walls at the same time be hypertrophied. The ureter has even been known to serve the purpose of a subsidiary bladder; in one well-known case it formed a tumour, easily recognized during life when distended, which reached from the pubes to the lower ribs, but which disappeared after catheterism. Dilatation of the pelvis of the kidney follows, and sometimes to such an extent that an organ so affected presents the appearance of a cyst, or congeries of cysts, a condition in which irreparable injury to the renal structure itself has been inflicted. Indeed, the kidney may ultimately almost disappear under the influence of such fluid pressure when long continued. Add to all these effects the noxious action of decomposed urine upon the whole extent of the mucous lining involved; the consequence of which is chronic inflammation, leading to ulcerative and gangrenous processes in various parts.

Such are the principal effects of prostatic obstruction, and the organic changes which take place in the organs behind its seat, as the results of large accumulations of urine, when not removed by artificial means.

CHAPTER X.

THE DIAGNOSIS OF PROSTATIC AND OTHER OBSTRUCTIONS AT THE NECK OF THE BLADDER.

Examination by Rectum.—Method of conducting it.—Points to be verified.—Examination by Urethra.—Sound adapted for the purpose.—Ordinary methods of using it.—Rigorous determination of Size and Form of Tumour often possible.—Diagnosis of Prostatic Enlargement from Stricture of the Urethra.—From Calculus of the Bladder.—From Tumours of the Bladder.—From uncomplicated Chronic Cystitis.—From Atony of the Bladder.—From Paralysis of the Bladder.

ALTHOUGH an observation of the ordinary symptoms of prostatic enlargement, when appearing in an elderly patient, affords good ground to the surgeon for entertaining a pretty correct surmise as to the nature of the complaint, still its existence cannot be asserted without a manipulative examination.

Just as in the case of suspected stone in the bladder, the sound must reveal the presence of the foreign body, so, in this case, must the sound and the finger recognize the existence of prostatic enlargement, irrespective of the evidence which may be derived from the existence of any symptoms whatever; and they should also define, as far as possible, what are its nature and extent.

The test which is chiefly depended on by the surgeon is an examination, by means of the finger, in the rectum. It is so familiarly known and commonly employed, that it might be deemed superfluous to enter into details respecting the method of its application. To the student, however, this will not be the case; neither will he, nor will any one unpractised in such explorations, learn much by employing it, unless he has attained some previous knowledge by experience, the accomplishment of which may, however, be greatly facilitated by some preliminary hints. Such is all I propose to offer here, in the belief that they will not be without their utility.

The patient should, as a rule, lie on his back upon a couch, the surgeon standing on the left-hand side, so that the fore-finger of his own left hand may be employed in the rectum, while the right hand is free to use a catheter if required, since by concerted movements of that instrument in the urethra, and of the finger in the rectum, more accurate information may sometimes be obtained than by either exploration conducted separately. The nail of the left index finger should be cut very short, and it is a good plan to fill the creases around it with a little common soap, and then to oil the whole finger thoroughly, as, by this means, fœcal matter is prevented from lodging in them. The patient's knees being drawn up and separated a little from each other, the finger should be made to glide slowly through the sphincter, and when introduced as far as possible, so that two phalanges are free to move in the bowel, the limits of the prostate may be defined. It is necessary to bear in mind, not less for the patient's comfort than in order to gain the opportunity of making a satisfactory examination for the operator, that the finger should be carried thus far, and through all subsequent movements, with the greatest possible gentleness, and with no rapidity or haste. The healthy prostate may now be at once defined; but first, situated immediately within the ring of the sphincter, in the median line of the anterior wall of the bowel, which recedes suddenly at this point, may sometimes be felt the termination of the bulb of the urethra, and, at all events, the membranous portion. Going higher, the apex of the prostate is distinguished, and gradually widening out, the body, which is firm to the touch, with an outline distinctly suggesting that of a flattened chestnut; the finger will readily appreciate a slight depression marking the line of the urethra, between its two lateral lobes, and may glide over the outer borders of the latter, into a hollow on either side; returning to the median line, it may pass upwards until the firm prostate is no longer felt, but the more yielding tissues of the bladder behind its base, especially in the centre, at the interlobular notch, which is readily reached in the natural state of the parts.

Once familiar with the normal conditions thus presented, deviations will easily be recognized; but without such previous knowledge, it is useless to expect the attainment of much information by the rectal exploration in the search for that which is abnormal; and quite impossible to appreciate any but very considerable developments of disease. It is desirable to pursue a methodical plan in conducting these examinations. For example: the first step of the inquiry should have for its object the deviations in size and form. The first-named is almost always in the direction of enlargement. Is it general, or partial? affecting one, or both lobes? and to what extent? Affecting breadth mainly, or forming a rounded protrusion into the bowel? I have found it so prominent sometimes, that the tip of the finger encounters the swelling the moment it enters the rectum, and has to be depressed very considerably before it can be carried beneath the tumid organ. Again, instead of finding the yielding coats of the bladder in the middle line, when the finger is carried up to its fullest extent, an increasing fulness and firmness may be encountered, due to an enlargement or outgrowth from the median portion ("middle lobe") occupying that situation, and defying all attempts to define it. Then the form of the enlargement may not be uniform or spheroidal; it may be irregular, knobbed, or mammilated. This also, it is of importance to note.

Secondly, the consistence of the parts is to be ascertained. Is the tumour soft, hard, or unequally so in places? Is there fluid in it? And—often a question of vital importance—can we appreciate fluctuation distinctly beyond it? In the latter case, the right hand should be applied to the hypogastric region, and firm pressure made there, with the view of ascertaining if a large body of fluid, such as a distended bladder, can be pressed down upon the apex of the finger, in the rectum below; then gentle but sudden taps should be made on the same region, for the purpose of imparting the wave-like impulse which, under such circumstances, will be communicated. This proceeding constitutes an important mode of verifying the condition of the bladder, and the proper position for the trocar, when the operation of puncture by the rectum is about to be performed, or its applicability to the case in question has to be determined. At the same time, the situation of arterial branches is ascertained, one or two of considerable size may generally be felt lying a little to the right or left of the middle line, and sometimes crossing it, branches of the hæmorrhoidal arteries. The presence of prostatic calculi may generally be thus ascertained, being usually felt with ease, when rather large or numerous, lying in one or more cavities of the prostate, with very little tissue intervening between them and the walls of the rectum.*

Thirdly. We seek for the degree and locality of tenderness on pressure. It is desirable to make, in a distinct manner, first giving the patient notice of our intention, firm pressure with the point of the finger in three different spots;—on the centre of the prostate, on the extreme right and left borders, consecutively, permitting an interval of a few seconds to elapse, that he may clearly distinguish each separate movement, the sensation occasioned by which he is required to describe. If inflammation is present, the pain will be extreme, and the mere introduction of the finger will be very distressing to the patient; in this case, heat and tension will be remarked also.

Lastly, supposing the catheter to have been previously introduced, we may, while holding it in the right hand, and communicating to it gentle movements downwards, gain an approximative

* I have a patient now under my care in the Marylebone Infirmary, in whose prostate a considerable number of calculi are embedded. The grating from these is very perceptible to a finger in the rectum, when a catheter lies in the urethra, and is gently pressed downwards.

idea as to the thickness of the tissue which intervenes between it and the finger in the bowel, and as to the situation and direction of the instrument, &c., in event of there being difficulty in introducing it. But its presence there is extremely useful, if only for the purpose of furnishing a solid body of known size and form over which the exploring finger may pass; and by the aid of which it is obvious that most of the above-mentioned inquiries can be prosecuted with additional facility and certainty.

Having learned, relative to these different points, all that can be attained through the rectum, the urethra is to be explored. A full-sized catheter, of the form ordinarily employed by the surgeon, should be first used, because any phenomena presented differing from those observed when the prostate is healthy are then at once made apparent. If by examination through the bowel we have found no variation in regard of size, and have now ascertained that the urine flows when the catheter has traversed not more than the ordinary distance, say from $6\frac{1}{2}$ to 8 inches, while the handle of the instrument itself has not required more than the ordinary amount of depression in order that its point may enter the bladder, we may be satisfied that prostatic enlargement does not exist, and we must seek for another cause of the symptoms complained of. But if the catheter had passed easily, say for 8 or 9 inches, which is at once known if the instrument is graduated (as all such ought to be) and still no urine flows; and if, in addition, while following its course, the handle has become more than usually depressed, approaching almost to the horizontal line (the patient being recumbent), there will be little doubt in respect of the existence of prostatic enlargement. The ordinary catheter being inadequate to reach the bladder, or doing so only when it has passed further than usual, and in the position described, another instrument may be employed. This is generally one which measures from two to four inches longer, and possesses a larger curve than the ordinary catheter; while some instruments describe also a larger arc, a third, for example, instead of a fourth, of the circle. If such pass readily, the increased length of the urethra is easily ascertained, and the direction of the prostatic canal is calculated from the position of the shaft noted at the moment that the point enters the bladder. A medium prostate catheter (see fig. 16 *a*, Chapter XII.) has its beak at right angles to the shaft, the recollection of which makes the direction and even the exact position of the beak obvious at once to the mind's eye, the axis of the shaft being, as it always is, in view. When the curve of the instrument is prolonged beyond this, the degree of incurvation being known may be allowed for, and its position is then ascertained without difficulty. In some few cases, while the beak passes through the prostatic part of the urethra, the handle will be distinctly deflected to the right or left, from which fact, if verified by two or three trials, a greater degree of enlargement may be suspected to exist on the side *towards* which the handle turns.

In this manner we may obtain approximatively correct views of the size and mode of development of the prostatic enlargement. The progress of the complaint may be noted from time to time, but I am not aware that any very considerable advantages can be obtained by the possession of more exact knowledge respecting the tumour, attainable, perhaps, at the expense of reiterated and more painful applications of the instrument than those at present alluded to; unless, indeed, there is any ground for advocating the adoption of some operative proceedings, such as the division of an obstructing bar or the like; a subject which is discussed at length in the seventeenth chapter. In view of any such undertaking, it is necessary to make an accurate diagnosis of the nature and size of the tumour; and this it is, within certain limits, in our power to accomplish, if ordinary care and the proper method be employed. But although we may not entertain any such intention, it is not the less desirable to be able to familiarize ourselves with the manner of accurately determining the condition of the prostate, of the bladder and its contents, especially in relation to the question of calculus or tumour, and for such a purpose the instruments with a large curve, already described, are wholly useless. Hence it is necessary to resort to one of different form, and that now usually adopted in sounding the bladder is well adapted for the purpose, viz., a sound with a very short curve at its extremity, or possessing a beak rather than a curve, which is much shorter and more angular than that of the ordinary catheter. Instruments somewhat resembling this description have long been employed, but their use for sounding the bladder, and more especially for examining the condition of the prostate, has, during the last thirty years, been more particularly advocated by those among the French surgeons who have bestowed special attention on maladies of the urinary organs, as Civiale, Heurteloup, Leroy D'Etiolles, and Aug. Mercier. The two latter have represented the forms which they employ; (see figs. 12 and 13); they differ little from the sound which is now generally used, and scarcely, if at all, from the form of the common lithotrite. The short beak is of course for the purpose of being carried into the bladder, in which cavity it can with care be turned freely in any direction, provided a sufficient quantity of urine or other fluid is retained. In this manner not only can every part of the bladder be searched for calculus, but information respecting the form and degree of obstruction at the neck, whatever may be its nature, can also be acquired. After the bladder has thus been traversed, the instrument should be gently withdrawn until the beak lies just within the urethro-vesical orifice, when by turning it round to the right and left, the natural condition, if it exist, of that part can be ascertained; or, on the other hand, the presence

FIG. 12.   FIG. 13.

Fig. 12, Sound of Leroy.
Fig. 13, Sound of Mercier.

of tumour, or of stone, the depth of the fossa behind the prostate (fig. 14), and other relative points, can be determined. It is

Fig. 14.

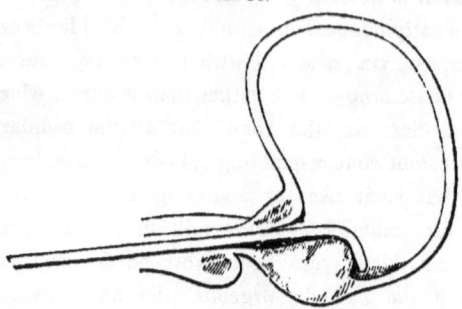

an advantage to use a hollow one, so that the quantity of fluid in the bladder can be augmented or diminished, without removing the instrument; and lest this should impair its solidity and weight I employ one made in steel, of which the extremity is a little bulbous and solid; the channel opening about half an inch or more short of the extremity on the convex surface. (See fig. 15.) Without giving my complete adhesion to a practice which requires the performance of the extremely numerous and varied manœuvres for the purpose of arriving at a diagnosis respecting the precise terms of a prostatic enlargement, which our French brethren habitually resort to, it may not be altogether unprofitable briefly to notice the methods by which the attainment of exact information respecting it is sought. I say sought, for I have good reason for believing that such manœuvres have sometimes failed of their object, even in the hands of the most expert disciples of the practice. And again, since I am compelled for the most part to express an objection to the practical end which these manœuvres have con-

Fig. 15.

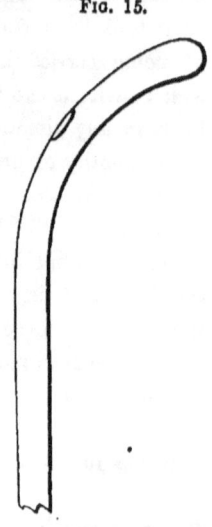

fessedly in view as employed in France, that is, the destruction by mechanical or chemical means of the obstructing portions of tissue, there is additional ground for doubting whether the employment of many of them can conduce to any beneficial result. Nevertheless, as some very useful hints may be derived from an observation of the manipulations referred to, I shall describe only one or two of the simpler means in use, for diagnosticating tumours of the neck of the bladder, premising that the instrument employed is always one similar to those delineated at figs. 13, 14, or 15. For the more-complicated proceedings of this kind, the reader is referred to the practice and writings of the authors themselves.

1st. Means of recognizing tumours which rise into the neck of the bladder.

The sound having been introduced, it is slowly and gently, but completely, rotated on its axis in the cavity of the bladder, and close to its neck. If the prostate is healthy, this is done without any elevation of the instrument, and the shaft retains an almost horizontal position (the patient of course being recumbent). But supposing that there is a tumour at the neck of the bladder, the beak will be arrested in the movement of rotation, and it will be necessary to elevate it proportionately to the height of the eminence, after which it will descend again, the movement of the handle indicating approximatively the size and form of it. If, in introducing the instrument through the prostatic part, the beak is found to rise gradually, the handle being depressed below the horizontal line, there is probably spheroidal enlargement of the median portion. (Pl. III.) But if the beak abuts upon an obstacle there, and has to be lifted over it in a direction upwards, entering the bladder with somewhat of a jerk, there is probably an enlargement of the same portion, affecting the form of a bar, with a deep sinus of the prostatic part of the urethra. (Pls. IX. and X.) But in this case the instrument can be rotated with facility, which, as has just been seen, cannot be accomplished when it is a case of tumour of the median portion of the prostate, and not a simple bar. In withdrawing the instrument from the bladder, the beak being turned downwards to the basfond, if the prostate is healthy, it will come back into the urethra easily, but if there is an enlargement there, it will hook against it, and not leave the bladder in that position.

2nd. In order to recognize an enlargement of the prostate projecting into the urethra, Mercier proceeds as follows : "After having explored the bladder, I draw the instrument gently back into the prostatic region of the urethra, pressing lightly upon it, at the root of the penis" (its upper aspect), "just under the pubic symphysis, so as to press the angle of the sound or salient part of its curve against the posterior wall of the prostatic urethra; then I draw it forward without elevating its shaft towards the abdomen as in ordinary catheterism, and without making it deviate much from the axis of the patient's body (15° to 25°). When there is a simple enlargement of the prostate in the antero-posterior diameter, the beak traverses it easily, without inclining either to the right or left. If, on the contrary, there is a projection of one of the lateral lobes, the beak, in passing the spot, inclines to the opposite side: the handle indicates this movement and the direction in which it is made."*

This portion of the subject may be appropriately closed with a few remarks on the diagnosis of prostatic enlargement from stricture of the urethra, vesical calculus, tumour of the bladder, simple atony or inertia of the coats of the bladder, and paralysis.

In Stricture of the urethra, the stream of urine is invariably small, in a confirmed case extremely so ; in the prostatic affection, though diminished in force, it is often less so in volume than in the previous case. The use of a full-sized sound, however, marks the distinction clearly. In stricture, obstruction is encountered almost invariably before six inches of the instrument have disappeared, always before it arrives at the prostatic urethra. In enlarged prostate, obstruction is not encountered until eight or nine inches have passed, and not necessarily then, for, provided

* Recherches Anat. &c., pp. 364, 365.

that the instrument be sufficiently long, it may pass into the bladder; but the handle has to be depressed between the patient's legs in a manner not required in the normal state. Lastly, stricture almost invariably makes its appearance before middle life, prostatic hypertrophy not until that period is passed.

In regard to Calculus, while many of the symptoms are common to both complaints, the occurrence of sudden cessation of the stream of urine, of severe pain at the close of micturition, the exacerbation of symptoms, especially of pain, and the appearance of a little blood after exercise, may be looked upon as strongly indicating the presence of stone in the bladder. But it may exist in the absence of most of these, the two first-named especially, from the circumstance that the calculus is usually situated behind the enlarged prostate, and does not approach the more sensitive region of the internal meatus. The fact of small quantities of florid and unmixed blood being occasionally passed after exercise, more closely approaches in value to a pathognomonic sign than any other. A persistent discharge of mucus, or of pus, in the urine, should also arouse suspicion. The use of the sound, however, can alone clear up this case also satisfactorily.

The existence of Tumour of the bladder is less easily affirmed. Compared with prostatic enlargement there is much more pain, and exquisite tenderness on the introduction of instruments, the urine is frequently or generally mingled with sanious discharge and flocculi, to which sabulous matter is often seen adhering. Examination of these under the microscope may reveal the peculiar structure of villous growth, or, which is almost equally significant, may demonstrate that they consist of organized structures, not of inorganic materials. Exploration by rectum sometimes detects a firm mass occupying some portion of the walls of the bladder, especially if malignant disease be present, of which other signs will also be ascertainable (see Chapter XIV.); but villous tumour is not appreciable by such, or by any other examination:

its most characteristic symptom is almost constant admixture of blood with the urine.

Simple uncomplicated chronic Cystitis, with catarrh, is by no means a common affection. The series of symptoms thus denoted is almost invariably due to the presence of a foreign body, to some form of obstruction, or to paralysis, depriving the patient of the power of expelling the contents of his bladder, a condition which is tantamount to obstruction. We may rely upon it that, in most of the obscurer cases, there is a material cause, most frequently calculus; the presence of which needs a more than ordinarily-searching examination to verify. It may be encysted, or otherwise rendered difficult of detection by the sound. The absence of all the physical signs of enlarged prostate, by rectal and vesical exploration, will, of course, prove the non-existence of that complaint as a cause.

Single or repeated acts of voluntary over-retention of urine are sometimes followed by Atony or inertia of the muscular parietes of the bladder, and a state of chronic retention follows from their consequent inability to expel the vesical contents. The condition resulting resembles much the retention produced by enlarged prostate, and requires frequent relief by the catheter in the same way, at least for a time. Here the absence of positive signs, the suddenness of the attack, its connection with a cause generally recognized by the patient, and the diminished power of discharging the urine *after a catheter has been placed in the bladder*, are quite sufficient to distinguish this affection. Particular attention should be paid to this last-named point. In enlarged prostate, the urine often flows with considerable force when the influence of the obstruction is removed by the introduction of a catheter, and the current can be accelerated materially by the will of the patient, unless there be atony also, as there may be from undue distension; however, it is not generally considerable, except in long-neglected cases. But when the cause of engorgement and retention is not obstruction, but complete atony of the bladder, the urine runs out of the catheter, and is not propelled, neither can the flow be much influenced by any efforts of the patient.

Lastly, there is Paralysis of the bladder, a condition in which its nervous supply is either impaired or destroyed. It is almost always associated with a similar condition of the lower extremities, and this may result either from disease or injury of the encephalon or spinal cord. There is no evidence of the existence of true paralysis, that is, a removal or impairment of nervous influence, *limited to the bladder;* nevertheless, the term paralysis is constantly applied, but most inappropriately, to denote inability of the viscus to expel its contents, whether the cause be obstruction at the neck, or over-stretching (atony) of its muscular walls. The bladder is not deprived of nervous force, and thus rendered paralytic, except when there is lesion of some nervous centre involving numerous other parts in the same predicament, any more than is the stomach, the intestines, or any other single viscus (see preceding chapter). There can then be no doubt respecting its presence; and the indication which catheterism presents, when it does exist, is also singularly characteristic. An instrument being introduced, the urine is propelled by the weight of the parts around, the will of the patient exerting no influence upon its flow unless the abdominal muscles should be in a normal condition, as in cases of injury (rare) occurring to the spinal cord between the sources of nervous supply to the muscles and to the bladder, in which case a slight influence is perceptible. Otherwise no impulse is noticeable, except through the agency of acts unassociated with micturition; such as deep inspiration, coughing, sneezing, and the like, by which a momentary pressure is communicated to the paralyzed bladder, and the stream is momentarily accelerated.

CHAPTER XI.

THE TREATMENT OF PROSTATIC ENLARGEMENT, FROM HYPERTROPHY AND SIMPLE TUMOUR.

The subject one of considerable importance.—May be treated under Three Heads.—1. TREATMENT for the purpose of obviating the results of Obstruction caused by Enlarged Prostate.—Necessity for removing retained Urine.—Question of patient relieving himself.—Instruments to be used.—Of permitting a Catheter to remain in the Bladder.—Evil results of not relieving the Bladder.—Treatment of ATONY OF THE BLADDER.—Of CHRONIC CYSTITIS.—Injections.—Counter-Irritation.—Baths.—Buchu.—Pareira brava.—Uva ursi.—Triticum repens.—Matico.—Lythrum Salicaria.—Alchimella Arvensis.—Epigœa repens.—Chimaphila.—Wild Carrot.—Copaiba.—Cubebs.—Benzoin.—The Demulcents.—Indications for use of the foregoing.—The Mineral Acids.—Alkalies.—Benzoic Acid.—A Case.—IRRITABILITY OF BLADDER.—Value of Opiates.—Injections.—HÆMORRHAGE.—Its Treatment.—INCONTINENCE of Urine.—Treatment.—Recurring attacks of Congestion.—2. THE GENERAL TREATMENT and Management of patients with Enlarged Prostate.—Dietetic, Regimenal, and Moral.—3. SPECIAL TREATMENT against Enlargement itself.—Hemlock.—Mercury.—Hydro-Chlorate of Ammonia.—Iodine.—Mr. Stafford's Method.—Bromine.—Kreuznach Waters.—Compression.—Depression.—Electricity.—Division and Excision.—Crushing, &c.

THE topic presented for consideration in this chapter is one of great interest and importance; and is well worthy to be the subject of prolonged and careful study. Its interest for the practical surgeon consists in the fact that, notwithstanding the generally-admitted intractability of the complaint, much may be done to palliate its most distressing symptoms, and to retard its progress; while, associated with this, is the knowledge that the attainment of any means capable of arresting that progress, or of curing the disease, would be one of the greatest boons ever bestowed by the science of medicine upon suffering humanity. The importance, therefore, of the subject is equally manifest. No wonder, then, that the search for remedial treatment should constitute one of the most alluring subjects of inquiry which can occupy the enthusiastic student of the healing art. Although by no means satisfied with our present achievements in this direction, there is no reason to esteem lightly the power which past experience and skill have enabled us to exert in prolonging the life, and ensuring the comfort of the patient. And I by no means agree with those writers who regard the enlarged prostate as one of the opprobria of medicine, and who urge, with little confidence and slender hope, the employment of treatment against it. I do not doubt that the day will come when the complete control of this evil will be in our power, adding another to the already numerous and splendid triumphs of scientific medicine. Meantime every addition to our appliances which accomplishes a desideratum, however small, not before attained, and every advance in the knowledge of the constitution of the organ itself, both in its healthy and diseased states, may be regarded as progressive steps by which the consummation of that expectation will be developed.

In considering this subject, it will be found convenient to make three distinct divisions of it, and to regard them separately, as follows:—

1. Treatment for the purpose of obviating the results of obstruction caused by enlarged prostate.

2. The general, or constitutional, treatment and management of patients with enlarged prostate.

3. Treatment directed against the enlargement itself.

First,—the subject of treatment as far as it relates to the means of obviating the results of urinary obstruction caused by an enlarged prostate.

The relief of obstruction is generally the chief indication presented by a patient so affected, when first applying to his surgeon for advice. Unaware of the nature of his complaint, he seeks to be relieved from some discomfort produced by an obstacle to the free course of urine, such as undue exertion necessary to perform micturition, the unnatural frequency with which the want is experienced, some degree of involuntary micturition, or, perhaps, an alteration of some kind in the characters of the urine passed.

A diagnosis having been made by the employment of the catheter and rectal exploration, the methods of arriving at which are fully considered in the preceding chapter, and the fact being discovered, as it usually is, that a certain quantity of urine is habitually retained in the bladder, however frequently or forcibly the efforts to evacuate it have been made, the first great principle to be followed out is this;—that it is necessary to ensure the complete removal of the urine from the viscus at least once a day. It may be very desirable to do this twice, or even three or more times daily, the necessity depending, in great measure, upon the degree of obstruction, and the consequent amount of residual urine. In general terms, if only two or three ounces are left after the act of micturition is performed, once a day may be sufficient; if it amounts to double this, twice will probably be better. If a much larger quantity of urine is retained than that passed by the natural powers, it is not unlikely that the instrument will be required three times in the 24 hours. And if the power of urinating is almost or quite lost, it will be necessary to employ it as often as a decided want to micturate is experienced. There are certain modifying circumstances which must be taken into account. Such are the facilities which exist for passing instruments, and the condition of the urethra itself. If the patient possesses the ability to pass a catheter easily for himself, and it is very rare indeed that he cannot attain it by tuition and practice, he complies with the demands of his case. But when the ability has not been attained, the difficulty may, in some circumstances, be serious, as when the operation is required to be performed every few hours, and the surgeon is unable always to afford the regular attendance which is desirable. It is, however, extremely imprudent to entrust this duty to any non-professional attendant; and the patient should be made to understand that, having his own sensations to guide him, he may soon attain considerable dexterity in the management of an instrument, at all events in the one passage with which he will become familiar. But, again, if the urethra is in an extremely irritable state, and this appears to be aggravated, as it must almost certainly be by the very frequent use of the catheter, however gently it may be introduced, it will be necessary to consider carefully the requirements both of the urethra and the bladder, and some compromise must be made between the conflicting interests of the two. So, also, the condition of the urine itself, and the amount of mucus with which it is charged, must be taken into account in determining the question: if the secretion is very acrid from decomposition, or much loaded with deposits, its more frequent removal is indicated; and an injection, as hereafter recommended, may be also desirable. In the majority of cases, however, in which these states are not present, the removal of the urine night and morning suffices to maintain the reservoir in a tolerably sound and healthy condition, and it is extremely undesirable to resort to artificial aid with greater frequency than is absolutely necessary to accomplish this.

With regard to educating a patient to the use of the catheter, it should be done systematically on the part of his attendant, as soon as the former has become somewhat familiarized to it by observation. The patient should be taught to note accurately how much of the catheter remains projecting from the urethra and its line of direction with regard to the body, at the exact moment when the urine flows, so as to accustom his eye to judge correctly as to the point at which he may expect the appearance of a stream when the instrument is in his own hands, a moment always of some anxiety to the inexperienced performer. In using the catheter for himself, he should generally stand with his back against a wall, the receiving vessel being placed conveniently, so that no unnecessary movement of the body be made during any part of the time at which the instrument is within the urethra. As to the kind of instrument to be recommended, I have seen silver and flexible instruments used with equal facility by different patients, the nature of the case, and the result of trials with both, deciding the question in each particular instance. Perhaps flexible instruments are, in the majority of cases, the better and safer kind. But where the senses are good, and sensibility acute, that instrument which the surgeon has found the best (and this will frequently be the silver one) will be most successful also in the patient's hand. The directions given will be precisely those which apply to the passing of instruments by the surgeon, detailed at length in the succeeding chapter, modified only by the fact of the operator having himself for the subject of his skill. This very fact ought to inspire him with confidence, for he possesses a most important advantage over his attendant in one respect, which will compensate largely at first, and altogether ultimately, for his want of varied experience on others which the surgeon possesses. It consists in his being able to regulate the movements of his instrument, not only by the sense of touch in common with any other operator, but by his consciousness of the sensations produced by it in the urethra, which the surgeon cannot possess. On this ground it is undoubtedly true that, after long practice, few men can pass a catheter for the patient so well as himself, provided that his senses are acute, and that he possesses ordinary intelligence. Accustomed also to follow but one track, he knows intimately every portion, and learns some little manœuvre adapted to meet any difficulty which may be presented.

When elastic instruments are employed, the curve is always required to be greater than they naturally possess as supplied fresh from the maker's hands. Hence the patient should be furnished with a good stock, and these should be kept for some months before they are wanted, each mounted upon a well-curved stilet (fig. 16, g), so that when required it may retain as much of the curve as may be desired after the stilet is withdrawn. It should then be oiled, but not warmed, and used without delay, as the curve soon lessens in the urethra. Nothing need be added here respecting modes of manipulation, which the patient learns from his surgeon, who will adapt his instruction to the necessities of each particular case. There is one little manœuvre, of trivial moment it may be thought, which was practised by a gentleman long a patient of my own, with enlarged prostate, and by whom it was adopted in order to obviate a degree of uncertainty produced in passing the gum catheter, by the yielding or bending of its shaft; and it certainly conduced materially to his success. He introduced into the instrument an iron stilet six inches long, which gave firmness to the stem or shaft, while the whole of the curved portion remained as flexible as ever. This ensured an amount of certainty in the manipulation of the instrument, which, as it was in his case assuredly not without utility, appears to me to be worthy of mention.

It has been recommended for these cases by some that a catheter should be permitted to remain in the bladder for some days at a time. There are two hypotheses on which the advice is given. The first is, that by permitting the bladder to remain empty, or nearly so during a considerable period of time, we encourage it to regain its contractility, assumed to be lost or impaired by overstretching. This, however, is not the true pathological condition which causes retention in these cases. There may be some loss in the muscular power of the vesical coats, but it is not much except in old and unrelieved cases. The material obstruction at the neck caused by the enlarged prostate, and not any "local paralysis," as it is commonly termed, is the sole, or almost sole, occasion of the urinary difficulty; as has been explained in the previous chapter. Consequently on this ground, the practice referred to gains little support. The second hypothesis is, that the constant pressure of the catheter promotes absorption of the substance of the tumour, and so tends to the material improvement of the patient's condition. But the fact is, that the tendency is much more to ulceration, than absorption, after the manner intended. Granting for argument's sake, that the desired action can be thus ensured, it has still to be proved that the object is gained, at an expense of pain and confinement, to say nothing of risk, sufficiently small to render the result an undoubted and valuable acquisition. Experience, however, negatives the supposition. It will, of course, be understood that there is no reference here to cases in which urgent urinary retention has existed, which has been relieved with difficulty by the catheter; or to those in which great pain is produced, or unusual obstacles are encountered in passing the instrument. In such circumstances, as we shall see when discussing that subject in the following chapter, we may be justified in permitting the catheter to remain in the canal for a considerable period.

The consequences of enlarged prostate already alluded to, viz. the increasing retention of urine and habitual distention of the bladder, which accrue from not completely emptying it daily, form only a portion, although a very important one, of the evils which can be obviated by this treatment. Associated with these, and entailed by the same cause, is that state of engorgement and overflow of urine, commonly termed, but inaptly, incontinence, which will assuredly follow gradually-increasing retention, when unrelieved, a state which necessarily exerts an injurious influence upon the ureters and kidneys, ultimately leading to their disorganization, and thus to a fatal termination. Short of these results are the scarcely minor ones of atony of the muscular coats of the bladder, chronic cystitis, catarrh, decomposed urine, hæmaturia, calculous deposit, and always accompanying them, impaired general health. All these can be, under ordinary conditions, avoided, and in most, kept in abeyance by the persevering use of the catheter. It is impossible to overrate the benefits arising to the patient from this means; and the responsibility which is incurred by overlooking or failing to impress his mind with the necessity which generally exists for its use, should be ever present to the surgeon's mind in dealing with any signs of irritability of bladder, and incompetence to retain perfectly the urine by patients in advanced years.

Supposing, however, that our attendance is required for a case in which, from neglect or otherwise, the urine has been long permitted to accumulate, and the residual amount after micturition has reached a pint, or even two. If we evacuate the bladder two or three times daily, great relief will undoubtedly be at once afforded to the patient; his frequent calls to pass water, the constant dribblings from a distended bladder will be removed, but the sudden change will frequently act with prejudice to the constitution. If we persist, some febrile symptoms will probably soon show themselves, the patient will become low, and even finally sink, his end being accelerated if the injunction to empty the bladder two or three times daily be regarded as invariably binding. Sir B. Brodie was the first I believe to point out the consequence of such treatment in these cases;* and observation has demonstrated to me the justice and the value of his remarks. Following the physiological law which forbids rapid and extreme change in the conditions and habits of the human economy, familiarly exemplified by the dangers known to accrue from sudden exposure to heat after extreme cold, or from appeasing, without stint, the cravings of the almost starved, the surgeon should commence his treatment by removing only a portion of the urine, not exceeding half the contents of the over-distended bladder, and in the course of two or three weeks, perhaps, by gradually withdrawing a somewhat larger proportion, and slowly accustoming the atonied coats of the viscus to their new condition, he may venture to empty it entirely. But in this proceeding he must exercise great caution, watching for symptoms, and doing all he can to invigorate the patient, by improving his digestive organs, and by administering as much nutritious food as he can digest, with as much tonic and stimulant as he may require.

In the management of other cases, however, of a kind less serious and advanced than those just referred to, and in which we can and ought to empty the bladder once or twice daily, there are other difficulties and emergencies which may be still encountered at some time or another. These it will be convenient to consider separately.

1. ATONY OF THE MUSCULAR COAT OF THE BLADDER.—The bladder having for a long period of time been unable to empty

* Lect. on the Urinary Organs. 4th ed. Pp. 203-5.

TREATMENT OF ATONY OF THE BLADDER.

itself, it is sometimes observed that the muscular fibres lose the power of perfectly contracting, even where the cavity is artificially emptied. Ordinarily, they partially recover their tone and power where this condition has been regularly attained once or twice a day for a few weeks. But this does not always happen. Partly from disuse, and over-distension, partly from inflammatory deposit in the vesical membranes impairing their contractility, the organ still continues in a state of atony. Thus it may be observed, when the catheter is passed into the bladder, the patient being in a recumbent position, that little or no urine issues unless pressure is made on the hypogastric region; or if he is standing, that the urine is propelled by the weight of the abdominal viscera acting on the bladder, and that in both cases the will exerts little or no influence on the stream, while an act of coughing or deep respiration affects it instantly.

This state is often partially remediable—sometimes entirely so; and it is extremely desirable, for many reasons, to resuscitate the lost function. Cold affusion suddenly made on the abdomen should be applied twice a-day. Cold-water injections into the bladder itself are often of great service, and may be employed on alternate days, and afterwards every day. These failing, a current of electricity may be passed by applying one pole to the lumbar region of the spinal column, and the other to the perineum and hypogastric regions alternately. A very mild current has been passed, it is said with advantage, after the failure of a less-direct route, by applying the latter pole to the neck of the bladder, through the urethra. The internal remedies are strychnia, in moderate doses, such as the $\frac{1}{30}$ or $\frac{1}{16}$ of a grain two or three times daily, but employed for a long period of time; tincture of steel, and ergot of rye: all combined with tonic regimen.

2. CHRONIC CYSTITIS—a consequence of prostatic obstruction—is generally encountered at some time or other in most cases. Its presence is manifested by the appearance to the naked eye of pus and mucus in the urine, by undue frequency of micturition, and some pain in the region of the bladder. It is generally very

amenable to treatment, and often disappears under the habitual use of the catheter above described. But when this is not the case, if the symptoms are not severe, and the discharge of viscid mucus which accompanies it is not mixed with blood, much benefit may be obtained by simply washing out the bladder once a day with warm water, either with or without a double current catheter; the temperature should be about 100° Fahrenheit. After the mucus and purulent matter have been removed, which may be assumed to be the case when the water flows clear and untinged from the cavity, decided benefit is in some cases obtained by using mild astringent injections, at intervals of two or three days—always commencing with extremely weak solutions—whether of the mineral acids, the acetate of lead, or the nitrate of silver. A small quantity of fluid only should be employed at first, and it should be thrown in very gently and gradually. This treatment will be further considered hereafter (see page 202)..

Depletion of any kind is, I believe, never admissible. Counter-irritation is difficult to employ, and often occasions much discomfort. I have at some times believed that benefit has been derived from the application of irritating plaisters over the pubes, and have occasionally used for this purpose a burgundy-pitch plaister, upon which 15 or 20 grains of tartar-emetic have been equably sprinkled. But some later experiences have appeared less favourable, since the eruption is apt to be sometimes more developed on the scrotum and perineum than on the locality of the plaister itself, to the great annoyance of the patient. Were it not for this circumstance I should hesitate less to employ this means. Blisters I have also used, but prefer a more chronic species of irritant. The best method of making permanent counter-irritation is that of rubbing a moistened stick of nitrate of silver on the skin above the pubes, first removing a portion of the hair. This is to be subsequently dressed with simple ointment. For more temporary purposes, nothing is better than a hot linseed poultice, the surface of which has been sprinkled with flour of mustard, and applied to the same region: this has often proved itself a means of great value.

Hot hip-baths, by which is intended a temperature ranging from 100° to 104°, or even higher, according to the patient's ability, natural or acquired, to sustain it, are among our most useful means. The patient should remain not more than from seven to ten minutes, the object being to make a smart impression on the skin, and fill its vessels, and not to cause congestion of the pelvic viscera, which most undesirable object would probably be attained by keeping him in the bath for 20 or 25 minutes. This distinction it appears to me important to maintain. When removed, he should be rapidly dried, wrapped in hot flannels, and placed in bed, or in a recumbent position.

Respecting the internal administration of medicines there is some difference of opinion, at least as to those which possess the most value in chronic cystitis. Those which take the first rank, and on which the most reliance is to be placed, are the following. A short notice will be devoted to each, from which the special indications which should determine the selection for any particular case may be inferred.

The Infusion of Buchu is somewhat stimulant and tonic; its

Buchu.—The leaves of the Barosma crenata and others; a rutaceous plant, from the Cape of Good Hope, introduced into practice in this country, about thirty years ago, by Dr. Reece.

Dr. Prout says, "In chronic affections of the mucous membranes of the urinary organs, . . among remedies of the balsamic class, the mildest, as well as one of the most efficient, is the buchu."—*Stomach and Renal Diseases.* 4th ed. 1848. p. 403.

Mr. Coulson writes, "In my experience, however, no medicine has been so generally successful in irritability of the bladder, as the infusion of buchu. I could cite several cases where it has succeeded after other medicines had failed."—*Diseases of the Bladder.* 5th ed. 1857. p. 85.

Sir B. Brodie believes the use of buchu to be mainly limited to that class of cases in which the bladder affection does not depend on obstruction at the outlet, but on renal disease, and states, "In these activity in regard of its action on the bladder seeming to be due to a volatile oil, which communicates its odour to the urine. From one to two or three ounces may be given from two to five times in the 24 hours; and it may be strengthened by 20 or 30 minims of the tincture, which is officinal in the Pharmacopeias of Edinburgh and Dublin, although not in that of London. I have no hesitation in testifying to its utility in many cases of irritable bladder, arising from stricture and prostatic obstruction. It is diuretic, but seems to exert a beneficial influence on the mucous membrane of the bladder, and to lessen mucous discharges from it. In a recent case of villous tumour of the bladder under my care, it appeared to alleviate the symptoms very materially. I have derived more benefit from it than from Pareira brava, which has often disappointed me. In consequence of the varied opinions which are expressed respecting the relative value and properties of these several agents by different authorities, I shall cite several of them in notes at the foot of the page, to present at one view the collective experience of those who should be among the best able to judge of this subject.

The decoction of Pareira brava is largely prescribed in this country, chiefly, it is probable, on account of the strong commendation it has received from Sir B. Brodie. He advised that

---

I cannot doubt that I have seen it productive of the most beneficial effects." He advises its continued use for a long period, if tried at all, and in doses of one and a half to two ounces three times a day. —*Lectures.* 4th ed. p. 151.

Dr. Gross thinks "its use is occasionally attended with benefit, but has never derived much advantage from it."—*Anatomy and Diseases of the Urinary Organs.* 2nd ed. p. 228.

Dr. Pereira says, " In chronic inflammation of the mucous membrane of the bladder, attended with copious discharge of mucus, it frequently checks secretion, and diminishes the irritable condition of the bladder, thereby enabling the patient to retain his urine for a long period; but I have several times seen it fail to relieve, and in some cases it appeared rather to add to the patient's sufferings."— *Elements of Mat. Med.* 3rd ed. 1853. p. 1913.

it should be made in a more concentrated form, and given in
larger quantities than officinally ordered in the late Pharmacopeia.
In the present edition accordingly the strength is largely increased, and the dose of this may be considered now as 2 to
3 ounces several times in the day. It may be rendered more
powerful by an addition of the extract; such as of from 10 to

---

PAREIRA BRAVA.—The root of this plant has been prized by
different nations for some centuries as an antidote to calculous
disorders.

Sir B. Brodie says, "I am satisfied that it has a great influence
over the disease (chronic cystitis) which is now under our consideration, lessening very materially the secretion of the ropy mucus,
which is in itself a very great evil, and, I believe, diminishing the
inflammation in the bladder also."—*Lectures on the Urinary Organs.*
4th ed. p. 112.

Dr. Prout states, "Of the remedies of a tonic and astringent
character, the Pereira brava is undoubtedly one of the best we possess in catarrhal affections of the bladder."—*Op. Cit.* p. 403.

Mr. Coulson speaks of it as adapted to irritability of the bladder
associated with pain, for which he regards it as "an excellent medicine," adding, that "it may be combined with nitric or nitro-muriatic
acid, or dilute phosphoric acid, to lessen the secretion of mucus."—
Op. Cit. p. 173.

Dr. Gross has never seen any good effect result from its use.—
Op. Cit. p. 227.

ARCTOSTAPHYLOS UVA URSI.—An ericaceous plant, the leaves of
which have long been employed against calculous diseases.

Dr. Prout writes, "When the kidneys and bladder are more than
usually irritable (in the early stages of organic renal disease), . .
I doubt if any remedy surpasses the Uva ursi when judiciously
directed."—*Op. Cit.* p. 159. Again: "In chronic affections of the
mucous membranes of the urinary organs, . . . next to the
Pareira brava, rank the Uva ursi, and the Lythrum salicaria. These
last are, however, more especially beneficial to those forms and
stages of the affection marked by irritative excitement, rather than
by vascular activity, or by organic disease."—p. 403.

Sir B. Brodie says, "The Uva ursi has the reputation of being

30 grains to each dose of the decoction, which then need not exceed a wine-glassfull. I have tried this preparation on numerous occasions, and in the fullest doses, and must confess that I have not obtained the amount of benefit from it I had hoped; generally, it has seemed to exert no influence of any kind. The indication for its use, according to those who advocate it, is not simple irritability of the bladder, but the presence of viscid mucus in large quantity.

The decoction of Uva ursi has also obtained a reputation for its power to check the muco-purulent discharge from the bladder, which is so commonly present as the result of irritation of its mucous coat by decomposing urine. It is said to be contra-

---

useful as a remedy for chronic inflammation of the bladder. I must say, however, that this remedy has generally disappointed me in these cases, and that I have not seen those advantages produced by it which the general reputation of the medicine had led me to expect."—*Op. Cit.* p. 112. Again, in cystitis, depending on renal disease, he states that "it may, in some instances, be employed with much advantage."—p. 150. But fuller doses than usual are advised.

Mr. Coulson speaks well of it, but thinks more highly of buchu. —*Op. Cit.* p. 96.

Dr. Gross says, "I have used it a good deal in the treatment of cystirrhœa, and have occasionally experienced the best effects from it. I have found it particularly serviceable in cases attended with excessive morbid sensibility of the neck of the bladder."—*Op. Cit.* p. 228.

Dr. Pereira writes, "My own experience of it amounts to this, that in some cases the relief obtained by the use of it was marked; whereas in other instances it was of no avail."—*Op. Cit.* p. 1544.

Dr. Wood, of Philadelphia, believes "the credit which it now enjoys is scarcely equal to its merits," and adds, that "in cases of cystirrhœa, persevered in for a long time continuously, for several months if necessary, I believe that it will occasionally effect cures even unaided, and will often prove a serviceable adjunct to other measures."—*Treatise on Therapeutics.* Phil. 1856. Vol. i. pp. 129, 130.

indicated when any degree of inflammatory action is present, being highly astringent from the large amount of tannic and gallic acid which it contains. The dose is from one to two ounces three or four times in the 24 hours. If it is desired to strengthen the preparation, 5 to 10 grains of the extract may be added to each dose of the decoction. I have found it occasionally to check mucous discharge and allay irritability, when the preparations previously noticed have failed. Of course its employment unmixed with any other agent is referred to.

The Decoction of Senega I have found to exercise a greater influence on the mucous secretion of the bladder than any other remedy. I was led to try it from its reputation for controlling the bronchial catarrh of elderly people. The dose should be one ounce, or one ounce and a half, two or three times a day.

An Infusion of the underground stem of the Triticum Repens (dog's grass, or couch grass *) is an agent from which I have derived much advantage. For 12 months I have used it largely, both in private and in hospital practice, in numerous cases of constant and severe irritation of the bladder from any cause; and in a certain proportion it has been of great value. It has long been prized in these circumstances in some country districts; and it is officinal as a diuretic, and as a " diet drink " among the French. My first knowledge of it was derived from a country patient suffering from severe stricture, and the frequent and painful micturition of which he was the subject was obviously more effectually subdued by this remedy, which he had long taken for the purpose, than by any medicinal means I could devise. Accordingly, I have tested its powers as above stated; and my

* TRITICUM REPENS.—James, in his Medicinal Dictionary, Lond. 1743, thus speaks of it:—"The decoction of the root drank is effectual against the gripes, difficulty of urine, and ulcers of the bladder, and breaks the stone. Vide Dioscorides, l. 4, c. 30. It is also moderately aperient and lenifying." The lithontriptic virtue of this plant has also been taken notice of by Boerhaave, and confirmed by abundance of experiments. John Gerard, in his famous Herball, Lond. 1623, speaks of it in similar terms. Among the ancient authors it enjoyed a great reputation ; Paulus, of Ægineta, names it more frequently than any other remedy in urinary diseases.

belief is, that it is of more value in the cases described, and in those of renal calculus especially, than in prostatic irritation. Nevertheless, in these cases also, where the indications are to lessen the frequency and the pain of micturition, it is unquestionably useful; often affording relief when buchu, pareira, uva ursi, and other infusions, have failed. The mode of administering it, which I have adopted, is the following. One ounce of the dried underground stem, popularly called root, is infused in a pint of boiling water for an hour, the liquid is strained off and taken when cool; from half a pint to a pint to be taken in 24 hours. It may be used as a vehicle for many other agents, but I have almost invariably given it alone, with the view of ascertaining its effects. It is somewhat diuretic, and perhaps slightly aperient.

Matico,* well known as a powerful astringent, appears to have been advantageously employed in a form of cystitis frequently accompanying prostatic enlargement. Dr. Neligan says, "I have found the tincture very useful in the treatment of catarrh of the bladder in the aged."† The infusion may also be employed, either with or without the tincture, in doses of from one to two ounces. It possesses a volatile principle, which appears to resemble that of cubebs, and to have a similar action.

A decoction of the Lythrum salicaria was employed by the late Dr. Prout, and considered by him to be very nearly allied in its properties to the Uva ursi (see note on page 186). It was formerly officinal in the Dublin Pharmacopeia. The dose is from one to two ounces.‡ On the same authority was recommended an infusion of the Alchimella arvensis, when the urine is alkaline and phosphatic, in which circumstances, says Dr. Prout, " a

* Matico belongs to the natural order, Piperaceæ, from which are obtained the cubebs and other peppers. The infusion, which is generally prescribed, is prepared with an ounce of matico to a pint of boiling water.

† Medicines, their Uses and Modes of Administration. 4th ed. Dublin, 1854. p. 76.

‡ Lythrum salicaria, the spiked purple Loosestrife. It has been employed chiefly in diarrhœa and dysentery, being mucilaginous and astringent. For the decoction, one ounce of the root is boiled in one pint of water.

strong infusion, taken frequently, sometimes gives great relief."*
The solvent power for phosphatic deposits, for which this plant
has enjoyed a credit, was believed by Dr. Prout to be due to the
malic acid which it contains.

Dr. Gross quotes cases illustrating the value of the Epigœa
repens † in chronic cystitis, with catarrh. It is administered in
the form of decoction, the dose being two ounces, frequently repeated. Similarly related to the Uva ursi, and possessing nearly
equal claims to utility in the complaint just named, is the Chimaphila, or winter green. It is employed in the form of decoction,
in one or two ounce doses, which may be strengthened, when
necessary, by the addition of extract.‡ The infusion of wild
carrot seeds§ must not be overlooked, as exercising a sedative
influence in some irritable conditions of the bladder. It has
been supposed to be chiefly useful in relieving the strangury
occasioned by blisters.

All the foregoing act probably, to some extent, by means of
their diuretic properties; inducing a flow of watery urine, and so
reducing the stimulating qualities of that fluid; while a certain
amount of astringent action may possibly be exerted directly
upon the mucous membrane, and so avail to repress the undue

* Alchimella arvensis: an indigenous plant. It is astringent, and has had an ancient popular reputation for the cure of gravel and calculus. The infusion is made of one ounce of the dried leaves with a pint of boiling water.—*The Nature and Treatment of Diabetes.* 2nd ed. p. 185.

† Op. Cit. 2nd ed. p. 223. Epigœa repens, the Trailing Arbutus—diuretic and astringent. The decoction is made with one ounce of the dried leaves boiled in a pint of water. This plant appears to be nearly related to the Uva ursi. Another American remedy is the Phytolacca decandra, or Virginian Pokeberry, employed in order to alleviate simple irritability of the bladder. Dr. Gross says, "Dr. Physick was in the habit of prescribing, with decided success in this affection, the saturated tincture of Pokeberries. He gave it in two-drachm doses, every seven or eight hours."—*Op. Cit.* p. 262.

‡ Chimaphila umbellata has long enjoyed a considerable reputation in America, both in urinary complaints and in scrofula, for which latter it has been considered a specific. It is diuretic and astringent. For the decoction, boil one ounce in a pint and a half of water to one pint, and strain.—*L. Ph.*

§ Daucus carota, common or wild carrot. An ounce of the seeds to a pint of boiling water are the quantities for infusion. Dose, two to three ounces every two or three hours.

discharge of its secretion. In some, of which the buchu is the most marked example, the volatile oil contained may, besides acting as a diuretic, exert some influence, so called specific, upon the vesical mucous coat, of a calmative kind, when that membrane is in that state of irritation or excitement, commonly known as chronic inflammation, with catarrh. That this is probable appears indicated by the known similar result of administering some of the balsams under the same circumstances, all testifying to the value of small doses of copaiba in chronic catarrh of the bladder. It is worthy a trial, and is often productive of amelioration of the symptoms. If beneficial at all, the result is soon apparent, and there is therefore nothing gained from pushing it, either by increased doses, or by long administration of the remedy. Indeed any increase of the dose beyond 5 or 7 minims appears to diminish the beneficial effect. It may be rubbed up with 15 or 20 minims of liquor potassæ, with a little acacia, and an ounce and a half of camphor mixture, or some bland vehicle; or if preferred, may be administered in a capsule. In whatever way it acts, the patient frequently experiences relief when micturition is difficult, and the urine is loaded with pus and mucus.

The Chios turpentine has been recommended, in doses of 4 or 5 grains, by Sir Benjamin Brodie. Both this and small doses of Cubebs pepper, 10 to 20 grains, have been found useful in checking inordinate catarrh. The volatile oil may be substituted for the powder if preferred, in doses of 10 minims, on sugar, or in mucilage; or better still, the preparation known as the "liquor cubebæ," which, when well made, is the most elegant and efficient form.

Mr. Coulson speaks well of another balsamic remedy, viz. the compound tincture of benzoin in drachm doses three times a day.*

Another class of useful agents is the Demulcents. These are, for the most part, simply mucilaginous, or starchy solutions, which form an agreeable means of diluting the renal secretion, at

* Op. Cit. p. 169.

the same time that they furnish some little nutriment to the body, and in some instances, perhaps, some special therapeutic influence. They are therefore often useful vehicles for the administration of the acids and the alkalies, when these are required; although originally given on account of certain soothing and sheathing qualities to the urinary tract, which, by virtue of their mucilaginous character, they were supposed to possess, there is no ground whatever for attributing their beneficial influence to that especial character, which must necessarily disappear in the process of digestion and assimilation.

Among the most useful are the decoction of marsh mallow, or of the common mallow in its absence; the decoction of caragreen, or Irish moss; the infusion of linseed; the decoction of barley, better known as barley-water, and a solution of gum arabic in water.*

Another remedy, which has obtained a considerable reputation in America for urinary diseases, and has recently been imported into this country, is the inner bark of the slippery elm, Ulmus fulva. It appears to possess little besides demulcent properties, but these certainly in a high degree. Tannic acid, in small quantity, is also present. The infusion forms an agreeable drink by itself, and offers a good vehicle for other agents. The following formula, which I have employed, answers extremely well. Macerate one ounce and a half of the bark in one pint of boiling water

---

* As good formulæ for these simple yet useful vehicles are often wanted, I have subjoined those which, having frequently employed, I know can be depended upon:—

DECOCTION OF MARSH MALLOWS.—Boil three ounces in three pints of water until reduced to two pints.

DECOCTION OF IRISH MOSS.—Clean and wash an ounce of the moss, afterwards boil in a pint and a half of water until it is reduced to one pint. Any proportion of the water may be replaced by milk, when required to be more nutritious.

DECOCTION OF LINSEED.—Boil an ounce of unbruised seed in a quart of water for an hour.

DECOCTION OF BARLEY.—Boil two ounces in a pint of water for five minutes, and throw away the liquor. Add two quarts of water to the barley, and reduce by boiling to one quart.

GUM WATER.—One ounce of pure gum dissolved in one pint of water.

for six hours; press thoroughly and strain. The United States Dispensatory orders one ounce to the pint of water, but the liquor is not sufficiently mucilaginous. By decoction the bark yields other matters which render the solution disagreeable, and indeed useless.

In regard to the efficiency of each one of the principal medicinal agents which have been enumerated, it is impossible to overlook the fact that the most opposite opinions are held by experienced practical surgeons. Nor is it possible to resist the conclusion, that either the virtues of these agents have been overrated, on the ground of estimates formed from the observation of some successful but exceptional cases—that is, of cases which have progressed very favourably under the use of the medicine, but which have been in reality examples of *post hoc*, rather than of *propter hoc* improvement—or that the selection of the remedies specially adapted to each form or phase of the complaint has often not been happily made; inasmuch as the appropriate and particular indications for the use of it have not been accurately defined, or, indeed, discovered; so that the successes have been due in part to chance, and are capable of being multiplied by more skilful adaptation.

The latter suspicion is not without foundation, and it is far from improbable that the three chief and most popular remedies in urinary difficulties are sometimes, perhaps not unfrequently, prescribed somewhat empirically and at random. This opinion is strengthened by the better success which is said to result from their employment in combination, a method which is more popular abroad than in this country. I had, some time since, an American patient with cystitis from obstruction, who had long experienced great benefit from the following mixture, and I may confess to having prescribed it with advantage in similar cases subsequently:—

R. Fol. Uvæ ursi.
Rad. Pareiræ bravæ a͞a ʒij.

Boil together in three pints of water until reduced to two, and strain. Two or three ounces to be taken from three to five times daily.

To this, when cold, tincture of buchu may be added if desired.*

The indications, then, which I believe will best guide us in the selection of the principal remedial agents enumerated may be briefly pointed out as follows:—

Chronic mucous discharge from the bladder in large quantity, associated with relaxation and debility, no inflammation being present, may be acted upon by the astringent tonic Uva ursi; or this may be conjoined with pareira, for which the latter, as well as the former, has been recommended. Chimaphila is equally appropriate in this form of complaint. In simple irritability of bladder—that is, when the desire to make water is frequent—in the absence of the causes or symptoms of any acute inflammation, the Triticum repens, and perhaps, Uva ursi, afford the best chance of relief; but the former is useful in inflammatory states also.

When there is some chronic inflammation present (not acute) as evidenced by irritability of bladder, some little pain above the pubes, and considerable tenderness experienced when a catheter is passed, certain kinds of volatile oil, which are excreted by the kidney and impregnate the urine are frequently beneficial. One of the mildest, safest, and most easily-digested forms is that found in the infusion of buchu. It may be used alone, or with the addition of 15 or 20 minims of tincture or liquor of cubebs,

---

* An illustration of the Transatlantic mode of prescribing these remedies is given by Dr. Gross, in his work already referred to:—" A combination of some of the articles above mentioned may often be advantageously employed. Indeed, the effect is usually much more conspicuous when they are given in this manner than when they are used separately. I have long been in the habit of administering, with the happiest effect, a 'combination of buchu, uva ursi, and cubebs, sometimes in the form of an infusion, but more generally in that of a tincture, given several times a day, in conjunction with a small quantity of bi-carbonate of soda. Occasionally a few drops of the balsam of copaiba, the muriated tincture of iron, or dilute nitric acid, may be advantageously added to each dose of these medicines."

Dr. Gross adds, naturally enough, "When thus combined, it is of course impossible to determine what merit is due to each respective article."—p. 229.

or this may be replaced by a few minims of turpentine or copaiba, but these are more prone to disagree with the stomach; yet they sometimes exercise a beneficial influence when simple buchu has failed, more especially in cases where much catarrh is present also.

An important point in the employment of the decoctions and infusions in question is to give them liberally. The ordinary doses by tablespoons are, I think, almost valueless. From 10 to 15 ounces must be given daily in order to obtain benefit in most cases. At least, I have obtained advantage thus after failing with smaller quantities. Formerly, they appear to have been so administered.*

In this place may be mentioned an excellent combination, for which I am indebted to Dr. Gross, useful in irritable, and even in some inflammatory states of the bladder. One ounce and a half of the leaves of Uva ursi, and half an ounce of hops, infused in two pints of boiling water in a closely-covered vessel for two hours: a wine-glass of the liquor to be taken several times a day. Dr. Gross writes—it "often acts like a charm; promptly allaying the pain and spasm at the neck of the bladder, and powerfully promoting resolution."†

But further, it cannot be overlooked that some of these infusions and decoctions have been generally administered in combination with other agents, and with the use of other means, which have, perhaps, really contributed largely to the favourable result, although from circumstances they have not been permitted to share the credit. The agents referred to are the acids and the alkalies. It has been by no means common to prescribe the vegetable solutions in question unmixed with one or other of the two very important classes of bodies just named; and much observation of

---

* Half-pint doses, two or three times a day, are recommended of a decoction of Uva ursi and Pareira brava combined, in Blackie's Disquisition on Medicines that dissolve the Stone, &c. . London, 1771. p. 182. This is but one illustration among several which might be referred to of similar date.

† Op. Cit. p. 186.

their effect when uncombined will be necessary before their specific properties are known more accurately than at present.

This remark brings us to the consideration of the influence of these chemical agents in chronic cystitis.

The mineral acids are constantly ordered when the urine is alkaline, and has a tendency to deposit earthy phosphates. But it is by no means in numerous cases that even full doses by the mouth can be depended upon to exert any marked influence over the chemical reaction of such urine. As general tonics to the system, in such circumstances often much needed, they are undoubtedly useful. But it is far otherwise with the opposite class —the alkalies. By their means it is in our power speedily and powerfully to act on the kidney secretion, and to change an acid to a strongly-alkaline urine if it be desired. They have long been regarded almost in the light of specific sedatives to the bladder under circumstances of inflammation or irritation, and are perhaps entitled to more uniform confidence in such cases than any other remedies known.* Such, at least, is, without hesitation, my own experience. Additional light has been thrown upon this subject by the observations of Dr. Owen Rees. He has been led to the belief, that even when the urine is alkaline, alkalies are often productive of a greater amount of benefit than any other remedy, allaying the irritation produced in the viscus by urine of that character, and tending to restore it to its normal acidity. It appeared to Dr. Rees, to quote his own language, "that an alkaline state of urine very frequently resulted from disease of the mucous surfaces over which the urine had to pass before excretion, and that urine which had been secreted of healthy acid character was, owing to this condition of the membrane, often passed of strongly-alkaline reaction, and containing a deposit of

* This view is taken by Mr. Adams. Anatomy and Diseases of the Prostate Gland. 2nd ed. 1853. pp. 42, 43. He compares their influence to that of quinine in neuralgia, giving the preference, among the ordinary alkaline salts, to carbonate of soda.

the earthy phosphates as a consequence. The patient, in fact, was secreting healthy urine, * * * * * the variation from the normal state consisting in the urine being rendered alkaline by disease of the mucous surface of the urinary passages. That the discharge from the urinary mucous membrane, when inflamed, was of a strongly-alkaline character, and sufficient in quantity to neutralize the acidity of healthy urine, I proved by an experiment on the inflamed surface presented by the fundus of an everted bladder which I examined in a case of deficient parietes of the abdomen, a congenital deformity not very uncommonly met with. In confirmation of the above views, I took the opportunity of adducing the fact that in several cases of alkaline urine I had succeeded in obtaining the secretion of healthy acid reaction by rendering the urine less acid on secretion, and therefore less irritating, and by perseverance in this plan till the inflammatory condition subsided, the normal acid reaction of the urine, as it passed from the bladder, was eventually obtained." *

Proceeding on this principle, Dr. Rees recommended those salts in which the alkali is combined with a vegetable acid, especially the citrate of potash, and the tartrate of potash and soda; the latter if the bowels require a laxative, and the former if this is not the case. Both exercise a powerful influence in neutralizing the acidity of the urine, notwithstanding the aperient action which is associated with one of them. I have had many opportunities of witnessing the good effects which have resulted from their employment in my own practice, particularly of the citrate of potash, which I have for some years past been in the habit of recommending as a habitual drink, so long as it is desirable to produce an alkaline effect upon the urine; and in a series of papers on Irritability of the Bladder, which appeared in the Lancet in 1854, I stated that it had proved in my hands often more useful for such a purpose than Vichy water (vol. i. p. 439).

---

* On the Pathology and Treatment of Alkaline Conditions of the Urine. By G. Owen Rees, M.D., F.R.S. Guy's Hospital Reports. Third Series. Vol. I. 1855. pp. 300, 301.

Its value in this respect was observed and pointed out not less than half a century ago.* But it is not invariably that alkalies exert a beneficial influence in these cases. On the contrary, I have seen the alkaline state of the urine decidedly increased by their moderate use. And it by no means unfrequently happens that alkalies in full doses augment the irritability of the bladder; while small doses, such as 10 or 15 grains of the citrate for example, frequently repeated, produce the desired effect.

Benzoic acid is a remedy which has succeeded with me after all other means have failed, in producing acid in the place of alkaline urine, and restoring the patient to comparative ease. A brief report of a case which I not long since watched very closely at the Marylebone Infirmary will illustrate its effects, of which I could subjoin other examples.

CASE No. V.

PROSTATIC ENLARGEMENT NOT CONSIDERABLE; OCCASIONAL ENGORGEMENT OF URINE AND OVERFLOW; CHRONIC CYSTITIS; ALKALINE URINE AND MUCH IRRITABILITY OF BLADDER. COMPLETE RELIEF.

J. B.—Aged 68.

Admitted to the Marylebone Infirmary under my care, Sept. 13, 1855.

For two years past some difficulty in passing urine, pain above pubes, irritability of bladder, and other signs of cystitis from obstruction, prostatic.

---

* It is interesting to observe that it is no new observation that the salts, formed by a combination of the vegetable acids with the alkaline bases, are capable of communicating, when taken by the mouth, an alkaline reaction to the urine; although it is, comparatively speaking, a recent achievement of chemistry to explain this. More than fifty years ago Sir Gilbert Blane was in the habit of prescribing citrate of potash for the express purpose of rendering the urine alkaline. See his paper "On the Effects of large Doses of mild Vegetable Alkali," read Nov. 1, 1808, to "the Society for the Improvement of Medical and Chirurgical Knowledge."—*Trans.*, vol. iii. p. 339.

Although easily prepared extemporaneously from citric acid and bi-carbonate of potash, it is more agreeably administered from bottles, as an aerated water. An excellent solution of half a drachm of the salt in each bottle is prepared by Messrs. Sandford and Blake, of Piccadilly.

ILLUSTRATIVE CASE.

Admission.—Passed a No. 10 catheter, drawing off 36 ounces of urine. General condition rather weak. Is distressed by loss of sleep from frequent micturition, and constant pain, not severe, in back, loins, &c. Urine ammoniacal, not fetid; cloudy; little tinged with blood; when cold deposits much tenacious mucus, pervading the whole of the fluid; strongly alkaline; no albumen. Under microscope—many pus corpuscles; a few blood discs; many large, round, granular-looking corpuscles; few triple phosphates.

Sept. 14.—Ordered—Ac. nitric. dil. ℞xx, ex. aquâ; 4tis horis.

18.—Urine has been daily removed by catheter from the bladder: residue not large: as alkaline as before.
To take acid every two hours.

20.—Urine the same. Substitute hydrochloric acid for nitric, every two hours.

23.—Urine perhaps slightly less alkaline. Sounded carefully for calculus, but found none.

25.—Apply—Emp. picis, cum antim. tart. ʒss.; to the hypogastrium.

27.—Altogether better. Urine less alkaline. Plaister has irritated him.

Oct. 1.—Urine neutral.— No suprapubic pain, except of skin. Pustulation limited to outline of plaister. To leave off acid.

4.—Urine strongly alkaline.
To resume—Ac. hydrochloric. dil. ℞xx, 2dis horis.

7.—Urine slightly alkaline.

9.—Urine faintly acid. Altogether greatly improved. To leave off acid.

13.—Urine alkaline. To recommence the acid.

18.—Urine still alkaline.

24.—As before.
Ordered—Pot. bicarb. Ʒj. four times a day in water. Alkaline treatment was continued, with variations in the quantity and form, for about three weeks, with no improvement.

Nov. 14.—Urine strongly alkaline.
Ordered—℞ Acidi Benzoici Ʒj.
Sp. Vini rect. ʒiss. Haust. ex aquâ— quaque 6tâ horâ.

18.—The urine faintly acid. All symptoms decreased.

22.—Has discontinued the benzoic acid, and the urine has again become alkaline. Recommence benzoic acid.

25.—Urine more decidedly acid than before observed.

Dec. 16.—The acid was again discontinued, and the urine became slightly alkaline. He returned to its use, the urine rapidly became acid, and he was soon much improved generally. The residual urine disappeared; he ceased to require the catheter, and was discharged without any bad symptom.

The improvement in the urine, especially at first, was doubtless due in great measure to rest and regimen; but it was clear that the mineral acids, in large and very frequently repeated doses, exercised slowly some power, not considerable, over its reaction. The alkalies were not only of no service, but appeared, indeed, rather to increase the evil. But the influence of the benzoic acid was much greater and more rapidly manifested on the condition of the urine than that of any other agent employed. The excellent final result was doubtless greatly due to the catheterism, and perhaps to the counter-irritation, in addition to the other means named.

Benzoic acid is extremely insoluble in water, which vehicle is, therefore, not well adapted for its administration. Twenty grains require a drachm and a half of rectified spirit for solution, which may be taken in a wine-glass of water, when it is precipitated in a very divided state, and should be instantly swallowed. Or it may be given in powders, rubbed up with a half or equal weight of white sugar; or it may be suspended in simple syrup, or in the mucilage of acacia.

3. Irritability of bladder, with pain increased during the repeated efforts to make water, is one of the most distressing accompaniments of the affection. The patient becomes worn and exhausted by loss of rest and sleep, if these are not ensured by sedatives and opiates; the latter, especially, being usually of great value. Generally speaking, opiates by rectum exert the best influence, and may be administered either in the form of suppository or enema. From one and a half to three grains of the extract of opium, or the same quantity of powdered opium, or half a grain to a grain of morphia, made up with six or eight grains of the

simple cerate of the London Pharmacopœia, are good forms for the purpose. A grain of extract of belladonna is sometimes a good addition when much spasmodic action is present. A combination with a few grains of hyoscyamus, conium, or lactucarium, may occasionally be made with advantage. A common form is ten or twelve grains of the pil. saponis cum opio; but a better vehicle for the active drug is the cerate just referred to, which possesses no irritating qualities, the effect of which in the soap is sometimes complained of. The butter of the cacao nut has also been of late employed, and for the same reason.*

When enemata are employed, they should be small in quantity and mucilaginous; as, for example, one or two ounces of starch or barley-water, containing from forty to sixty minims of laudanum. The fluid form ensures a more rapid action, when such is required.

At the same time we may administer by mouth morphia, or some other preparation of opium, in quantity appropriate to the patient's condition: and it is usually well borne, and of great value in these cases. I have found in some cases small doses, such as one-sixth of a grain of morphia nightly, produce a very admirable effect in checking irritability of the bladder. Where,

---

* The following are excellent formulæ for suppositories:—
℞ Morphiæ hydrochloratis, gr. ⅓ to j.
Butyri cacaonis,
vel Cerati, gr. x.—Misce bene; fiat suppositorium.
℞ Ext. Belladonnæ, gr. j.
Pulv. opii, gr. ij.— iij.
Butyri cacaonis,
vel Cerati, gr. vij.—Misce bene.

These are not only unirritating, but of good consistence for use, and are readily soluble at the temperature of the body.

The following is one recently recommended by Dr. Simpson:—

"Take of acetate of morphia, six grains; sugar of milk, one drachm; simple cerate, half a drachm, or as much as may be sufficient to make a proper consistence, and divide the mass into twelve suppositories; then dip each suppository into the following mixture, to form a coating: take of white wax one part, lard plaister two parts; melt together. The best way is to insert a needle into the apex of the suppository, dip it into the melted wax and lard, and immediately afterwards into cold water, to harden it before it loses its shape. The shape is conical, like a pastile."—*Medical Times*, Feb. 7, 1857.

unfortunately, it does not agree, we must have recourse to conium, belladonna, or Indian hemp, or to full doses of hyoscyamus, or extract of hop. Sometimes large doses of camphor produce a powerfully-sedative effect; one or two teaspoonfuls of the ordinary spirits of camphor may be given in three or four ounces of water, or of infusion of hop, for this purpose.

4. When there is phosphatic deposit in the urine, while this is loaded with mucus and is ammoniacal or fetid, much benefit may often be derived from injecting the bladder. Indeed, these are the circumstances in which this treatment is especially applicable. The formula most generally employed is the dilute nitric acid, in the proportion of half a drachm to the pint of water, which is sufficiently strong to commence with, afterwards gradually increasing to, but not exceeding, two drachms. It is scarcely necessary to say that where there is much purulent secretion, and more especially if there is sanious discharge, and great pain is experienced, this remedy is contra-indicated. Sir Benjamin Brodie, who long ago advocated this practice in the treatment of cystitis, says:—"It is better to begin with washing out the bladder with a little tepid water; then to inject the acid solution, allowing it to remain for not more than thirty seconds in the bladder. At first the operation should not be repeated oftener than once in every two days, afterwards it may be repeated once daily, but never more frequently than this."* The quantity thrown in must depend upon the capacity of the bladder, and will vary from two to four ounces. It may be added, on the authority just quoted, that injections "are especially to be avoided where the mucus deposited by the urine is highly tinged with blood."*

I have often witnessed the good effect of injecting very dilute solutions of the acetate of lead in cases of large mucous secretion from the coats of the bladder. The best mode of proceeding is the following: a gum-elastic catheter having been passed into the bladder, the external orifice is to be connected with an india-rubber bottle or with any injecting apparatus which may be pre-

* Lectures on the Urinary Organs, pp. 114, 115.

ferred, and a few ounces of plain warm water are to be injected, but not sufficient to produce a sense of uneasiness, much less of pain, from distension of the viscus, and after remaining about half a minute should be permitted to flow out. A solution of acetate of lead having been prepared, of the strength of one grain to the ounce of distilled water, a small quantity of this should be added to the next quantity injected, say in a proportion not exceeding one of the solution to five or six of warm water; this is to be slowly injected, allowed to remain from 30 to 50 seconds and then to run out. A trial of one part to four of water may be next made; or be employed on the next occasion, after 24 or 48 hours' interval. In no case is it desirable to exceed one-third of a grain of the acetate to the ounce of water, unless the object is to act on a phosphatic calculus, in which case one grain to the ounce may be employed; for which see Chapter XIX. at the close of this volume.

5. A very frequent accompaniment of enlarged prostate is the appearance of blood in the urine. It may be simple, unprovoked hæmorrhage from some part of its surface, or it may be the result of passing an instrument. In the latter case, it will sometimes take place although passed with the greatest care and without pain, so ready to bleed is the organ in some states when enlarged; but this is an exceptional occurrence. It may be caused by the patient moving about after the catheter has been passed, while it still remains in the canal; supposing, for example, that it has been passed in the recumbent position, and that the patient rises to evacuate the fluid. Some cases there are in which although the instrument passes without obstruction, or even causing any painful sensation, it is nevertheless exceptional not to see a little blood at the time or soon afterwards. *A fortiori*, hæmorrhage may occur, and to a large extent when the employment of the catheter is difficult, or occasions pain. This circumstance is said to indicate ulceration of the organ. It may be so in rare cases, but I know that it often happens when no breach of surface can be found after death. It is not improbable

that the hæmorrhage takes place from enlarged and congested capillaries in the mucous membrane of the bladder, the veins from which are often pressed upon and probably obstructed when the prostate is enlarged.

When hæmorrhage is very slight, it requires no special treatment beyond an attention to increased care in the use of instruments, and the observance of perfect quiet in the recumbent posture for a short time after passing one. A change from an elastic to a silver instrument, or the reverse, will sometimes cause it to disappear; or the employment of one that is one or two sizes smaller than that which has been habitually used. If the amount lost is sufficient to threaten or produce any appreciable effect on the patient's general condition, it will be necessary to use internal remedies. The surgeon may then select from the following, or try them, if necessary, in succession, if the first attempt is unsuccessful. Gallic acid, five to seven grains, with or without a few minims of the liquor opii, three times a-day, or more frequently if necessary; sulphuric acid in infusion of roses; acetate of lead and opium; or ten to fifteen minims of turpentine, suspended in mucilage, and frequently repeated. Sometimes preparations of iron are indicated, especially the sesquichloride, and may be given in the form best adapted to agree with the patient's stomach. Mr. Adams prefers the use of alkalies in these circumstances, stating that—" of internal remedies, the simple salts of soda and potash, as the carbonates, in small and repeated doses, are decidedly preferable to acids."* If the hæmorrhage is considerable, the bladder becoming distended, which sometimes happens to an enormous extent, a large tumour being definable, extending midway towards the umbilicus, or higher, other means must be adopted. Bladders containing ice should be applied to the perineum and hypogastrium, the patient maintaining the most perfect quiet in bed, his person being lightly covered, and the bed-clothes elevated from the body by means of a cradle. Two or three ounces of ice-cold water may be frequently injected

* Op. Cit. p. 116.

into the rectum, if it can be done without disturbing the patient too much by producing action of the bowel, which in this quantity it is not likely to do. It is recommended that a catheter should be passed into the bladder and the clot broken up; and that efforts to remove portions of this should be made by applying an exhausted gum-bottle, or syringe, to the instrument. I must venture to dissent from the injunction to interfere, unless it is rendered absolutely necessary by retention of urine, which I can scarcely believe can be occasioned by the presence of the clot. I am sure I have seen fresh hæmorrhage excited by this practice. The best results in my own experience have been those in which, after first passing the catheter to verify the state of matters, I declined all mechanical interference. In one of the first so treated the bladder was so full as to resemble the uterus of a pregnant woman. The patient was very exsanguine, and had a small and very feeble pulse. I ordered gallic acid and opium every hour, the latter until it allayed the painful and spasmodic straining to evacuate the contents of the bladder, which is usually present, and causes great distress; and nutriment in teaspoonfuls constantly; ice locally; and absolute quiet. Fresh blood ceased to ooze by urethra, and in a few hours, urine so thick and deeply coloured as to resemble grumous blood, passed. In the course of 48 hours it gradually became lighter in colour, the bladder smaller, and ultimately the whole of the clot was dissolved, and came away in solution. The patient is at this moment perfectly well. The indications appear to me to be exceedingly plain not to interfere, provided there is not absolute retention. The breaking-up or otherwise disturbing the clot is liable to provoke fresh bleeding; the hasty removal of the vesical contents is extremely likely to open orifices of vessels now closed by plugs of clot within them, and pressed upon by the contained mass, besides affording a fresh cavity into which more blood may be poured. Besides, there is no ground for regarding the clot as a great evil, which must be got rid of at all hazards; much less for adopting such means, recommended on no mean authority, as injecting several ounces of a strong mixture of acetic acid and water into the bladder in order to dissolve it. The solvent power of the urine, possessing as it does a temperature of about 100° Fahr., has appeared to me remarkably great, and is probably not only one of the most efficient, but the very safest of agents in effecting the purpose.

It does happen in a few cases that some of the mechanical means above referred to must be adopted to remove firmly-adhering coagula, and the bladder has even been opened above the pubes for that purpose;* but such necessities are happily extremely rare, and I am persuaded that we shall, in the majority of cases, conduce more certainly to a favourable result by permitting Nature to do her own work, without any undue haste on our part to be officious in offering her assistance. It is true that the distress occasioned to the patient by the urgent sensation of the want to micturate and desire to empty the bladder, besides the pain and spasm sometimes, although not always, associated with this state, indicate that something should be done for the removal of the coagula. In these circumstances, especially if the patient is much reduced, the use of opium, either by the mouth or rectum, will relieve much of the distress, and we may still trust somewhat to the course of events for relief, without resorting to instruments. I have seen several cases since the foregoing remarks appeared in the first edition, and I am more than ever convinced that they indicate the proper course to be pursued in the circumstances described. But when it appears absolutely necessary to adopt mechanical interference, it must be done with the greatest caution, since the neck of the bladder, which is probably the source of the hæmorrhage, must be exposed to some disturbance and pressure by all movements of an instrument made

---

* Mr. A. Copland Hutchinson, with the concurrence of Sir A. Cooper, opened the bladder, from above the pubes, and removed a pint of coagulum, only twelve hours after the occurrence of hæmorrhage. He felt enlargement of the prostate from the cavity of the bladder. The patient died three days after the operation, and no *post mortem* was made.—*Lond. Med. Repository*, vol. xxii. p. 123. 1824.

within its cavity. Nevertheless, the necessity existing, a full-sized catheter might bo passed, and a syringe, or stomach-pump, adapted to it, by means of which a considerable portion of clot may be withdrawn. In relation to the subject of internal remedies in cases of extreme urgency, when the most powerful astringents we possess must be freely administered, it should not be forgotten that Sir B. Brodie records a case in which, all other remedies having failed, a dose of Ruspini's styptic, repeated two or three times in the course of twelve hours, was followed by complete cessation of the bleeding.* The tincture of matico has been also recommended, and, if employed, should be frequently given, and in not less than drachm doses. It contains, however, no recognized astringent principle: its use in other affections of the bladder has been already referred to.

6. A distressing result of enlarged prostate is sometimes incontinence of urine. By this I do not refer to the overflow from engorgement relievable by the catheter, usually termed incontinence, which has been already fully considered, but a real inability on the part of the bladder to retain more than, at all events, a small quantity of urine—a condition, therefore, in which the calls to make water are necessarily frequent, and cannot be re-resisted. As soon as the bladder has received some one, two, or three ounces of urine, it begins to flow, in the absence of any voluntary efforts on the part of the patient. This may occur, although rarely, as has been already shown in the Ninth Chapter, from peculiarity in the form of the enlargement, preternaturally opening the vesico-urethral orifice. If, then, the frequent micturition does not depend on a cause relievable by medicine, such as that occasioned by irritability or inflammation of the mucous membrane of the bladder, but on some organic source of the kind described, a proper receptacle (often useful in the latter case also) is the principal remedy, and should be almost constantly worn. Such are manufactured of india-rubber, and they most

* Op. Cit. p. 201.

efficiently provide for the necessity. They are, however, too well known to render any description necessary here.

7. Another result of obstruction at the neck of the bladder from enlarged prostate is a susceptibility to congestion and inflammation in the organ from very slight causes, and a few hints respecting its management will close this section of the subject of treatment.

Various circumstances, of which the most common are sitting on a damp or cold seat, especially out of doors, general exposure to cold, the movements encountered in a long journey, horse exercise, indulgence in alcoholic stimulants, and sexual excitement, will produce in a sudden manner an attack of increased difficulty in making water, and occasionally a profuse discharge from the urethra of a muco-purulent character, sometimes tinged with blood. These symptoms are generally accompanied by others of ordinary fever, although of a mild form. Such attacks sometimes produce a good deal of alarm; but they usually subside rapidly with rest in bed, hot fomentations, hot hip-baths, a mild aperient, and sedative treatment: warm enemata may also give relief. The bladder must be relieved in the usual manner, but generally with a smaller elastic catheter than might be otherwise employed, if the urethra is swollen and tender, as it commonly is. Occasionally, also, a few leeches to the perineum, or around the anus, are useful. Dry cupping to the perineum is sometimes equally efficacious, and may render the loss of blood unnecessary. The treatment may not be otherwise antiphlogistic, as the age and constitution usually met with in these patients indicate a full amount of support, in the form of good nutritive food, sometimes without the ordinary amount of stimulant; but more commonly, it is undesirable to reduce it materially, if at all. Such attacks are to be considered as congestive in their character, or at most as subacute inflammatory affections of the urethra and prostate, and are widely different from the affection commonly known as acute prostatitis, already considered in the Third Chapter.

II. The general treatment and management of patients with enlarged prostate is next to be considered.

It is of great importance to maintain all the functions of the body in health since any slight derangement is liable to augment the urinary symptoms. A fit of catarrh, indigestion, or constipation, are apt to produce increased obstruction and irritability of the bladder. The principles, then, on which the diet, regimen, and general management of the patient are founded, it is by no means difficult to understand and apply. Modifications will be necessary for every individual case; but a general plan may be sketched here, from which material deviations will not probably be often necessary.

First, in the matter of diet, the patient should restrict himself to such plain, simple, but nutritious food as his experience has shown that he can easily digest. Moderately-cooked, tender and juicy meat, of which the best in most circumstances is mutton, should be taken at least once a-day, varied by poultry and game, and occasionally fish, with well-cooked vegetables, or fruit, in moderate quantity—the former habitually, the latter occasionally; home-made bread, not less than thirty-six hours old; fresh milk if it agrees with the stomach, eggs in moderation, and farinaceous puddings, furnish the principal varieties of food from which the diet should be selected. All that tends to derange the stomach and bowels, to tax unduly the digestive powers, or to over-excite the circulation, must be avoided. Pork, and salted or dried meats and fish, highly-seasoned dishes, rich soups or gravies, pastry, cheese, dessert in all its indigestible variety, strong tea and coffee, all unripe, and most uncooked fruit, all raw vegetables and pickles, should be unhesitatingly denied. The question of alcoholic stimulants is often entirely an individual one, for which express rules cannot be given, except that in no case is more than a strictly-moderate use of such to be permitted. There are many, however, I firmly believe, of these cases in which the withdrawal of the accustomed two, three, or four glasses of sherry per diem would be a positive injury: where, during a considerable number of years, this moderate indulgence has been allowed, and often, à *fortiori*, where it has been exceeded, we shall inflict an injury by withholding it. In most cases, a light and dry sherry of the first class is the best kind of stimulant. Port-wine is not generally admissible, often wholly to be avoided, as are spirits also. The substitution of the lighter wines of the Rhine or of Bordeaux is in some cases permissible, but in advising this we must be guided by the previous habits, experience, and the constitutional tendencies of the patient. There are other cases, perhaps exceptional, in which light bitter ales agree better than any other dietetic stimulant. Whatever is selected of this kind should be regarded as the single article of use, and variety should never be indulged in. During an access of inflammatory symptoms, the stimulant should be withdrawn, if necessary; and, lastly, it may be affirmed that where the patient has not been accustomed to it, or if he feels certainly as well, if not better, without it, its disuse is undoubtedly indicated.

The clothing should be such as encourages and maintains a due action of the skin. Flannel or woollen garments should cover the trunk and the limbs, and all changes of temperature should be efficiently provided for. The patient often suffers, as the season of autumn approaches, from a foolish prejudice against unnecessary wrapping. The lighter summer flannels should, early in this season, be exchanged for a heavier description. Any chill or check to the transpiration by the skin is attended with danger of internal congestion of the prostate in the subject of enlargement. Damp must be sedulously avoided, or removed after exposure, especially from the feet, without delay: the lower limbs should be kept habitually dry and warm, a habit of the first importance, as freedom of circulation and healthy vascular action here is one safeguard against the recurrence of congestion elsewhere, and particularly at the point which demands especial protection.

The healthy functions of the skin must be promoted by habitual tepid sponging, and occasional warm-baths, both always followed by well-applied friction with the rough towel or horsehair gloves.

The warm foot-bath should be frequently employed, on the principle just alluded to.

The question of exercise is one of importance. The subject of enlarged prostate must not be encouraged to believe himself too much an invalid, but must exert his physical powers, as far as they exist, in daily exercise in the open air, of which walking is decidedly the best form. Riding is generally out of the question; the movement of trotting is undoubtedly prejudicial, and I have several times seen bleeding caused both by it and by a long drive over rough roads, or indeed after a long railway journey; and, at the same time, increased difficulty in micturition. On ordinarily smooth surfaces, carriage exercise may be added to, but should not supersede, that of walking, where the latter can be taken; if not, the former must be substituted. No exercise should be carried to undue fatigue, and after it, rest in the horizontal position is desirable. Neither should the patient withdraw too much from intercourse with cheerful society, as some, under a morbid sense of the gravity of the complaint, are very apt to do. This, often encountered at the onset of the symptoms, gradually gives way in some cases to a very remarkable extent; while in others the complaint becomes a mental incubus, which depresses the patient for the remainder of his days, and probably tends to shorten them materially. The patient should be reminded how many men there are who have long been the subjects of prostatic enlargement to an extent rendering it requisite that they should remove from the bladder the whole of their urine, by means of the catheter, from inability to pass any by their own efforts, and who, nevertheless, are so actively engaged in the pursuits of life, whether those of business, or merely of pleasurable occupation, that their daily associates are often wholly ignorant of their infirmity. The habitual tone or temper of mind exhibited by the patient is to be noted, since there is no manner of doubt that an unhappy and desponding one is prejudicial to its victim; will tend to encourage the steady progress of disease, and unnerve his constitution for resisting its casual attacks. The opposite state must be sedulously encouraged, not merely as an important therapeutical means, which it nevertheless is, but as the legitimate result of a proper estimate of his complaint, so susceptible of palliation, so slow in progress as it is in the majority of cases, and so little prone to shorten life (pp. 135-37), where judicious care and management are exercised. To assist in producing this healthy and natural state of mind, the mere assertion of the propriety of cherishing it is not sufficient; occupation of a cheerful character, suited to engross the thoughts and energies of the patient, should be found, if possible. I have observed that those whose lives are little, if at all, shortened by the complaint, are almost uniformly men who are interested and engaged in the daily pursuits of business or professional occupation; men who take the brightest view of their trial, and who by strangers may be regarded as persons of average health. It is far otherwise with those whose time hangs heavily on their hands, who are oppressed with ennui, and who, naturally enough, acquire a habit of brooding over their complaints, and permitting their thoughts ever to be in contact with the painful subject, until it acquires exaggerated proportions, and exercises a mastery over the mind, which it is intensely difficult to shake off. The mental and moral management of such is sufficiently indicated by these few remarks, which a sense of their importance has impelled me to make. Before leaving this subject, the necessity for reminding the patient of the evil effects which may result from great excitement of all kinds should not be overlooked. Mental disquiet and anxiety, when excessive, I have seen exerting most unhappy influence upon the urinary function in these cases. Undue indulgence in sexual excitement must be similarly guarded against; it may be even necessary occasionally to advise abstinence from intercourse, when obviously followed by the mischief which in some cases results.

III. The special treatment to be directed against the enlargement itself.

This resolves itself into two distinct parts:—The medical and the mechanical.

The medical treatment has been hitherto marked by uncertainty, not to say inefficiency, as to any power exhibited in effecting a reduction in the size of the enlarged organ. Nevertheless many agents have been administered for this purpose. One of the earliest on record is hemlock, which probably gained its reputation, or rather its introduction into practice against enlarged prostate, independently of its ancient celebrity even from the time of Pliny for the "reducing of all tumours," from a remark of John Hunter to the following effect: " I have seen hemlock of service in several cases. It was given upon a supposition of a scrofulous habit. On the same principle I have recommended sea-bathing, and have seen considerable advantages from it, and, in two cases, a cure of some standing." * There can be no doubt from the tenor of this passage, that Hunter alludes to the enlargement, which occurs from effusion after acute inflammation in young subjects, and not to the totally different affection now under consideration—a distinction which even now does not seem to be always sufficiently maintained. On the next page we learn that Hunter had heard of the virtues of burnt sponge in a single case; and in another, of benefit derived from a seton in the perineum, as long as it continued open; the age of the last case, twenty years, being given. It is very evident that these have no bearing whatever upon our subject, and there is no reason therefore for founding any treatment on the practice referred to. Mr. Coulson speaks of hemlock in relation to the hypertrophied prostate, and "believes it may prove beneficial in some cases." He "frequently combines it with iodide of potassium,"† which fact certainly cannot be regarded as additional testimony in its favour. The reputed efficacy of this plant in the resolution of enlargements, particularly of the lymphatic and of the mammary glands, seems to rest on good evidence, but I have not yet been able to obtain any proof of its utility for the cases before us, although it may be worthy a trial. One thing is certain, there is no preparation

* On the Venereal Disease. 2nd ed. London, 1788, p. 174.
† Diseases of the Bladder and Prostate Gland. 4th ed., p. 438.

in the materia medica respecting which more care is necessary in order to obtain it pure, than the extract of hemlock; and this should never be forgotten when it is intended to employ it. Further, in order to ascertain its real value, it should be always given alone, and for a considerable period of time. The dose is from two to five grains, two or three times a-day.

Mercury has been written about as a remedy, and although never highly recommended, has been spoken of as " worthy of a trial." If there is one agent more than another, for the administration of which no indication in this complaint exists, it is, I should think, mercury. I cannot conceive an elderly patient with hypertrophied prostate being otherwise than injured by a course of this indiscriminately-prescribed drug, and must protest against its admission into a catalogue of medicines, having the slightest claim to influence favourably the disease in question.

The hydrochlorate of ammonia has enjoyed a considerable reputation in Germany since Dr. Fischer of Dresden first claimed for it, in 1821, the power of reducing senile enlargement, when taken in sufficiently large doses. Several have subsequently reported its efficacy; among the most recent of whom is M. Vanoye, who relates two cases in which he employed it successfully.* He commenced by giving, in numerous doses, four grammes (about one drachm) daily, increasing the dose to double, and at last to three times that amount. Larger quantities than this produce unpleasant symptoms. No result was observable until it had been taken four or six weeks; after an interval it was again administered for six or eight weeks longer: the full dose of two or three drachms daily being employed during the greater part of that time.

I believe the remedy has not been tried to any extent in this country. Further experience does not show that it possesses any value in hypertrophied prostate.

Iodine and its combinations have been put in requisition pretty

* Annales Méd. de la Flandre Occid. April, 1852. From the Bull. Gén. de Thérap. 1852, p. 521.

extensively, with a view to the removal of prostatic enlargement, and a considerable degree of success has been claimed as the result. The known power of this substance in effecting the resolution of numerous swellings of the lymphatic glands, of the thyroid gland, and other tumours, naturally induced a trial of it in this affection also. The late Mr. Stafford first called the attention of the profession to it with this view in 1840, and in 1845 published a second edition of his work, giving the result of further experience; and the opinion which he formed therefrom was expressed in the preface in the following terms:—" I have very little hesitation in saying, in the general cases of enlargement we meet with, success will attend the treatment, if it be properly effected and persevered in for a sufficient length of time." *

Mr. Stafford's plan consisted in administering iodine internally by suppositories in the rectum, occasionally by the mouth, and in applying it to the urethral surface of the prostate in the form of a weak ointment; commencing with one grain of the iodide of potassium to the drachm of simple cerate, and increasing it to ten or twenty grains to the drachm, sometimes even adding to this a small quantity of the pure iodine. He states that his success was complete with numerous cases of the complaint in advanced age, as well as in that form to which young men are subject.

On analyzing the cases detailed, 27 in number, 11 only can be regarded as examples of the enlargement of old age, and in these the inference depends on the evidence afforded by the author's examination, by exploration of the rectum. In several of the others, an enlargement confined to the " middle lobe " was diagnosed, because there had been difficulty in introducing an instrument near to the neck of the bladder, no enlargement being presented in the rectum. These, of course, are rejected, first, because existence of the difficulty described is no proof that

* An Essay on the Treatment of some Affections of the Prostate Gland. By R. A. Stafford. 2nd ed. London, 1845. Preface, p. vii.

enlarged "middle lobe" exists; and secondly, because it is not very common to find this form of enlargement unaccompanied by augmentation of the lateral lobes also. In other cases, confirmed stricture was present; when, if difficulty in drawing off the urine was experienced after the catheter had been passed through the contraction, enlargement of the "middle lobe" was in consequence affirmed to exist. But the co-existence of senile enlargement with stricture is not common, while difficulties in traversing the posterior or prostatic part of the urethra after a stricture has been passed, no hypertrophy existing, are familiar to every surgeon of experience from other and well-recognized causes. Lastly, the remaining cases ranged between 21 and 40 years of age; the affection generally followed a severe gonorrhœa; and they were examples of enlargement from prostatitis in early life.

But 11 cases of success, in individuals ranging between 50 and 80 years of age, are recorded without hesitation, and must be accepted or rejected according to the view which is taken of the author's powers of diagnosis, and accuracy in reporting. It is almost unnecessary to call to mind the fact, that a very careful and discriminating exercise of both these faculties is necessary in observing the increase or decrease in size of a tumour, the chief evidence respecting which is derived from rectal examination, especially if the progress be slow. The reader cannot, however, but be struck with the ease and confidence with which the author notes the effect of only two or three suppositories, as perceptible to his finger in the diminution of the enlarged prostate; with the general completeness of the cures vouched for, and the rapidity with which they appear to have taken place. From two to four months is affirmed to have sufficed, in several instances, to reduce the prostate of an aged person from "the size of a hen's egg" to its natural dimensions (*Op. cit.* p. 101 and elsewhere). All that can be said further is, that the success was marvellous, and that other surgeons have been less fortunate, notwithstanding that Mr. Stafford's experience certainly induced numerous trials of his remedies by others. And thus it has come to pass that the use of iodine in these cases has been almost relinquished in practice.

Regarding this substance, nevertheless, as one of the most powerful and useful agents in the materia medica, in accomplishing the removal of enlargement when affecting various organs of the body, a view which is entertained of it by the profession at large, does it not appear to be possible that we may err in pursuing an opposite extreme to the practice just described, and in banishing it altogether from the list of remedial means? With our present views of the disease and knowledge of remedies, there appear to be none which afford so fair a hope of possessing a power to influence it, as iodine, and its congener bromine. Much must depend upon the manner of its administration, which requires the exercise of judgment and watchful care. The influence of these agents over simple enlargements of the uterus is undoubted, and a certain analogy between prostatic and uterine enlargements has already been pointed out (page 113, *et seq.*).

To affirm that there may be good grounds for believing that those remedies which promote absorption or discussion in the former complaint may be similarly effectual in the latter, is only a legitimate inference. The highly-charged iodine and bromine springs of Kreuznach in Rhenish Germany have obtained a deserved celebrity in the uterine cases referred to, as many of the highest authorities in London and Edinburgh testify. A knowledge of this fact, and a hope that some new source of efficient treatment might be found for the prostatic affection also, impelled me, several years ago, to pay a visit to Kreuznach. And what I then learned still confirmed this hope. Dr. Prieger, the well-known physician there, assured me that in the treatment of the chronic enlargement of the prostate of age, he believed he had seen some valuable results from the employment of the waters by bathing and clysters. Since that time the Kreuznach waters, and their saline constituents, have been imported in different forms, and employed in various complaints in this country, although I do not know that they have been used by any but myself for the disease in question.

Now, without being able to report very confidently at present, for there is no one subject more difficult to form accurate opinions respecting, than the effects and actions of medicinal remedies, and none concerning which it is more necessary to be sceptical in regarding evidence, and slow to judge, I do think that there is some ground for anticipating a beneficial result from the plan of using these means, which I have, with considerable variation, adopted for numerous cases. The difficulty of arriving at exact conclusions is also singularly great in the cases before us, because it is not easy to determine slight changes in the size of the prostate by tactile examination at different periods. None but a considerable and material change can possibly be verified in this, the only completely satisfactory manner we have of observing the effect of an agent in diminishing its volume. The symptoms may greatly improve; the habitually-retained urine may decrease in quantity, the expulsive efforts of the bladder may become more powerful, but all this does not prove the specific effect of the remedy in diminishing volume. It affords collateral proof, or presumptive evidence; but the absolute proof of the material fact is difficult to realize accurately for one's self, and still more difficult to communicate to another.

But then it must not be forgotten that the embedded or isolated tumour of the prostate, like that of the uterus, cannot fairly be regarded as amenable, at all events to any great extent, to the influence of this treatment. I do not know that there is any conclusive proof to be adduced of the disappearance or even of the considerable diminution in volume of such a tumour in either organ, as a result of the action of any medicinal agent whatever. That general hypertrophy may be thus influenced it is not unreasonable to suppose; continuous outgrowths are, probably, less amenable to therapeutic action, although perhaps not to be regarded as so intractable as the isolated tumours just referred to. Although the latter form is common, yet ordinarily there is associated with it more or less of general hypertrophy as well.

I am of opinion, then, that it is certainly worth while to attempt the reduction of enlarged prostate, especially if it be an example which is ascertained to constitute a pretty uniform tumour in the rectum. If the patient enjoys a fair share of health, there is nothing to contra-indicate it; the treatment may be pursued without exhausting the constitution, or deranging the digestive functions.

The plan which I have pursued in several cases is the following:—

Tepid hip-baths, daily, of water to which the bittern or mother-lye, of the Kreuznach springs has been added, in varied proportions, beginning with half a pint, or pound, according to the form in which it is obtained (see note below*), to four gallons of plain water, at a temperature of 90° to 94°, or warmer, if pre-

---

* The principal spring at Kreuznach employed for medical purposes is the Elizabeth-Quelle. Its temperature is $54\frac{1}{4}°$ Fahr., and it contains about 90 grains of solids in the 16 ounces, with about 5 cubic inches of carbonic acid gas.

The following is one of the most recent analyses, by Bauer :—

Chloride of Sodium	72·92
„ Potassium	0·97
„ Magnesium	0·25
„ Calcium	12·98
Carbonate of Magnesia	1·57
„ Lime	0·27
„ Iron	0·20
Bromide of Sodium	0·30
Minute quantities of Iodine, Manganese, and some earthy bases with chlorine	1·47
	90·93 grs. in 16 ounces.

In this form it is administered internally in small quantities.

But for topical applications, the water of this spring is strengthened in saline constituents by the addition of the mother-lye after the elimination of the chloride of sodium at the Salt Works, which exist on a very large scale close by the town, at other saline springs. This mother-lye, of which the specific gravity is between 1·3 and 1·4, contains no less than between 2000 and 3000 grains of solids in 16 ounces. A late analysis gave the following result.

ferred; in this the patient should be seated for twenty minutes every morning.

Local application may be made either by enema or suppository; if by the former method, the following formula may be depended upon as not too irritating to the rectum. It should be retained there as long as the patient can conveniently do so. The best instrument for injecting it is an india-rubber bottle with ivory tube, as the constituents of the Kreuznach water will rapidly injure metallic apparatus.

> ℞ Potass. iodidi, gr. v.
> Kreuznacher Bittern, ʒj
> Dec. Hordei vel Lini, ʒiij
>
> Misce pro enema, quotidie utendum.

To this a little opium may be added if necessary, in order to enable the bowel to retain it.

The suppository, which I now decidedly prefer, and which is certainly more easily administered and borne than the enema, may be used after the following form:—

In 16 ounces of the mother-lye there were 2484·16 grains of salts, constituting about a third part of the mixture.

Chloride of Calcium	.	.	1789·97
„ Sodium	.	.	226·37
„ Potassium	.	.	168·31
„ Magnesium	.	.	230·81
„ Aluminium	.	.	1·56
„ Lithium	.	.	7·95
Bromide of Sodium	.	.	59·14
Iodide of Sodium	.	.	0·05
			2484·16 grains.

This fluid being evaporated, the saline matters have for some time past been imported to this country for medicinal use; but the result is considered somewhat inferior as a therapeutic agent both here and at Kreuznach, to the original mother-lye, while it is certainly somewhat less convenient for use. More recently this has been imported by Messrs. Schacht, of 38, Hounsditch; from whom my patients have obtained it in any quantity required, at a very reasonable rate.

℞   Potass. iodidi, gr. ii.—v.
        vel
    Potass. iodidi.
    Potass. bromidi, ā ā, gr. ii.—iii.
    Cerati, gr. viii.
    Misce, fiat suppositorium.

This should be employed at the time of going to bed, and may be repeated every night for a considerable period.

The Kreuznach water itself, from the Elizabeth-Quelle, is now obtainable in this country, but is probably less useful than the bromide and iodide of potassium, given internally. I must confess I am disposed to believe less in the value of internal remedies than in that of the topical means described. At most, I would employ only small doses of the bromide and iodide of potassium conjoined, in some suitable vehicle; these are more likely to be useful, and are much better borne by the stomach than the natural water, with its large proportion of chloride of sodium. From three up to ten grains of the bromide with, at most, two or three of the iodide, twice a-day, is the quantity I have employed. It is scarcely necessary to say that this course must be persevered in for a considerable period of time, namely, from two to four months, during which the dose may be gradually increased.

Of the application of these irritants in any form or degree to the surface of the sensitive mucous membrane of the urethra I wholly disapprove. Nothing is easier than to pass down to the prostatic part a small portion of ointment impregnated with some chemical agent, and project it into the urethra there. But that it can remain there in any quantity, or for any time, adequate to the absorption of a part of the salt introduced, I do not believe: the greater part, if not the whole, is speedily removed to the bladder, and the utmost which can be expected to result is an amount of irritation corresponding to the quantity of the agent employed—an effect which, in any degree, is positively injurious.

Now, although I think we may by perseverance in this line of treatment, aided by those other appliances, and by an appropriate regimen, which are necessary, and have been already described, attain some improvement in the condition of the prostate, or, perhaps, be able to retard its increase, I do not think we are warranted in expecting to reduce the bulk of a considerably-enlarged prostate of long standing by the means described. They are, however, simple, easily employed, and unattended with any danger to the patient: and they certainly hold out more promise than any other therapeutic agents with which we are acquainted. They can be tried by the patient himself for a considerable period of time, (and without perseverance for some few months it would not be desirable to commence their use,) with but the occasional superintendence of his medical attendant, when once instructed at the outset. All the advantages which the natural springs possess are now attainable at home, since the treatment, mainly consisting as it does of external applications, is pursued with precisely the same elements as at Kreuznach, in no respect altered by their transmission here; while for internal remedies, the artificial product of the chemist is preferable to the crude salt water of the native springs.

Under these circumstances, I believe that we shall act judiciously in advising most patients whose health is good, who do not suffer from inflammatory symptoms, and in whom the complaint is not considerably advanced, to make trial of the treatment in question.

COMPRESSION.—The influence of compression in retarding the progress of morbid growths and enlargements has long been recognized, and is constantly employed with success. Thus tumours have been discussed, and inflammatory products are absorbed, of which instances are familiar to all. The question has therefore arisen, can reduction of the enlarged prostate be effected by the same agency? Its solution has been attempted in various ways, but never, it would seem, with any great degree of success, the practical application of adequate pressure having been generally found either impracticable, or, which amounts almost to the same thing, excessively difficult. It has been supposed, moreover, that the benefit accruing from the use of large catheters in this complaint is in part due to the compression which they exercise on the prostate; and there is probably some amount of beneficial influence thus exerted. It does not appear, however, to act to an extent sufficient to retard development appreciably, although it doubtless tends to maintain a more patent state of urethra than would otherwise be found. It is evident that a much more powerful act of compression, than that which can be exercised by a catheter even of the largest size, is required, if any result is to be anticipated in the way of reducing the size of the organ, or of rendering more patent the partially-obstructed neck of the bladder.

Physick, the American surgeon, attempted to accomplish this object by distending with fluid a small bag of gold-beater's skin, previously rolled up and introduced on the end of a catheter, into the bladder; and by then attempting to withdraw the dilated sac through or into the vesical orifice; and a successful result is reported by Parrish.* Subsequently, M. Leroy D'Etiolles attempted by means of metallic instruments to compress the posterior border of the neck of the bladder. He first of all employed straight, or nearly straight, sounds, by means of which, when carried into the bladder, it is not difficult, as may be readily understood, to make pressure with the hand, after a somewhat rude fashion, directly upon the floor of the vesical neck. As a modification of this method, and owing to the difficulty of introducing straight instruments into the bladder when the prostate is enlarged, he devised an instrument having the form of an ordinary prostatic catheter, the curved portion of which could be rendered straight at will, after its introduction into the bladder, by means of a stilet made of articulated portions capable of being brought into a right line through the agency of a screw in the handle of the instrument; and in this manner he succeeded in exerting a certain degree of

* Surgical Observations, p. 258.

pressure on the same limited portion of the organ.* After this M. Leroy proposed to dilate the vesical neck more equably in various directions, by employing an instrument with three expanding metal blades, and again also with the lithotrite of that period (1831-2). Other French surgeons have proposed to introduce a large curved gum elastic catheter, and afterwards to forcibly straighten it by introducing a strong straight stilet. Mercier speaks of having employed this method, and states that a flexible extremity to the steel stilet enables it more easily to traverse the catheter. But he also has devised a special instrument for dilating the prostatic portion of the urethra, or the neck of the bladder. It possesses the form of his catheter with two handles, "bicoudée" (fig. 16, f, page 231), but so contrived that pressure in an antero-posterior direction can be exerted by means of a second portion, which, being glided along the shaft, and continuing its direction, furnishes two divergent blades, by which forcible dilatation can be made.† He has recorded good results from the proceeding, and believes that these arise not from any absorption of tissue produced by the pressure, but by a mechanical depression of the obstacle which exists at the neck of the bladder. In one of his cases, he has witnessed marked benefit from six applications of the instrument of about five minutes each.‡ Some other apparatus have also been designed in France, but of so complicated a nature, and so obviously inefficient for any practical purpose, that no description need be given in these pages. The foregoing attempts are only detailed here for the sake of illustrating the views which have at different times been held by various surgeons with regard to compression as a therapeutical agent in enlarged prostate, as well as the means which have been employed in order to accomplish it.

After some experience of the employment of compression in

* Exposé des Procédés pour Guérir de la Pierre. Paris, 1825, p. 180, *et seq.* Par Dr. Leroy D'Etiolles.
† Recherches, &c., 1856, pp. 174, 175.
‡ Etude sur divers Points d'Anat. et de Path. des Organes Génito-urinaires. Paris, 1860, p. 24.

its simplest and most innocuous form, viz., by expanding an india-rubber tube with water, so as to exert a considerable influence on the neck of the bladder, I am of opinion that the risk encountered of irritating the parts does not compensate for any little benefit attained, and which is mostly only temporary in its character. Hence I do not advise adoption of any known method, believing that in most cases the progress of the disease is more retarded by good general management of the case, and the avoidance of irritation in every form, than by any specific attempt of a mechanical nature to reduce the tumour or dilate the neck of the bladder.

Several times during the last 20 or 30 years, attempts have been made to reduce the volume of a hypertrophied prostate, by passing through it a current of electricity. It is not difficult to make application of this agent; but, supposing that any advantage could be expected to result from it, the sittings must be very numerous and extend over a long period of time, in order to afford it a fair trial in any given case. Were any marked benefit likely to arise, there would be good grounds for advising the patient to submit to the inconvenience even of such a course. Experience, however, does not warrant us in drawing any favourable conclusions respecting its employment. Very recently, a case has been reported in Paris * of marked success following the use of Faradisation to a case of "hypertrophied prostate." One conductor was passed into the rectum to the rectal face of the organ; the other was passed to the prostatic urethra. On examining the details of the case, there cannot be a moment's doubt that it was one of mere chronic inflammatory enlargement of the prostate, succeeding to gonorrhœal discharge: the age of the patient was 44 years, a period of life at which neither dissection nor careful clinical observation has ever detected a case of hypertrophy of the prostate. The number of sittings was 70, and extended over a period of 0½ months. In this time, under any appropriate treatment, it is more than probable, that the reduction of such an organ would readily have taken place.

* Gazette Medicale, 1861. No. 70, p. 309.

Division of the obstructing portion at the neck of the bladder has been performed. Other operations have been also attempted for effecting a similar purpose, such as the excision or the crushing of a protruding portion ; and even the ligaturing of a polypoid outgrowth. Respecting the division of an obstruction, barlike in its form, elevated from the posterior border of the neck of the bladder, it is no doubt a proceeding to be accomplished without much difficulty, with the exercise of ordinary care. In most cases, although not invariably, the bar is a prostatic development, and when well marked may perhaps in some cases be incised with advantage, and without danger to the patient. Such was the opinion of the late Mr. Guthrie. As, however, he introduced the consideration of this subject to the profession in connection with his views of another affection occasionally met with, altogether distinct from enlarged prostate, and to which he gave the name of "bar at the neck of the bladder," I shall defer any further remarks respecting the operative proceedings (which must be the same, or nearly so, whatever be the constitution of the obstruction in question) to the seventeenth chapter, which is devoted to an examination of that subject. In that place the various cutting operations which have been applied to the prostate will be considered at length. A very few words will suffice for the notice of crushing and the ligature. Some of our French brethren have performed on the living body these procedures, the first named not unfrequently. A portion, supposed to be the protruding one, is seized between the blades of a lithotomy forceps, or an instrument very similar, and is torn away, if possible, or crushed, so as to ensure a state of sphacelus in the part attacked. Jacobson's lithotrite has been also used, and is preferred for the purpose.* M. Leroy has also described an ingeniously-contrived apparatus for carrying a ligature round the pedicle of a polypoid tumour, springing from the median portion of the prostate. It is engraved in a late

---

* The mode in which M. Leroy adapts the instrument of Jacobson to the purpose is explained and illustrated by a drawing in the Gaz. des Hôpitaux, January 27, 1849.

work, where he states also that he has used it with success.* No details of the operation are given, although, it must be confessed, they would have been exceedingly interesting, both in respect to the difficulties overcome, and the subsequent effect of leaving a putrid slough in the bladder, as a result of the process. Very recently, the same surgeon has designed an ecraseur, contained in a canula of the form of a catheter, for the purpose of removing these outgrowths.†

In estimating these proposals, I think most English surgeons will be content with awaiting further experiences in the hands of those who have hitherto seen fit to adopt them. For my own part, I have no expectation that any benefit will be conferred on the patient by such methods of accomplishing the ends proposed, even granting that no doubt existed as to the possibility of carrying them into execution. Mention is made of them here, solely because there exists no good reason for ignoring the practice which is followed by well-known surgeons in the great continental capital. Let it, however, by no means be supposed that such mention implies approval.

* Thérapeutique des Rétrécissements, des Engorgements de la Prostate, &c. Paris, 1849, pp. 75 and 77.
† Bull. de la Soc. Anat. Paris, 1856, p. 420.

CHAPTER XII.

THE TREATMENT OF RETENTION OF URINE FROM ENLARGED PROSTATE.

Urinary Retention from Enlarged Prostate, generally due to Congestion of the Organ; first indication, to relieve Congestion—Baths; second, to allay Pain and Spasm—Opium; third, instrumental relief—Catheters, various.—Comparative Advantages of.—Modes of passing solid, flexible, &c.—False passages the great cause of difficulty.—Modes of avoiding them.—Mercier's Instrument. —Should the Bladder be emptied at once?—A case.—Should the Catheter be retained ?—Catheterism unsuccessful, what Means are to be employed ?—Perforation of Prostate,—Puncture of Bladder above Pubes; by Rectum; through Symphysis Pubis. — Comparative Merits of. — Case. — Perforator. — Perineal Operation.—Conclusions.

It has been already premised that the term "complete retention of urine," here used, does not include or designate that chronic retention of urine so frequently present as a result of enlarged prostate, and so familiarly known; but is intended to apply alone to that urgent condition in which, from this cause, the patient is unable to pass any urine at all, or, at all events, only in quantity so small, as not to equal the amount of excretion naturally produced; a condition in which he is therefore in a state of hourly-increasing difficulty and danger, and from which it is not merely expedient, but necessary, in order to save his life, that he should be relieved.

Some external circumstances generally give rise to that exacerbation of the habitual symptoms which constitutes the state in question. In by far the greater number of cases, exposure of the surface of the body to cold or wet, or to both combined, is the agent, which, augmenting the distribution of blood to internal organs, produces congestion in the already-enlarged prostate, and an engorgement of its vessels, which temporarily increasing its volume, occludes the already narrowed urethro-vesical orifice. Whatever the cause, however, by which prostatic congestion is thus suddenly favoured, this is, in almost all instances, the essential nature of the obstruction. Hence may be inferred the first indication by which to direct our treatment, viz., to overcome or dissipate as much as possible internal congestion. The second is to allay pain, and quiet those involuntary but unavailing efforts to pass water, which the patient almost invariably suffers from to a distressing extent; and thirdly, and mainly, to give exit to the urine, and unless in very exceptional cases, by the natural passage of the urethra.

We shall fulfil the first indication in the most efficient manner, by employing the hot bath for the whole body. This may always be ordered with advantage, unless the presence of the surgeon is required at a very late period in the course of affairs. It sometimes is alone successful, and if so, ensures the desired end with the smallest amount of suffering or risk. If not, it at least places the parts in a better condition for the subsequent use of the catheter. The temperature should be high, from 100° to 104°. Patients who have resorted to it frequently require a much higher degree of heat than those who are not so accustomed; and the duration of the bath may be about fifteen or twenty minutes, the heat being rather increased than diminished during that time. Before the first-named period has elapsed, it is more than probable that a full effect will have been produced on the skin, that its vessels are filled, and a considerable derivation from the internal viscera must have been accomplished. Supposing, then, that the patient has not been able to relieve himself freely by this time, and is not unduly faint, an attempt may be made at once, while he is still in the bath, to introduce a catheter. If, on the other hand, he is becoming faint, and this is more likely than otherwise to happen in patients somewhat advanced in years, as the subjects of these attacks always are, it will be better to wrap him in warm blankets and move him at once to bed, before taking the next step to give effectual relief to the bladder. Meantime, however, if there be signs of much suffering—and it is almost certain that such will be the case, and particularly if involuntary paroxysms of straining to pass urine are present and uncontrollable—some sedative should be freely administered. Opium is one of the best; and no better form need be desired for the purpose, than the "liquor opii sedativus," of which thirty to fifty minims may be given, according to the judgment of the medical attendant.

With the third and last indication comes the question of the catheter; not a question as to the propriety of using it forthwith, or of delaying; for of the former course there can be no question, but that of the kind of instrument to be employed, and of the best manner of overcoming the varied difficulties which may present. Let me premise, then, that with the exercise of great care, and of ordinary skill and judgment, there should rarely, very rarely indeed, be failure in the attempt to reach the bladder and remove its contents, by introducing an instrument safely through the track of the urethral canal. There are certain circumstances under which this accomplishment would be impossible, but such are fortunately of very unfrequent occurrence. The surgeon should be provided with silver and gum-elastic catheters of the ordinary prostatic length and curves. First, respecting silver instruments, a prostate-catheter should not be less in size than No. 9 or 10; it should be from 12 to 14 inches long from the rings on the handle to the end of its beak or point; and the curved portion should comprise about a fourth

---

EXPLANATION OF FIG. 16.

a Prostatic Catheter, No. 1, adapted for most ordinary cases.
b Prostatic Catheter, No. 2, the curve and size usually sold by instrument makers as that of Sir B. Brodie.
c Prostatic Catheter, No. 3, the largest size, the curve commonly called "Liston's."
d A Prostatic Catheter strongly recommended by the late Mr. Guthrie. "This particular curvature I obtained after many trials with a flexible metallic instrument, and I believe it to be the best."—*Anat. and Dis. of Urinary Organs.* 3rd ed. p. 34.
e The Prostatic Catheter of Mercier—"sonde coudée."
f Another, recommended in some cases of difficulty—"sonde bicoudée."
g An Elastic Gum Catheter, mounted on an over-curved iron stilet, for the purpose of ensuring a suitable curve in the instrument after the stilet is withdrawn.

Fig. 16.

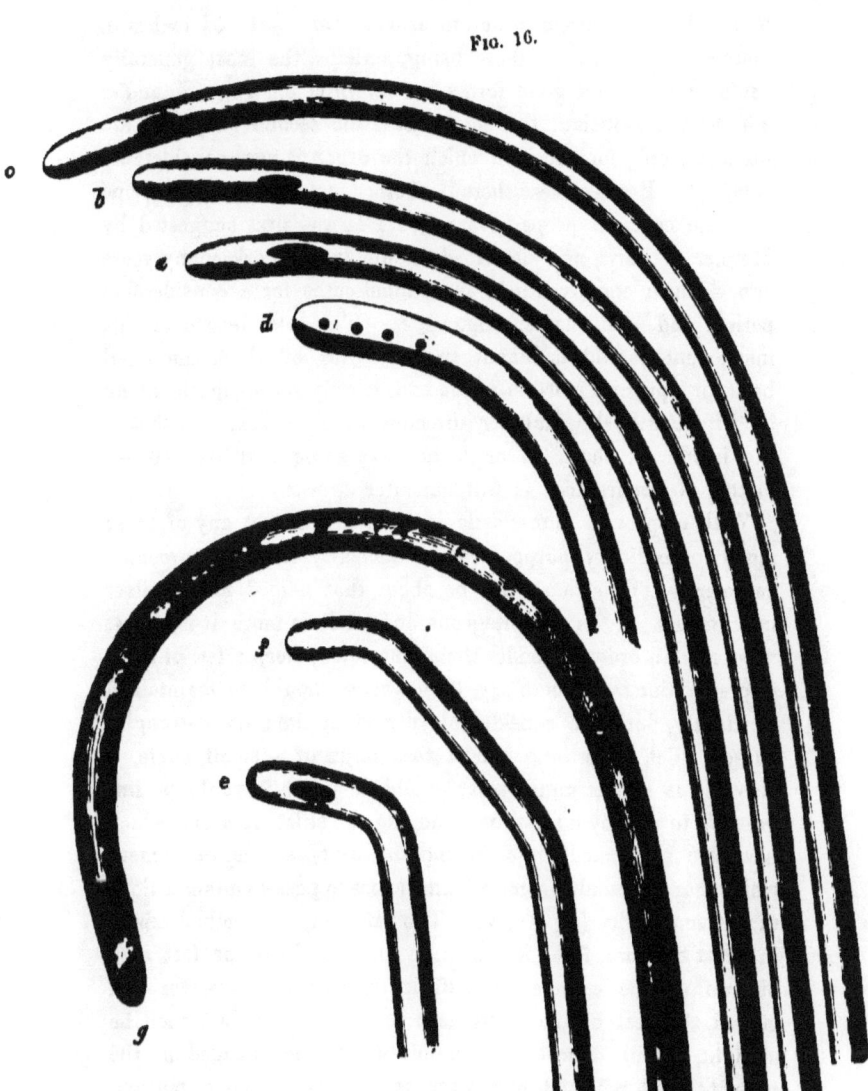

to a third of a circle, which measures from 4½ to 5½ inches in diameter; the mean of these being, perhaps, the most generally useful size. Three good forms are shown at fig. 16, *a, b,* and *c.* The first is sufficient for most cases, the second and third are necessary only for those in which the organ is very considerably enlarged. Besides these there is another instrument, a useful one in some cases, if properly managed. It was first suggested by Mercier of Paris, and described by him in his work many years ago. I have employed it in exceptional cases for a considerable period, and with success (fig. 16, *e*). The total length of this instrument should be about twelve inches, of which the small beak, or upturned portion at the end, is only seven-eighths of an inch in length; this takes a direction which makes, with that of the handle or shaft, an angle not exceeding 100° or 110°—a matter of importance, as will hereafter appear.

With respect to gum-elastic instruments, almost any of those made for ordinary purposes are sufficiently long for prostate-catheters. The size should be about that named for the silver instruments. It is advantageous, indeed it is more, it is almost necessary, in order to render them efficient, to keep a few of these in preparation; that is to say, the catheter should be maintained constantly, during a considerable period of time, on a strongly-curved stilet, describing almost two-thirds of a small circle, a curve, it is almost unnecessary to add, in which it would be impossible to employ it; but on removing the stilet from one which has been so treated for a few months, we possess an instrument which may be found in some circumstances to possess qualifications of extreme utility (fig. 16, *g*). The value of this method arises, in great measure, from its ensuring that the beak, or last inch or two of the catheter, is sufficiently curved. However well curved the rest of the instrument may be, if the last inch be straight, it will, almost to a certainty, become engaged in the prostate, and will not pass over an enlarged median portion. This should never be forgotten in giving the intended curve to the iron stilet, which cannot be done with the fingers; the last inch can be well curved only with a pair of pliers. A gum catheter also may be used, either with or without a stilet; in one case being a flexible, in the other an inflexible instrument; so that it possesses sometimes an advantage over metallic instruments in its capability of being adapted in form to any curve which the peculiarity of the case may appear to demand.

It will, I think, answer no practical end to refer to the usage and recommendations of acknowledged authorities on the debatable point as to whether the flexible or the metallic instrument is to be preferred in catheterism for retention from enlarged prostate. We should, by doing so, but place in juxta-position the most opposite opinions and practice, and that from men of large experience and sound judgment. By some the silver catheter is exclusively used; others believe the elastic instrument infinitely superior. Now, although for cases of stricture of the urethra I advocate the use of inflexible instruments as the rule beyond all question, and one which admits of very few exceptions (a subject fully discussed in my work on that subject), I have as little hesitation in regarding the two varieties as possessed of almost equal utility in cases of prostatic disease. And there is nothing paradoxical in this. The object in stricture is to introduce an instrument of small size into a narrow opening, situated usually in the straight part of the canal, or, more accurately speaking, in that part of the canal which has little or no natural curve, and which is maintained almost as readily in the straight as in any other direction. The direction in which the catheter is employed may be completely determined by the operator, and his success depends much upon the control which he exercises over it. But in prostatic enlargement the case is different; the ability to determine the precise direction of the catheter, although considerable, is less complete. First of all, the obstruction is situated in or beyond that part of the canal, which naturally possesses a curve; secondly, the curve is liable to be indefinitely increased by the presence of the disease in question; and thirdly, the canal may deviate irregularly either to right or left, as an effect of the same condition. Under these circumstances, while confessing at the same time a natural predilection to a silver catheter, as generally a more certain and satisfactory instrument, I nevertheless have met with many cases in which the gum-elastic catheter has proved the most successful instrument, and has been productive of less pain to the patient. We may consider instruments of both kinds necessary to a properly-furnished armamentarium to meet the emergencies under consideration.

It will be easily inferred from the foregoing that I prefer, as a rule, to use a well-oiled and warmed silver catheter at the outset; it furnishes us with more certain means of making correct diagnosis as to the precise situation of the obstruction, can be better felt through the perineum or from the rectum, and can better reveal the condition of the canal to an experienced hand than the flexible instrument. In very many cases it will enable us to afford relief to the patient "cito, tuto, et jucunde." As it is desirable to follow a uniform manner in introducing it, and to adopt one of the best, I may describe in detail the various steps which should be followed by the pupil in his endeavour to acquire facility in the practice. The surgeon, standing at the left side of his patient's bed, who should lie on his back in an easy position, takes the catheter lightly between the thumb and fore and middle finger of his right hand, which occupies the supine position, the former (the thumb) being, therefore, applied to the upper surface of the handle, close to the rings; the two latter supporting it below, and in a horizontal direction. The penis may be either held indifferently between the thumb and fingers of the left hand, or uniformly according to the following method, which is not without a certain convenience. The left hand is in this case applied, the palm being upwards, so that the middle and ring fingers hold the penis just behind the corona glandis; the index finger and thumb are then at liberty to be applied for the purpose of retracting the prepuce if necessary. The point is then introduced into the urethra, the direction of the shaft being parallel with the line of the left groin, and the instrument carried down as far as it will go without elevating the handle from the horizontal line. This is now gently carried to the median line of the body, and at the same time a little raised, so that the point enters the sub-pubic curve. The penis being now untouched, the shaft of the catheter is brought to the perpendicular, and moved slowly downwards towards the interval between the patient's thighs, while at the same time slight traction is made upon it, so as to keep the beak of the instrument closely along the roof of the urethra and enable it to slide closely under the pubic arch; after which, unless the obstruction be considerable, it will soon enter the bladder, gliding over any little prominence at the floor of the vesical neck. Supposing that the difficulty is not overcome by this simple means, there remain but two modes of manœuvring which afford a reasonable chance of success with the silver instrument of the ordinary prostatic curve. The first is the attempt to follow more closely, or accurately, the upper aspect, or roof, of the urethra, either by withdrawing and sooner depressing the handle, or by employing an instrument with a longer and more strongly-curved extremity, so as to override, if possible, a large tumour of the median portion, or other prominence of that part. The second is to incline the beak, when arrived at the prostate, either to the right or left, so as to pass through the sinus or hollow which, to a greater or less extent, exists on each side of such median projection, entering the bladder not over, but laterally, as regards the obstruction. These are the points which experience, as well as our knowledge of the pathological anatomy of the organ, indicate to be borne in mind.

Failing in attempts by this manner, a gum-elastic instrument, which has been kept for some time on an over-curved stilet, as before described, should be well oiled, not warmed, as this will destroy or weaken the curve, and then removed from the stilet immediately before it is required to be introduced. It is to be slowly carried down the canal, the pendant part of the penis being drawn over the left thigh or groin, and accommodated to the curve of the instrument, so as not to unbend the latter more than is necessary; and the general direction already indicated in the rules just given for the use of the silver catheter is to be followed for the remainder of its course. Owing to its flexibility and a well-curved point, it not unfrequently glides in without a hitch; but if it be otherwise, the lateral direction of its point may be varied, by twisting the handle slightly to either side. By withdrawing and again pushing forwards, by communicating a little screwing motion as it is thus made to advance, by making pressure on its convexity with the point of the index finger on the perineum, or placed within the rectum for that purpose, or by pressing closely on the root of the penis in front, it is probable that the end will be attained.

The use of the stilet is twofold. First, if the silver instruments are insufficiently curved, we possess the means of employing an inflexible instrument, although constructed of gum, of any curve we desire, by first communicating it to the stilet, which ought to be stouter and stronger than a mere wire. Secondly, it enables us to put in practice a manœuvre of considerable utility, well known as having been originated by the late William Hey, of Leeds. Many have derived an advantage from adopting it in circumstances of difficulty. It may be thus described: the catheter, mounted on its stilet, having been introduced as far as to the obstacle, the stilet is then withdrawn about an inch, which has the effect of increasing the curve and elevating the point of the catheter, so as often to carry it over the enlarged portion in a manner less easily accomplished in any other way.* The stilet should be so large that it cannot issue from the eye of the catheter in any backward or forward movement which may be communicated to it.

Another method, which has occurred to myself, is to employ a full-sized silver catheter (No. 10 or 11) of the No. 1 pro-

* Pract. Obs. in Surgery. By Wm. Hey, F.R.S. Lond., 1814, 3rd ed. pp 399, 400.

static curve (fig. 17, *a*), but completely open at its extremity, as if the point were cut off. A flexible gum-elastic catheter, of a

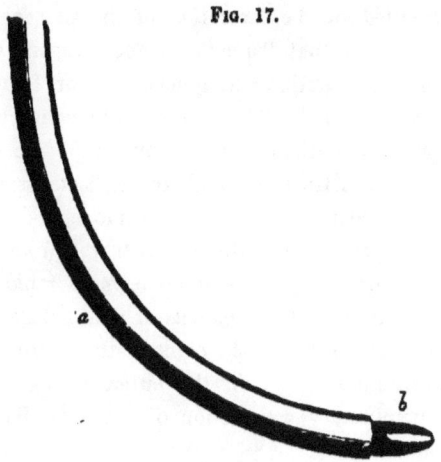

Fig. 17.

size adapted to fill pretty accurately the channel of the silver catheter, is to be passed through this, just so far that the point of the flexible instrument protrudes, forming an obturator and point (*b*), this apparatus being passed down to the obstruction, and the silver catheter being held by the left hand, the flexible one may be gently pushed onwards through the former, and may find its way into the bladder, when neither the silver nor the flexible instrument would pass alone.

In employing any manipulation hitherto mentioned, the instrument should be held with extreme lightness, and used only with gentleness and delicacy. By no other means can the operator learn to appreciate the kind of resistance which its point encounters; by the gum-elastic instrument especially he may be easily deceived if he uses an improper degree of pressure, which, instead of advancing, is, perhaps, merely driving it into some lacuna, or against some fold of mucous membrane, while the instrument, nevertheless, continues to disappear under the hand, by becoming bent within the urethra. No practised hand, however, can mistake the sense of resistance offered under these circumstances; nevertheless, it has ofttimes been a source of deception, and frequently has resulted in the laceration of the passage. It is a trite remark of Civiale that, "a catheter goes properly only when it is *swallowed* by the urethra; no quicker or more forcible movement is *allowable* there." There is a truth conveyed by this remarkably appropriate simile, which should never be forgotten; the instrument should travel through the delicate, sensitive, and contracted mucous canal by a slow, continuous, and easy movement, resembling that by which the morsel travels from the fauces to the stomach. Anything more forcible, more rapid, does but excite resistance, either of a mechanical or vital nature; if, in the former case, by carrying before the point of the instrument a fold of mucous membrane; or if in the latter, by exciting either voluntary or involuntary contraction of some of the muscular structures around.

Sir B. Brodie recommends, in cases where "the urethra is irritable and liable to spasm at the membranous part," that the gum catheter should be passed to the neck of the bladder without a stilet, after which this may be introduced in order to enable the operator to surmount the obstruction there. Less opposition is encountered, he believes, by the instrument when flexible than when made rigid by the stilet.*

The method of employing the angular instrument of Mercier comes next to be described. It is not to be passed in the manner of the ordinary catheter, nor even exactly as the lithotrite, the form of which it strictly resembles. Success in its use in difficult cases depends upon the management of it after a particular method, otherwise no advantages can arise from its employment.

The idea of an instrument with a beak nearly at right angles with the shaft, and not more than three-fourths or seven-eighths of an inch in length, takes its origin from the peculiar form and size of the urethra, known to be assumed in largely-developed

* Op. cit., p. 191.

swelling of the lateral lobes of the prostate. As has been shown in the fifth chapter relating to the anatomy of hypertrophy, the urethra within the prostate becomes increased considerably in measurement from its floor to its roof, and the opening into the bladder elongated vertically, so that the canal forms a long oval in its recto-pubic diameter, instead of a spheroidal section, when distended by an instrument; at the same time a sudden elevation generally exists on the floor at the vesical neck, a kind of elevated step at the entrance to the bladder. The adaptation of the instrument to the formation described will therefore be seen. The following manner of using it, in order to carry out the idea suggested by the morbid anatomical condition referred to, is derived from Mercier, its originator, although not given here precisely in his own words. The operator stands indifferently on the right or left side of the patient, who is lying on a couch or bed; the right being more convenient at the last step of the process, in which position we will therefore now consider him. Taking the penis in his left hand, and holding the catheter in his right, he introduces it in the ordinary manner as far as to the bulbous portion of the urethra, when the shaft is to be raised nearly to a right angle with the patient's body; it is not now to be simply depressed between the patient's thighs, as this would merely tilt up the beak against the roof of the urethra, and not cause its advance along the canal, as in the case of the curved catheter; but there is to be a combination of the two actions of pressing in the line of the shaft and depressing, the degree of the latter to be regulated according to the advance which the beak is felt to be making as it glides through the deep portions of the urethra to the prostate. Thus far we have introduced it precisely according to the mode in which we should pass an ordinary lithotrite. But if there is obstruction of the kind now under consideration, as soon as the instrument has obviously arrived at it, and will not advance further by the the means directed, its shaft is to be depressed until it nearly reaches the horizontal line of the body along the interval between the thighs, in which direction it is to be gently pressed forward, so that the whole of its short beak, and not its point, advances along the prostatic urethra, the form of which in advanced cases of enlargement, as we have seen, admits of this action, until arrived at the step or bar at the neck of the badder, over which the same gentle pressure, slightly varying the direction, in order to "humour" its course, may cause it to glide directly upwards into the bladder. The adaptation of the instrument to this form of the urethra is seen by examining figs 3, 4, 5, and 6, at pages 87, 88 ; as well as Plates V., VI., IX. and X.

I have not had any experience of this instrument for cases in which I have previously failed by other means. I have, however, introduced it in the manner described with ease, for some of those to which it has been adapted, for the purpose of trial. I cannot, therefore, speak of its superiority as a final resort, that is to say, after the failure of the curved instruments, flexible and inflexible, for which emergency chiefly its author recommends it ; and it is necessary to add that he regards it under these circumstances as often superseding the necessity for forcible catheterism through the median portion ("third lobe"), or the puncture of the bladder. I certainly should not fail to try it where such are encountered, and confess that I should do so with some degree of confidence. I have but once been compelled to puncture the bladder in retention from enlarged prostate, but in that case under circumstances which would have rendered Mercier's catheter as useless as all others were under the conditions then present. This case, which was one of considerable interest, will be reported when the subject of puncturing the bladder comes before us. (*Vide* page 250.)

The real cause of difficulty in these cases is the existence of false passages. It often happens that in previously-made attempts the catheter has perforated the obstruction, has passed out of the urethra, and has been carried onwards a considerable distance in the belief that it was still progressing towards the bladder, when it has really been boring a false route by the side of the rectum or the bladder. After this unfortunate occurrence, it follows that the instrument takes the wrong course much more readily than the right; in all cases it is difficult, in some almost impossible, to avoid it. Under these circumstances, a metallic instrument, the movements of which are susceptible of more complete control than the elastic one, should be employed. It should be as large as the urethra will admit; its point should be blunt, not tapering, and this should be kept throughout its course close to the upper aspect of the urethra by gentle pressure of the point upon it, so as to avoid the lower part where the obstruction is generally encountered, and the walls of the urethra are the most yielding; hence it is that the false passage is generally found in that situation. Arrived at the pubic curve the handle should be depressed in order to elevate the point, and here the fingers placed on the perineum to make gentle pressure on the convexity of the instrument, felt through the tissues, often aids its transit. But if, after passing some seven or eight inches, it is arrested in its course, the forefinger should be introduced into the rectum to ascertain the situation of the catheter, and if it is felt too distinctly, the mucous membrane only appearing to intervene between the finger and the instrument, we learn that it has left the urethra, probably by its floor or one of its sides, and has passed between the bladder and the rectum. In these circumstances the catheter is to be withdrawn three or four inches, the finger being still maintained in the bowel in order to be used as a fulcrum upon which to tilt upwards the instrument, in the next attempt to introduce it, a method which often proves successful. If still failing, the instrument is withdrawn to the same extent, and its point, as it is again carried onwards, is to be pressed gently on the right side of the urethra; failing in that, it is to be introduced while pressing it on the left; since one side must be unbroken by the orifice of the false passage, and guided by the entire side it is probable that the catheter may be slipped into the bladder. These manœuvres, carefully and systematically employed, rarely fail in overcoming the difficulty presented by the conditions in question. Mercier proposes an ingenious method in order to avoid a false passage of the kind described. He uses a metallic catheter with an eye, situated in the convexity, two or three inches from the termination of the instrument, which is solid beyond the eye. Having passed this first, the end of which, as usual, finds its way into the false passage and fills it, he slides down the hollow of the catheter an elastic instrument which, issuing at the eye in the convexity, above described, avoids the false passage, and has a good chance of following the course of the urethra into the bladder. (See fig. 18.)

Fig. 18.

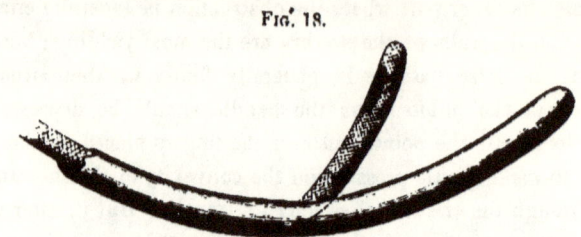

Fig. 18. Mercier's instrument for avoiding false passages. It is a silver catheter, hollow up to the dotted line; beyond this is a solid portion which enters the false passage. A small gum catheter is passed along the instrument and issues at the orifice in its concavity to traverse the urethra in front of the false passage occupied by the extremity of the instrument.

The question occasionally arises, is it desirable at once to evacuate the entire contents of the bladder when retention has existed for a considerable period of time? In very rare instances the removal of a large quantity of urine, amounting to several pints, has been followed by fainting and depression, from which the patient has never rallied. When the extent of vesical dullness is very considerable, it is therefore prudent to afford relief in a gradual manner, and, supposing that the catheter is retained, this may be easily accomplished. The removal of some 30 or 40 ounces will probably afford complete ease, and after the lapse of half an hour or an hour another portion may be withdrawn; in this manner the bladder may be gradually brought to adapt itself to the normal condition of contraction, which, subsequently, as a rule, must be insured at least once or twice a-day.

Since the publication of the first edition, in which the preceding paragraph appeared, a striking illustration of the necessity of adopting the prudent course there indicated has presented itself. The case became the subject of a legal inquiry in the country, and an action at law resulted, at which my opinion was desired. The circumstances were these. An aged pauper had suffered from "incontinence of urine" for some time, this condition being, in reality, as it so commonly is (see page 156), the sign of a greatly-distended bladder. The amount of urine daily passed by continual dribbling, was equal, or nearly so, to the normal quantity, and the medical officer had concluded that it was unnecessary to pass a catheter. Circumstances, however, brought about the decision to employ an instrument; the old man was placed upright against a wall, the catheter introduced, and six pints of urine were withdrawn in full stream; but the water had no sooner ceased to flow than the patient fell, dead, at the surgeon's feet. Fatal syncope had taken place, doubtless, in consequence of the rapid removal of so large a body, which had previously pressed on the abdominal veins and viscera; the patient being, unfortunately, in the worst possible position for meeting the result. Had the precaution above recommended been taken, undoubtedly the sudden catastrophe would not have occurred.

A point of some importance remains. A catheter having been introduced with some difficulty for the relief of retention, should it be permitted to remain? This question is answered negatively and affirmatively by different authorities. In support of the negative it is said that the parts are already in a state of considerable irritation, and that it is therefore undesirable to permit any chances of adding to it, of which the presence of a catheter may be one. On the other hand, it is urged that the bladder, after long retention, will very soon fill again, that the obstruction may again act as before, and that less hazard is incurred by the presence of the instrument than by a probable repetition of the efforts to place it there, the argument receiving additional force if more than ordinary difficulty was experienced in passing it in the first instance. I confess I have no hesitation in coinciding with the latter view, as the rule, reserving the right to make exceptions under peculiar circumstances. I have seen great danger incurred by the too early removal of a catheter which had relieved an urgent attack of retention. The indication for tying in is strengthened, if the catheter, which has been introduced with difficulty, is of gum elastic, and not a metallic one, the former being much better tolerated than the latter. In fixing it in its place, it is necessary to be careful that its extremity only just reaches the bladder, and does not project far into it, since the latter position is prone to cause irritation. A plug is fitted to the external orifice, and the water is drawn off from time to time as may be necessary. In some cases, however, it is better, especially when the water is acrid and offensive, to substitute for the plug a piece of light and flexible india-rubber tubing of sufficient length to reach a vessel at some distance, under the bed for example. Or, if it conduces to the patient's comfort, the tube may be used long enough to reach to an adjacent room, a plan I have frequently adopted in private practice, with good results in these conditions, the whole of the excretion passing off as easily into a vessel at 12 feet distance, as into one placed in the bed itself. It is only necessary to be careful that the tube is slender and light, and that it is fixed at two or three points in its course, commencing at the side of the patient's bed, so that no weight or traction may be made upon the catheter by it. All this is very easily managed by a little contrivance. The employment of the catheter in this manner is of course to be dispensed with as soon as the patient's condition permits.

The next question presenting itself for solution is, what step is to be taken when it is imperatively necessary to afford speedy relief to retention, all attempts by gentle catheterism having failed?

Rarely as such a condition of things is or ought to be realized in practice, it must occasionally present itself. We must, therefore, be prepared for the best method of encountering it. An artificial opening must be made into the bladder; this may be accomplished by perforating the obstructing portion of the prostate, or by puncturing the bladder from some external part. The different operations involved by these proceedings shall be first described, and their applicability to the various phases of prostatic retention discussed hereafter.

Perforation of the obstructing portion of the prostate—in other words, forcible catheterism—is usually performed with a strong silver catheter, about No. 9 or 10 in size, of somewhat conical form at its point, rather longer than the ordinary catheter, but not possessing the large curve of the full prostatic instrument. . The operator introduces this to the seat of obstruction, and satisfies himself by means of a finger in the rectum that it lies fairly in the urethra, and is engaged in the prostate: he then steadily carries it onward towards the cavity of the bladder, by pressing the point firmly forwards, and at the same time slowly depressing the handle; and he desists on feeling that the point is free in a cavity, and on finding that the urine flows through the instrument. This, according to Chopart, was done by Lafaye on the person of Astruc; in more recent times it has been done in this country by Home * and Brodie.† Liston accomplished it with a cutting stilet "carried through a slightly-curved and long canula," and "practised the operation a few times successfully."‡ Whatever be the instrument employed the surgeon must be particularly careful to maintain it in the middle line of the body, and also to aim at making the point emerge just behind the neck of the bladder, neither too near the pubes on the one extreme, nor the posterior wall of the bladder on the other. The catheter or canula, as the case may be, should be retained not less than forty-eight hours afterwards in the bladder, that the tissues around may consolidate, and no difficulty be experienced in replacing it by a catheter when withdrawn.

Puncture of the bladder is performed in three ways; above the pubes, through the rectum, and through the pubic symphysis.

* Practical Observations on the Treatment of the Diseases of the Prostate Gland. By E. Home. London, 1811. Vol. i. p. 163.
† Lectures on the Diseases of the Urinary Organs. By Sir B. C. Brodie. 4th ed. London, 1849, p. 195.
‡ Practical Surgery. By Robert Liston. 4th ed. London, 1846 p 48f.

The suprapubic, at one time regarded as the only possible mode of reaching the bladder from the surface in cases of enlarged prostate, is performed as follows:—The patient being placed in a half-sitting, half-reclining position, and the pubes shaved, a vertical incision of the integument is made directly above the symphysis pubis, about an inch and a half or two inches in length at the surface; this is to be carried downwards through the linea alba, so as just to admit the tip of the finger to recognize the distended bladder. Meantime an assistant, standing behind the patient, should press one of his hands firmly on either side, against the abdominal walls, in such a position as to steady the bladder. A straight, or a slightly-curved, trocar (if the latter, the convexity of the curve should be upwards) is then to be carried, with a very little inclination downwards into the bladder. The puncture should be made, not quite close to the pubes, or as the bladder contracts the opening will tend to recede downwards behind the symphysis: nor should the distance from it at all exceed an inch, else the peritoneum will be endangered. When the distension is considerable, this membrane is carried two inches or more above the margin of the symphysis. In this case also it is better not to empty the viscus immediately when very large, in order to avoid the evils already described, as resulting from the sudden removal of pressure from the abdomen. After the operation the canula should be exchanged for a silver tube specially adapted to slide through it, secured by tapes and a T bandage, which may remain a variable length of time, at all events until lymph has been effused upon the edges of the wound, when it may be withdrawn, and an elastic gum catheter worn in its place, an instrument which is generally better tolerated by the bladder than one made of metal. In a case in which I performed this operation in University College Hospital recently, I passed a gum catheter through the tube and withdrew the latter on the operating table.

The puncture by rectum, commonly adopted in stricture, but which has also been resorted to in a few cases of retention with enlarged prostate, may be performed in the following manner.

The rectum having been emptied, if necessary, by means of an enema, the patient is to be placed on his back, in the position for lithotomy, and firmly held by two assistants, not tied. The surgeon is then to introduce the forefinger of his left hand into the bowel, and ascertain the limits of the prostate, defining its boundaries, if possible, particularly the posterior one. *Fluctuation should be felt there,* communicated through the contents of the bladder, from a tap, or from momentary pressure made on the hypogastric region; and the point at which it is most distinctly perceived in the median line selected. Any spot within fair reach of the finger, under the circumstances of retention and consequent distention of the parts, may be considered safe as regards the peritoneum. Having directed an assistant to support firmly the lower part of the abdomen with both hands, so as to press down and steady the bladder towards the rectum, a well-curved trocar, seven or eight inches long, should be carried along the finger, and carefully directed to the part indicated; the handle is then to be depressed, and the point carried upwards through the coats of the rectum and bladder, until it is felt free in the cavity of the latter. The canula is to be carefully kept in its place while the trocar is withdrawn, and afterwards retained there by means of a bandage and tapes. In order to prevent the liability to slip from the bladder, which the ordinary canula is found to exhibit, Mr. Cock, of Guy's Hospital, has contrived one, the extremity of which can be made to expand somewhat after its introduction into the bladder, and with which there is therefore less danger of the occurrence of that accident. The form is that of the trocar generally employed, but increased in length and thickness.* I have three times performed the operation with Mr. Cock's instrument, and have no hesitation in saying that I prefer it to the original one.

Lastly, there is the puncture of the bladder through the symphysis pubis.

This operation appears to have been first proposed by Dr. J. M.

* Medico-Chirurgical Transactions. Vol. xxxv., p. 186.

Brander, now of Jersey, in the year 1825, when a student in Paris, where he read a paper on the subject to a medical society, advocating its employment on theoretical grounds alone, derived from the supposed advantages of the situation regarded anatomically.* Subsequently he presented a paper on the same subject to the Royal Medical and Physical Society of Edinburgh, and afterwards to the Medical and Physical Society of Calcutta, in whose Transactions it is published, an account of a case being appended.†

Several successful cases have since occurred in the practice of Dr. Brander and others. One in that of a man aged seventy-two, in whom the retention occurred from prostatic enlargement, has been performed by Dr. Leasure, of Newcastle, Pa., and is recorded at length in the American Journal of Medical Science, April, 1854.

Dr. Brander appears to have employed in the second, if not in the first, of the two cases reported, an ordinary hydrocele trocar of middle size, although he speaks of one flattened in form in the original paper. The cylindrical instrument seems to offer an advantage alluded to by Dr. Brander, from its form admitting of a rotary motion as well as of direct impulsion. The patient should recline, and the trocar should be introduced, whether after a small preliminary division of the integuments or without it, appears to be immaterial, about the centre of the symphysis, reckoning from above downwards, and in a direction at about right angles to the vertical axis of the body. Dr. Brander says, "somewhat obliquely downward and backward toward the sacrum, varying the direction according to circumstances; a piece of flexible catheter is then to be introduced through the canula," and retained by a tape.

In reviewing the comparative merits of these proceedings, it

* Séances de l'Athenée de Médicine. 1825.
† Transactions, 1842. Vol. viii., part 2, pp. 208–239. A Paper on the subject, and one case read December, 1839. In the Appendix to this volume is a second case, which occurred in 1841. The first patient died in a few hours, the second in about nine days, after the operation.

must be admitted that each of the two modes of puncturing the bladder possesses in certain cases some special advantages. Most surgeons have regarded the puncture above the pubes as more dangerous than the puncture by rectum, and perhaps it may be so, since, although often successful and affording a relief which has continued for several years,* it is attended with some risk of extravasation and suppuration behind the pubic symphysis, or beneath the peritoneum. If the relief is intended or expected to be only temporary, the rectal is the simpler and the safer operation, and the suprapubic should be employed when an enormous prostate forbids the application of the former. But, on the other hand, if it is reasonable to believe that the new opening must be the patient's only resource for a considerable period of time, as in cases where he has lost all power of passing urine, and depends on the catheter solely, and moreover, where the urethra will probably be difficult to traverse with a catheter in future, it is incontestible that an opening above the pubes forms a much more convenient situation for artificial relief than an opening within the rectum.

The puncture by rectum, which has been employed occasionally to afford relief in these cases, as our museums testify, is almost universally stated by authors to be inadmissible when the prostate is enlarged, an occasional exception being made in favour of the operation when the enlargement is not considerable. It is, perhaps, a question whether its applicability to a large proportion of prostatic cases, has not been overlooked by many. There are certainly exceptional examples of prostatic enlargement in which the

* There is a preparation in the museum of the Royal College of Surgeons exhibiting the bladder and prostate of a man aged 66, on whom the operation was done for retention from enlargement of the latter organ, who lived four years subsequently, resuming his former business habits. No. 2043. My friend Mr. Paget, of Leicester, has a patient now living, whom I have seen, who wears a short flexible tube above his pubes, upon whom he performed this operation fifteen years ago; and during the whole of this period he has evacuated all his urine by means of this tube. He has another patient who has done the same thing for two years. Neither will consent to part with their tubes, from the great relief afforded: both follow their occupations with comfort and regularity.

finger is unable to reach the tumid bladder behind, and detect fluctuation there. It is the more desirable to make the opening through the rectum, if there be any doubt about the existence of a distended bladder above the pubes; for there may be imminent danger from retention without the presence of this condition. And further, in a corpulent patient, its detection, if present, is not always satisfactorily to be accomplished, while in such a person the suprapubic operation is less easy, or, at all events, less advisable. I have only on one occasion been compelled to puncture for prostatic retention, and on this performed it with ease and success by the rectum. This case is an instructive one, and is subjoined; it also exhibits circumstances which must compel the surgeon to demur to the soundness of the well-known proposition of Desault, that puncture of the bladder ought in no circumstances to be performed.

Case No. VI.

In December, 1853, I was summoned at night by his medical attendant to see a gentleman in the Regent's Park, who was suffering from complete retention of urine. He was 73 years of age; had experienced more or less difficulty for 10 or 12 years past, but had not required medical aid until the last three weeks. He had passed no water since the preceding night, and was suffering greatly. Repeated and prolonged attempts to pass a catheter had been made without success, followed by bleeding. On examination I found vesical dulness midway to umbilicus; and through the rectum, a considerably-enlarged prostate. I first introduced a long and well-curved silver catheter, but not succeeding, mounted a No. 10 gum elastic one, upon a strongly-curved stilet, and introduced it without any difficulty into the bladder. It had evidently glided over an obstruction at the neck. Three pints of high-coloured urine were drawn off. I left the instrument in, and advised it to be retained at least 48 hours. Next day I heard the patient was extremely comfortable.

At midnight following the second day, I was again sent for. I learned that the catheter had been removed, contrary to my wish, in the morning at ten, about 36 hours after my first visit, that he had passed no water since, and that, prior to this summons, prolonged and painful attempts to replace it were unavailingly made, and had been attended with very considerable bleeding. I again employed the catheter, but now found, despite all my efforts, the beak of the instrument invariably slipping near to the rectum, where it could be felt with too much distinctness. The patient was weak and suffering severely, straining violently, and begging for relief. Medicinal agents had also been fully administered. I proposed, therefore, puncture of the bladder, and, being able to detect fluctuation beyond the prostate as far as my finger could reach, decided upon performing it there, which I did without any difficulty in the manner described above—drawing off again a large quantity of high-coloured urine. The canula was tied in, and retained some time, as it gave little inconvenience. I saw this patient no more after the first fortnight, as he was exceedingly comfortable, and exhibited extreme repugnance to the introduction of a catheter, the necessity for which I endeavoured to insist upon. I subsequently learned that he lived more than two months after this, gradually sinking, with no other symptoms than those of increasing debility; that the canula was retained during that period; and that disease of the prostate, bladder, ureters, and kidneys was found after death, but that the urethra was not opened. I believe he never permitted any attempt to be made to pass a catheter by the urethra.

High authority, both by example and precept, exists, for the practice of puncturing the obstruction itself, whatever it may be, in most cases undoubtedly an enlarged median portion of the prostate, by means of a silver catheter, or cutting instrument. And the inconvenience which a canula in the rectum generally occasions, if retained there for any longer period than a few days, has been admitted as an objection to the rectal operation. But the cases will be very few indeed in which the urethral canal cannot be rendered pervious to the catheter after a little rest and withdrawal of the urine by another channel, when the canula may be removed, and the opening between the bladder and the bowel heals rapidly enough.* Where the operator has not much hesita-

* For further observation respecting this matter, and in relation to the whole subject of the puncture of the bladder per rectum, see the Author's work on the Pathology and Treatment of Stricture, chapter xi.

tion in assuring himself that the point of his instrument is not lodging in a false passage, but is placed in the prostatic urethra and impinges against an otherwise impassable obstruction at the vesical neck, I believe the best proceeding will be to introduce an instrument similar to a trocar and canula; but sufficiently long and shaped nearly like a middle-sized prostate catheter (although with less curve), and perforate to a necessary extent for relieving the bladder. I have proposed a slight modification of this instrument, which was used by Liston, which I think is likely to be useful. Instead of the trocar being made to fill completely the canula, a small groove should be hollowed out in its substance from the side, so that the moment of its entry into the bladder may be announced by a small stream of urine issuing through the groove. Otherwise, the arrival of the instrument at the bladder can only be learned by withdrawing the trocar (see fig. 19).

There is yet another mode of reaching the bladder already described, which, as far as I am acquainted, has not been yet performed in this country, but which, undoubtedly, possesses claims meriting our regard—viz. the puncture through the pubic symphysis. I can add nothing respecting it from my own experience on the living subject. On the dead I have repeatedly performed the operation; and in elderly subjects have found a good deal of

Fig. 19.

EXPLANATORY NOTE.—*a, a,* Canula; *b,* Point of cutting stilet. The upper asterisk points to the groove described, and the lower one to the aperture for escape of urine.

force necessary in order to introduce the trocar through the symphysis. The interosseous cartilage is very narrow in such subjects —indeed sometimes nearly ossified—a condition which I have verified by dissecting and macerating six examples between sixty and seventy years of age. Nevertheless, I have never failed to puncture the viscus safely with a trocar in these trials. The theoretical objections which have been made to it are, the possible occurrence of urinary infiltration into the cellular interval between the bladder and pubic bones, and of peritoneal inflammation by extension from the wound. The advantages claimed for it are, on the other hand, the ease with which the bladder may be reached in this situation, even should the organ be contracted in size, or the prostate be greatly enlarged, and the facility with which the catheter may be used and retained in the wound, or replaced after removal if necessary; add to which, it may be regarded as a more desirable situation than the rectum in relation to the comfort of the patient himself. As far as the objections referred to are considered, it is but fair to state that they appear not to have been encountered in practice, and hence they cannot be regarded as possessing any very great weight. It is a proceeding which doubtless well deserves the consideration of practical men, and experience appears to warrant a trial of its merits when opportunities offer for testing them.

An operation in the perineum, which has for its object the opening of the urethra at the membranous portion, or near to the apex of the prostate, has been recommended, and in one case practised, by Dr. Lawrie, of Glasgow, in retention from enlarged prostate. He introduces into the urethra a grooved lithotomy staff, preferring the angular staff of Dr. Buchanan with a short beak, and opens the canal upon it for an extent just sufficient to admit the finger at the point named. From this opening he passes an almost straight metallic catheter, twelve inches long, and open at each end, into the bladder. Through this open tube a probe-pointed wire can be introduced, upon which, after withdrawing the former, an elastic tube will glide into its place, should such an one be preferred for the purpose of retaining in the bladder.* This proceeding, which is by no means of recent origin, although it has fallen into disuse,† appears to me calculated to be successful in those cases in which the urethra has not received any serious injury from violent catheterism. It sometimes happens, however, that from this cause false passages have been made below the prostate, constituting, in fact, the difficulty for the solution of which the question of employing a knife or trocar is entertained (as in Case VI., p. 250). In such circumstances, an incision upon a grooved staff might, perhaps, lead only to embarrassment, as the urethra might still not be opened or even found.

Provided, however, that this lies safely in the urethra, there is not much doubt that an instrument may be passed without great difficulty from the perineal opening, upwards into the bladder. I fear, however, that a urinary fistula would always remain. Sinuses of this kind in the perineum, associated with enlarged prostate, are particularly rebellious to treatment; indeed, do not generally close so long as the patient is troubled with frequent micturition and unhealthy urine. However simple and effectual this mode of relieving the bladder may in some cases be, the probability of the result alluded to should, I think, not be overlooked.

Our present experience of the various procedures detailed may be stated in the form of conclusions, which, if not absolutely established, may be regarded as fairly deducible from the available data, and therefore as approximatively correct.

1. That the cases of urinary retention from prostatic enlargement which cannot be relieved by the introduction of a catheter, over and beyond the obstruction at the vesical neck, are extremely rare.

* Glasgow Medical Journal, July, 1854, p. 211.
† It is described by Heister as employed before his time; but he recommends as preferable the direct perineal puncture by means of a trocar plunged into the body of the bladder behind the neck: a dangerous procedure, now obsolete, but much in vogue during the 17th and 18th centuries.

2. That when it is necessary to make some artificial opening into the bladder in order to afford relief to such a case;—if the beak of the catheter can be carried along the urethra to the obstruction in its prostatic part, perforation may be made, either with that instrument, or with one of similar form, containing a cutting stilet, expressly adapted to the purpose.

3. That if the condition of the urethra, from the existence of false passages, laceration, or other lesion, be such that the operator cannot be certain of his ability to carry a catheter fairly along the canal to the obstructing portion of the prostate, some other mode of entering the bladder must be adopted; in which case, if fluctuation can be distinctly felt within easy reach of the finger, in the median line behind the prostate from the rectum, the puncture through that bowel is an easy and safe mode of giving exit to the urine.

4. That if, on the contrary, fluctuation cannot be distinctly felt by the rectum, the puncture through the bowel is a proceeding of doubtful propriety; and the question of opening the bladder, either above or through the pubic symphysis, must be solved for the case in question. That the former operation may be regarded as somewhat less hazardous, when the patient is not corpulent, and vesical dulness can be clearly defined in or above the hypogastric region. Further, if the circumstances of the case indicate that the artificial opening may be required for a long period of time, the suprapubic puncture must be preferred as more safe and convenient than the rectal.

5. Lastly, that the experience of the operation of puncture through the pubic symphysis is not extended enough at present to permit of a comparison being made in regard to its results with other modes; but that it is sufficient, coupled with the apparent advantages derived from anatomical considerations, to recommend the operation to the test of practice, in order that its merits may be duly ascertained.

CHAPTER XIII.

ATROPHY OF THE PROSTATE.

Pathological signification of Atrophy.—Kinds of.—From Exhausting Disease.—Senile Atrophy, degree of.—Frequency of.—Nature.—From Mechanical Pressure.—From Local Disease in the Prostate.—Congenital Atrophy.—Symptoms and Treatment.

By Atrophy of the Prostate is to be understood a diminution in the bulk and weight of the organ resulting from a gradual disappearance of some of its constituent structures. Regarding solely the results of this action, it may be considered as the converse of Hypertrophy.

Senile Hypertrophy, however, in its popular signification as applied to the prostate, is not, strictly speaking, the converse of Atrophy, since the enlargement of the organ which occurs in the later periods of life is not the augmentation of structure due to increased function, analogous, for example, to the augmentation of a muscle by increased use, but is a process essentially morbid, having no conservative, or compensating design and action; *e.g.* hypertrophy of the heart for the purpose of overcoming the increased resistance offered by a calcified valve, &c. Atrophic change cannot, however, be regarded as resulting from any active pathological influences, exerted in the organ itself. It is a passive condition rather, and consists in simple wasting of the organ, in the gradual disappearance of elementary structures.

What is the precise physiological action by which Atrophy is determined? Is it some active process of absorption removing the constructive elements of the prostate, in the same manner, but more rapidly, than that ordinary process, which results from the effete tissues throughout the whole body, in order that they may be, as constantly and gradually, replaced by new material? I think not, but believe it to be rather the result of failing power on the part of the body to replace, by new material, the effete tissues removed by the natural process of absorption. It is not that the process of degradation is much more rapid, but that the powers of supply and re-formation are less vigorous than heretofore. When the resources of the body are inadequate to supply the plastic material and the formative power, in an equal ratio with the expenditure, general Atrophy must result.

Nevertheless, there are undoubtedly various forms of Atrophy liable to affect the Prostate, which must be considered separately.

I. The Atrophy of exhausting general disease.
II. The Atrophy of old age.
III. The Atrophy caused by pressure.
IV. The Atrophy caused by disease in the prostate itself.
V. The Congenital Atrophy.

The first is the Atrophy which occurs from exhausting constitutional disease, particularly in Phthisis. The extent to which Atrophic change occurs from this cause is sometimes very remarkable, as I have several times had opportunities of noting. I have recently had occasion to dissect a specimen from a man 21 years of age, who died of phthisis, in whom the organ weighs only 54 grains, or less than one drachm. It is not unlikely that from long-continued disease in this case the sexual organs had not been so fully developed as would have been the case at this age in health. Another example is No. 151 from a man aged 78, who died of phthisis, the prostate weighed only 2 drs. 45 grs. In all cases of this kind there is very considerable wasting of all the structures of the body, and herein this form of atrophy often differs from the second form. But the proportion of diminution affecting the prostate appears to be larger than that suffered by most other organs in the body.

There are other diseases in which the prostate becomes atrophied; all wasting diseases produce this condition to a greater or less extent, but in none is it so marked as in tubercular and scrofulous disease. A good example is seen in No. 150 of the table of preparations, the prostate of a man aged 90, who died exhausted by carbuncle; here the organ weighed 2 drs. 50 grs. Now in this form of atrophy all the tissues of the organ seem to be about equally affected: one component tissue does not seem to have been diminished in greater proportion than another, judging from the appearances presented on making sections in different parts of the organ.

The second kind of Atrophy is that which occurs during old age. There is frequently a general diminution in weight and bulk of the solids as individuals advance in life beyond a certain age. A mere participation in this condition is not what is intended by Senile Atrophy. A prostate, which is the subject of this affection, is one in which the diminution is relatively greater than that which affects the rest of the body. It has been said, but it appears without sufficient foundation, that when hypertrophy of the prostate is not present in the aged, atrophy will always be found to exist. This, certainly, is not borne out by facts. Thus, by analysis of the table in Chapter II., we find no less than 50 individuals at and above the age of 70 years (70 to 94 years), whose prostates ranged in weight between $3\frac{3}{4}$ and $5\frac{1}{2}$ drachms, a great majority of them being between 4 and 5 drachms. These must be regarded, almost without exception, as cases in which neither hypertrophy nor atrophy could possibly be present. In none of them was there any sign of urinary derangement during life. The number of prostates in this condition is much larger, viz., upwards of 90, if all the individuals between 60 and 70 years are also reckoned; the latter age was chosen for this calculation to place beyond a doubt the statement that a normal condition of the prostate is common at very advanced ages.

Among the 164 examples at and above 60 years, 11 only had prostates weighing less than $3\frac{1}{4}$ drachms. Organs weighing upwards of $3\frac{1}{4}$ and under 4 drachms cannot be considered the subjects of Senile atrophy from the mere fact of weight; some of them are certainly normal; but a knowledge of their structure and of the size of the individual, are data which it is necessary to possess, since, as in other organs, it is not the absolute weight, but, as just observed, that which is relative to the weight of the body, which must be known in order to decide the existence, or the degree, of atrophy present in a doubtful case. Accepting the number of undoubted examples of atrophy as 11, and deducting two as occurring in individuals who died of exhausting disease (phthisis and carbuncle), there remain nine, which may be classified here. Consequently, on this calculation, senile atrophy occurs in rather more than 5½ per cent. of individuals at and over 60 years of age; and with the wanting data supplied, it is probable that this percentage might be a little increased. This fact appears to have no practical bearing; nevertheless, possessing the materials, it is desirable to eliminate it, for it is not altogether without a scientific interest.

Senile atrophy, that is the atrophy which has its origin, not in exhausting disease, but which occurs independently of it in some elderly subjects, is of a somewhat different character, histologically regarded, from the atrophy of exhausting disease. It has been seen that, in the latter, all the constituent tissues are about equally diminished, as far as dissection and careful comparison enable us to judge. In senile atrophy, on the other hand, the glandular tissue seems more diminished than the fibro-muscular stroma of the organ. This latter is often hard and tough, and may even have some small tumours of the same material existing in it. But the glandular element is in smaller proportion to the fibro-muscular or stromal than in the healthy organ. This form of atrophy is, perhaps, also less extreme in its degree than that of exhausting disease. While in the latter case we have seen an adult who died from phthisis, with a prostate weighing only 54 grains: we have never seen a prostate affected by senile atrophy weighing less than two drachms and a few grains.

The third form of Atrophy is that produced by mechanical pressure. No peculiarity can be affirmed to characterize this form of atrophy. It is that species of diminution, of disappearance of elementary constituents, which is observed in all the tissues of the body under the influence of continued mechanical pressure. Thus the prostate is sometimes observed to be very considerably diminished in weight and volume from the action of pressure exerted by adjacent tumours, which may have fluid contents, as abscess and hydatid; or by bony and other solid tumours; by calculi in the bladder or embedded in the prostate itself. Not unfrequently, also, from long-continued and extreme distension of the bladder with urine, and of the prostatic urethra itself, when very confirmed structure exists, a considerable pressure is sometimes exerted, which in like manner seems to result in marked atrophy of the prostate. In these cases the structures are sometimes thinned to a very considerable degree, the natural ducts and cavities are dilated, and nearly one half of the organ, estimating it by weight, may disappear in the course of long-standing and unrelieved stricture of the urethra.

The fourth form of Atrophy is that produced by some local diseases of the prostate itself. Abscess in the prostate will produce disintegration of a considerable portion of the prostatic tissue by impairing the local nutritious supply of the adjacent structures through the agency both of ulceration and pressure; and thus atrophy may be said to occur. A deposit of tubercle will produce similar consequences, and sometimes to a considerable extent. In like manner, malignant disease will cause the proper structure of the prostate to disappear, superseding it by its own morbid growth; and in this sense also the prostate has been said to be atrophied.

Fifthly. Congenital Atrophy. — The prostate is sometimes found in a condition at birth, in which it may be said never to have been naturally developed. This condition is usually associated with congenital malformation of other portions of the genito-urinary apparatus,—in extroversion of the bladder for example. It is in conformity with custom alone, a custom scarcely warranted by philosophical precision in the employment of terms, that the condition referred to is named under this head, since atrophy supposes, strictly speaking, a pre-existing normal condition; no further notice of it therefore will appear here.

SYMPTOMS AND TREATMENT.—It is not known that simple uncomplicated atrophy of the prostate declares itself by any signs or symptoms. Neither is it conceivable that it should do so, except by the diminution in size ascertainable by rectal examination and by diminished secretion. As regards the latter, we know so little of its specific amount in health that no conclusions can be adopted respecting the change which atrophy produces. No doubt it is associated with a decline of the sexual powers in age. Neither does the condition demand any notice in relation to treatment. There is no reason to believe that we possess any means of restoring an organ affected with senile atrophy. When atrophy results from other causes, such as stricture, abscess, &c., their alleviation or removal will doubtless be a means, and the only means, of beneficially effecting the prostatic affection.

CHAPTER XIV.

CANCER OF THE PROSTATE.

A Rare Affection; but probably less so than generally supposed.—Why it is so.—Analysis of Tanchou's Tables.—Malignant Disease of Prostate almost invariably Encephaloid.—Examination of reported Cases of Schirrus.—Melanotic Deposit.—Ages at which Disease appears.—Duration.—Course.—Morbid Anatomy.— Symptoms.— Hæmorrhage. — The Urine. — Treatment. — Eighteen Cases.—Tabular View.

THE prostate may be the subject of cancer either as its primary seat, or when secondarily implicated after the primary appearance of the disease in some other part of the body. In either case, prostatic cancer is a rare affection. It is a question, however, whether its rarity has not been in some respects exaggerated, judging from the general impression which appears to prevail respecting it. I am inclined to believe that a certain small proportion of instances is lost sight of among the very large number of cases assigned to senile hypertrophy. The course of malignant disease, when well marked, it is impossible, with ordinary care, not to diagnose from the last-named affection; but in more chronic forms, perhaps sometimes occurring, but most especially in those cases in which a malignant growth arises in a prostate previously the subject of senile enlargement, the cancerous character is sometimes, I believe, overlooked. For a good example of the co-existence of these two affections, which occurred under my own care, I beg to refer to a case at the close of this chapter. The morbid parts illustrating the course of disease referred to, I exhibited at the Pathological Society of London, in 1854.

In reference to this question of frequency, the statistical researches of M. Tanchou are commonly quoted by writers on this subject, and in the form comprised in the following brief statement. Among 8289 fatal cases of cancer, he met with only five affecting the prostate. It will be desirable, however, to pursue our inquiries a little further with these figures, for the simple statement is calculated at first sight to produce an incorrect impression respecting the fact. The manner in which the data in question and the statistical summary were arrived at is as follows: M. Tanchou made an abstract from the registers of deaths for Paris and its suburbs, of every case in which the fatal result was attributed to cancer, during the years 1830 to 1840 inclusive; reporting *what was believed to be the primary seat* of disease in each case, naming one organ only, and classifying the whole accordingly, with a view, among other objects, to form a numerical estimate of the primary seats of malignant disease throughout the body. The total number comprised 6957 females and 2161 males=9118. In 829 cases, the seat of disease was not originally reported, leaving 8289 cases. Of these 1904 were males, and among these the disease was recognized as a primary lesion, five times in the prostate, and all in adults.* But 72 cases are given of cancer of *the bladder* without distinction of sex. As a primary lesion, this is, I believe, more frequent in the male than in the female. In the latter sex vesical cancer is almost invariably due to extension from a uterine growth. Supposing, then, that fully, or more than, one-half of these may be claimed for the male sex, it is not unreasonable to suppose that some of them may have been prostatic in their origin. The fully-developed prostatic encephaloma soon becomes vesical, and may often not be distinguishable except by careful examination. Thus in two of our own cases, viz., Nos. IX. and XXIV., *the bladder was almost filled by* a tumour which, nevertheless, had its origin in the prostate.

Regarding the source from which these figures were derived, viz., from an ordinary register of deaths, for the purposes of which no special examination is required, it appears to me ex-

* Recherches sur le Traitement Médical des Tumeurs Cancéreuses du Sein. Par S. Tanchou. Paris, 1844, pp. 256-261.

ceedingly probable that a source of error, so difficult to guard against, must have vitiated the result in relation to this question. And we are entitled, I think, to believe that the proportion of five cases of primary disease in the prostate, out of 1904 male cases of cancer, is very much smaller than the true number.

The occurrence of secondary is much more rare than that of primary cancer in the prostate. It has most commonly been observed that the invasion has then taken place by extension from the bladder. I have, once only, seen it succeed encephaloid cancer of the penis, itself a very rare affection.

As to the form of cancerous disease which most commonly affects the prostate, no doubt exists that it is encephaloid. In children it appears to be always so. In adults, an exception, to say the least, is extremely rare. Indeed, the occurrence of schirrus there has been denied. After a close examination of all the cases reported, I believe the opinion expressed by Dr. Walshe in 1846, resulting at that time from an examination of necessarily fewer data than we now possess, was correct, viz. that "the evidence of the occurrence of true schirrus of the prostate is defective."* I have tabulated 18 carefully-reported cases at the end of this chapter, having rejected those respecting which the evidence is insufficient, and among them at least two well-known cases commonly accepted by contemporary authors as examples of schirrus. Granting its existence, the extreme rarity of this form renders minute details absolutely necessary to establish an example. Wanting these, we must not hesitate to deny the admission to a category so designated, of any case not accurately observed. Moreover, it is necessary to bear in mind that, among the older authors, the term schirrus has been used with the intention of implying merely the presence of an indurated structure. The precision of meaning which modern pathology has assigned to it must not, therefore, be understood when referring to their records.

* The Nature and Treatment of Cancer. By W. H. Walshe, Professor of Medicine in University College. London, 1846, p. 414.

In the two cases just referred to, little beyond the single physical character of extreme hardness of the prostatic substance existed, to support the allegation of schirrus. The presence of encephaloid deposit in other parts of the body was noted in one of them, and to a degree which would indicate that, whatever the prostatic affection might be, it was, at all events, not the primary lesion.

The cases are those by Mr. Howship and Mr. Travers. Mr. Howship's patient, at seventy, had a prostate of normal size, but unusually firm in texture; the disease causing death consisting of large encephaloid masses, involving the vessels, nerves, and viscera in the pelvis and abdomen, and occupying more than half the space of the latter cavity.* Mr. Travers briefly states, respecting a case of his own, "I found the prostate occupied by a tubercle, possessing all the characters of schirrus, upon section, in an old nobleman, long subject to retention."† Wanting the evidence derivable from the condition of the adjacent glands, from microscopical examination of the growth, and knowing the prevalence of firm fibrous tumour in the part affected, it is impossible to affirm the applicability of the term schirrus, as designating a peculiar species of cancer to these two examples. Yet they are adduced as instances by most writers on the subject, excepting, it must be added, by Dr. Walshe in his work above referred to.

I am aware of but one recorded case of schirrus, in which the term has been employed by an observer of high standing, experience, and competency to judge, with the express intention of indicating a specific form of the genus carcinoma. Mr. John Adams, of the London Hospital, gives the following account of the case. A gentleman was the subject of urinary derangement for the first time, at 59 years of age, and he died three years afterwards from the advance of the complaint. At the post-mortem, the lumbar and iliac glands were affected with schirrus, " the prostate gland was enlarged to nearly twice its natural size; an ovoid mass, *distinctly schirrous*, the size of a small nut pro-

* Med.-Chir. Trans., vol. xix. p. 35. † Ibid., vol. xvii. p. 346.

jected into the bladder from the upper surface. The left lobe was occupied by a long schirrous mass: the right lobe appeared healthy." Mr. Adams further added, that "the tumours had been examined by an experienced microscopist, who had pronounced it to be true schirrus in every particular."*

Melanotic deposit is said to be occasionally found associated with encephaloid of the prostate. Its presence is reported in two cases, one at adult age, the other in childhood. It should not be forgotten, in the examination of these cases, that interstitially-effused blood in a fungous growth may be mistaken for true melanotic deposit, which to the naked eye it sometimes resembles.

Regarding other forms of carcinoma, such as colloid, or epithelial disease, frequently ranked as cancer, there is no example known of their occurence in the prostate.

Malignant disease has at present been observed only in childhood and at advancing age. No authenticated cases are on record between the ages of 8 and 41. The duration of the disease, from the first appearance of symptoms to the fatal result, appears to vary from one and a half to five years in adults, and from three to nine months in children. It should not be forgotten that in the former class we may encounter a source of error, which, if not pointed out, tends to produce an over-estimate of the duration of the disease. It has been shown that encephaloid deposit may sometimes take place in a prostate previously hypertrophied, and already the cause of obvious symptoms of urinary obstruction. In such a case it is clear that the period which elapses between the first appearance of urinary difficulties and the fatal termination, is not to be regarded as the duration of the malignant disease. Such a case has been already referred to. This condition of things it may not be always easy to verify during life; nevertheless, if, the existence of enlarged prostate having been ascertained some years ago, exacerbation of symptoms rather rapidly occurs, with manifest increase in the size of the

* Report of the Meeting of the Royal Med. and Chir. Soc., April 12, 1853. —*Lancet*, 1853, vol. i.

tumour, attended by cachexia, above all by enlargement of the lymphatic glands in the neighbourhood, we may pretty safely conclude that malignant action has supervened. Undoubtedly these are very rare cases, still the existence of such may be regarded as established.

In referring to the table of cases, it may be observed that in childhood the encephaloid deposit was limited to the single organ, or to it and to the adjacent lymphatics, although the latter has not been positively reported, and that it ran a very rapid course. Whereas in the adult cases, the development of disease was slower, and other viscera were usually effected besides the prostate. These facts harmonize with those presented by the course of encephaloid generally. Always progressing rapidly, its growth seems active in proportion to the youth of the individual. The function of growth generally is much more active in early life than at any subsequent period; a fact which may not improbably account for the result noted. On the other hand, the appearance of the deposit in several organs, observed in adult age, may, perhaps, be regarded as an effect of the slower rate of progress manifested by the disease affording time for numerous local developments to occur.

MORBID ANATOMY.—Little can be positively affirmed respecting the morbid anatomy of encephaloid in its early stage in the adult subject. Anatomical examination can very rarely be brought to bear on any specimen which has not already arrived at a late period of its fatal course, the death of the patient having been almost invariably caused by the prostatic disease in an advanced stage. There is reason to believe, judging by inferences drawn from rectal examination of these cases in the earlier stages, some months before death, that a deposit takes place at one point, from which, as from a centre, the disease extends; this point being usually in one of the lateral lobes. In the first stage, probably, the smallest amount of active increase is displayed; but after the swelling has reached a certain size it rapidly enlarges, and often a very considerable mass is formed before, through the death of the patient, it can become amenable to close anatomical analysis.

The conditions then commonly found are as follows:—The prostatic mass is of irregular form; portions project from different aspects of the organ, usually from its upper and posterior ones, and direct their course into the cavity of the bladder. These are of very unequal consistence; some parts are hard and tense, appearing to be firmly bound by the enveloping capsule; but the most projecting parts are often soft and sprouting, as encephaloid of rapid growth is observed to be in other parts of the body. The colour varies: usually brown and reddish tints prevail, apparently produced by great vascularity, or by internal hæmorrhage, the latter being sometimes more clearly evidenced by the existence of dark blood-clots, enveloped by the mass and disclosed by making sections. Some portions are becoming gangrenous, and manifest gray and slaty tints, especially in the urethral mucous membrane. On making incisions into the tumour, if in the later stage, examination often shows that most of the prostatic structure has disappeared before the invading action of the cancerous product. The most recent portions of the growth are seen to be soft to the touch, slightly jelly-like, opalline, faintly transmitting light, and of a pale buff tint, almost white; others are yellowish, and some reddish up to deep blackish-red. Some parts are seen to be softened, pulpy, and disintegrated; from some a creamy juice exudes—from others sanious fluid, or fluid containing flaky masses of dead structure, and here and there are found small collections of purulent matter.

On searching carefully for prostatic structure among the deeper parts (in the sprouting fungus of course there is none), and if we discover the confines of the morbid growth where it runs insensibly into still-remaining portions of the organ, we may infer from all appearances that the gland structures are first involved in the disease, and that the stroma is less readily invaded; the pale muscular fibres appearing to continue free from cancerous deposit to a later period than the gland-follicles. Occasionally, but rarely, the prostate is affected rather by a general infiltration of all the parts than by a distinct localization of the growth as a tumour; in one such case examined, the condition of the stroma just alluded to was well marked.

The mucous membrane of the urethra is generally not entire; beneath its surface nodules of the diseased growth may be seen, of which one or more have perhaps penetrated, and are commencing to form a fungous growth; and around these nodules there is an arborescent arrangement of minute blood-vessels.

In some examples there is no outburst of fungus from any part of the organ, the disease being contained within its limits, albeit these are greatly extended. In other instances this is far from being the case; the capsule gives way, there is not only fungous growth, but ulceration, or even cavity, from which diseased structure has been thrown off, and from which hæmorrhage also has probably taken place from opened blood-vessels.

Under the mucous lining of the bladder adjacent there are also, sometimes, several nodules of cancerous disease; and finally, if the prostatic affection is advanced, the neighbouring lymphatic vessels and glands are always diseased. Even the veins of the prostate have been seen to contain cancerous matter.

Such are the appearances I have witnessed in cases of encephaloid affecting the adult. In the child, encephaloid of the prostate attains a larger size as compared with the bulk of the individual. Here we witness all the results of more rapid formation, a softer structure, more succulent and juicy, of brighter tints; often more vascular. Otherwise the general characters in the adult and in the child are the same.

SYMPTOMS.—The symptoms of the malignant affection are those common to prostatic obstruction of any form, but generally declaring themselves with greater rapidity than in the cases of senile hypertrophy. These need not be repeated. But besides them, there are other and distinctive characters, such as more severe pain, often very intense; occasional, often frequent, hæmorrhages; and more or less constitutional cachexia. The pain is felt in the rectum, or in the region of the sacrum, and shooting down the thighs, either the anterior or posterior aspect. In one case of my own, recorded here (No. VII.), the suffering experienced during the early and middle term of the complaint was very slight, apart from that produced by retention of urine, which occasionally happened, and was relieved by the catheter. During the last few months of his life the patient lost the powers of sensation and motion in the lower half of his body from encephaloid deposit in the upper part of the spinal column, otherwise doubtless his sufferings might have been severe. The prostate is not invariably tender to the touch in these cases, at least not notably so; an observation which is supported by the foregoing instance, before the paralysis supervened; and especially by another case very carefully watched by my friend Dr. Armitage, to whom I am indebted for the history forming Case No. VIII.

Hæmorrhage is a common occurrence both at an early and late period in the course of the disease, being almost universally present at one time or another, and sometimes to an alarming extent. The blood is usually voided almost pure or unmixed, and frequently appears with or after some attempt to urinate, which, from some circumstance, has been attended with greater exertion than usual. Much less commonly is the hæmorrhage observed to be continuous for any length of time, or constantly communicating a bloody tint to the urine, as happens with some tumours of the bladder; unless, indeed, that organ should also be implicated, or the disease have assumed the form of a fungoid growth into its cavity. Nor is the urine so liable to be mixed with pus or mucus as in the vesical affection.

The enlargement formed by the prostate itself, when examined by the rectum, is always hard at first, and may or not be irregular in outline or consistence. Softening may in the later stages be felt, but the patient's powers do not always sustain him to so late a period as that in which the growth either softens or fungates. The deposit may be ascertained to affect the whole organ, and most commonly does so, but it may affect one portion more than another. Frequently, especially in the somewhat more chronic form of the disease which is met with in adults, as compared with that which affects children, other organs are affected, but by no means invariably. But there are always diseased lymphatic glands adjacent, and sometimes the infection reaches more distant groups. The existence of such swellings in the course of the iliac vessels, and sometimes in the inguinal region, may frequently be verified by examination of the abdomen, and constitutes a most valuable sign, when present, in relation to the diagnosis, and, consequently, to the prognosis of the case.

The urine should be closely examined, in cases of a doubtful nature, for the presence of cells which may be regarded as malignant, from inspection of their forms and constitution. Some observers state that they have verified cancer-cells in the urine. Others have failed to do so after close and repeated examinations. A good deal of débris may usually be seen in advanced cases; its presence appears to indicate that the growth has fungated, and throws off more or less of its elements in the condition of sloughy detritus. I have searched for characteristic forms of cancer-cell in the urine of malignant disease of the bladder, but unsuccessfully, and am not disposed to think that much reliance can be placed upon the appearance of the cells met with. The urinary passages yield epithelium cells of all forms and sizes abundantly, and these, I suspect, have been mistaken sometimes for the "cancer-cell." I remember a case in which a good observer diagnosed a cancerous tumour of the bladder from the discovery of what he considered undoubted cancer-cells, taken in connection with other signs. The patient died soon after with villous tumour, a form of disease miscalled cancer, and having no structural resemblance to the malignant affection. The presence of enlarged glands, the cachexia, and the history of the case, afford better means, I think, of establishing the diagnosis than any which are derived from the observation of the so-called cancer-cell in the urine.

Regarding the treatment of malignant disease of the prostate, nothing more can be offered here in relation to the constitutional affection than applies to it when occurring in any other part of the body. The treatment is palliative, and must be regulated according to the various necessities which may arise in the progress of the case.

Thus accumulation of urine must be provided against, at the smallest possible risk of irritating, much less of injuring, the part. If catheterism can be dispensed with altogether, so much the better. In no circumstances is it of more importance to be extremely gentle in the manipulation of instruments. The pain must be relieved by anodynes administered both by mouth and rectum. The addition of conium to opium, by enema or suppository, is often particularly useful (see page 201); and by mouth belladonna is sometimes a valuable auxiliary in mitigating pain, given in doses of from one-fourth to three-fourths of a grain twice or three times a-day. Hæmorrhage must be treated on principles already fully illustrated in a preceding chapter (page 203). The powers of life are to be supported by every means in our power. Nutritious food, both in the solid and fluid form, with a due proportion of alcoholic stimulant, must be supplied in accordance with the digestive powers of the patient.

Case No. VII.

J. A. aged 60. Came under my care in the end of 1851, for frequent and difficult micturition. The cause of this was an obvious general enlargement of the prostate, not attended with tenderness on rectal examination. After a short time he was greatly relieved by the use of the catheter and medicines, and I lost sight of him. Subsequently he attended frequently at the St. Marylebone Dispensary that his bladder might be relieved. The urine was alkaline. He suffered little pain or inconvenience, except at periods when the retention became almost complete.

His symptoms becoming worse, he was admitted to the St. Marylebone Infirmary, Dec. 9, 1853.

I then learned that he had gradually become paraplegic about one month previous to admission. He had now entire loss of motion and sensation as high as the hips, and soon after sensation disappeared up to the arm-pits. The fæces passed involuntarily, and the urine during sleep. There was no pain in the back; no history of any injury; nor were there any facial signs of cerebral lesion.

Under the employment of the catheter, &c., he improved at first slightly, but in Feb., 1854, the urine became bloody, contained much phosphatic deposit, and he died, greatly emaciated, on the 23rd.

P.M. The bladder was not large, somewhat thickened, and its mucous membrane was corrugated, exhibiting dark red and slaty tints, and adhering calculous matter. Several little nodules of a light colour, isolated from each other, were seen beneath it. The prostate was uniformly enlarged, and about the size of an orange.

The adjacent iliac glands were enlarged. The right kidney atrophied: its pelvis dilated, and containing three or four ounces of pus. The lower end of the corresponding ureter was completely obstructed by an enlarged gland pressing upon it. The left kidney enlarged, and containing a little calcareous deposit.

On removing the posterior arches of the spinal column, a flattened patch of encephaloid matter was found adhering to that of the first dorsal vertebra, and smaller portions were seen lower down. The cord itself exhibited no marks of change, except, perhaps, some undue injection of its substance. There was no cerebral lesion, nor was there any malignant tumour discovered in any other part of the body.

On making section of the prostate, the enlargement appeared to be mainly due to the displacement of the true prostatic tissue by deposit of matter, which, to the naked eye and microscopic examination, was obviously encephaloid. The same appearance was found in the glands, in the interspinal mass, and in the nodular deposits beneath the mucous membrane of the bladder. I exhibited this specimen at the Pathological Society, March, 1854.[*]

CASE No. VIII.

J. B. Aged 65. A sawyer by occupation. Enjoyed good health until one year and ten months before his death. He then began to

[*] Transactions, vol. v. p. 204, which see for further particulars, and a detailed report on it by Mr. Hutchinson.

suffer from pain in the sacral region, which gradually increased, but he did not give up work entirely until 14 months before death. At that time he passed urine reported to have the "colour of elder wine," for some days.

May and June, 1856, he passed in St. George's Hospital. Soon after leaving, he voided a considerable quantity of blood by urethra during a period of six days. The total amount was estimated by himself at "not less than ten or twelve pints."

In July he twice passed blood, but in smaller quantity; and on two occasions since, slightly.

About the middle of September, he came under Dr. Armitage's care. His state at that time was as follows: "Complexion sallow; body emaciated. Complains of very severe pain in the region of the sacrum, and passing down the back of the thighs. Passes urine twice or three times in the night, rather more frequently in the day; occasionally with pain, when much ropy mucus is present and appears to block up the urethra."

Urine is generally alkaline, with more or less tenacious mucus and earthy phosphates: no blood.

A full-sized catheter being introduced encountered no opposition. The bladder did not completely empty itself, and the patient passed the instrument for himself twice a-day for relief.

Examination by rectum verified the existence of a large and very hard tumour in the situation of the prostate, apparently about the size of a large duck's-egg, but more prominent on the left than on the right side of the median line. Examination not painful; *no obvious tenderness on pressure.*

In the right inguinal region, there were several enlarged and hard glands, and on making deep pressure similar swellings were felt in the course of the iliac vessels. The same was observed, but to a less extent, on the left side.

These conditions altered very little before death, which occurred on the 17th March, 1857. The patient had become a little jaundiced, more emaciated, and weaker. The glands exhibited no manifest change. The tumour continued much the same in size, and as hard as ever, until about a month before death, when it became much softer, especially on the left side.

Post-mortem examination. The prostate was found enlarged, and evidently the subject of malignant disease, as were the lymphatic glands described. No appearance of any similar disease in any other part of the body, which was closely examined. Dr. Armitage having kindly presented me with the prostate and adjacent parts in the fresh state, I carefully dissected it, and add the following notes made at the time.

Bladder enlarged, thickened, and fasciculated; mucous lining shows marks of inflammation. Several small bodies, isolated and about the size of hemp-seed, are seen underlying the mucous membrane near its neck.

Between the orifice of the left ureter and the prostate is a mass of oblong form about the size of a large almond, evidently formed by some deposit in the muscular substance of the bladder.

The prostate is four times its natural size. The general outline is somewhat uneven, and slightly nodulated; the left lobe larger than the right. From its rectal surface, a soft fluctuating swelling extends backwards, nearly in the course of the left vesicula seminalis, which, on being opened, gave exit to yellowish fluid of a thick consistence, containing débris of tissues. The vesicula is much thickened, and loaded with deposit in its structure; the right vesicula also in a less degree. A section of the mass showed the tissues to be infiltrated with soft pale matter, from which a creamy juice abundantly exuded when pressure was applied. This under the microscope exhibited numerous nucleated and binucleated cells, large and small, characteristic of malignant growth; so also did the other deposits described.

Case No. IX.

MALIGNANT TUMOUR OF THE PROSTATE.

I.B. Aged 68. Had symptoms of diseased bladder five years. He suffered excessive pain in the back, bladder, and rectum, especially during the last six months of life. Rectal examination showed the prostate to be diseased; and catheterism was extremely difficult. There was at last very frequent and painful micturition, and on one occasion considerable hæmorrhage. He died comatose.

P.M. The right ureter had sloughed and given way, admitting a large quantity of urine to be effused between the peritoneum; "three pints of fluid being measured. The bladder was almost filled with a fungous tumour of loose coagulated blood, mixed with a white pulpy substance; which derived its origin from the back part of the prostate," and had plugged up both ureters as well as the urethra. There were several small masses of the same matter in the lungs and liver. — *Trans. of Med. Chir. Soc.* vol. viii. 1817, p. 279. By G. Langstaff, Esq.

Case No. X.

A man, aged 45, had suffered five years with symptoms of diseased bladder, from obstruction. His urine was loaded with mucus, and sometimes was tinged with blood. During the last six months of life he had voided pus and blood per anum.

At the post-mortem, the bladder was found to be contracted and hypertrophied. On opening it a large tumour was seen "produced by the prostate gland, which was converted into a fungoid and carcinomatous tumour resembling melanosis." The kidneys were inflamed. "The absorbent glands in the pelvis and abdomen were converted into medullary and carcinomatous matter, and the veins in the pelvis were filled with fungoid deposits." There was cancer of the stomach also.—*Catalogue of Preparations in the Museum of Geo. Langstaff, Esq.* Lond. 1842. p. 352.*

Case No. XI.

A gentlemen, aged 59, in whose case Mr. Adams was consulted. Enlargement of the left lobe of the prostate was detected nearly at the outset; irritability of the bladder, and inability to empty the viscus, being the earliest symptoms. For the last year of life complete control over the bladder was regained. The main symptoms were, excessive pain in the lumbar and pelvic regions, and in the thighs and legs, with pain and swelling of the testicles. Duration of disease about three years and a quarter.

P.M. Prostate gland twice its usual size, the left lobe being occupied with a carcinomatous tubercle, right healthy. The adjacent glands much enlarged and similarly diseased, but no other abdo-

---

* A man, aged 63, "had been afflicted with disease in the urinary organs and prostate for several years." The symptoms increased in severity considerably for some time before death. At the post-mortem the prostate gland was found to be "affected with melanosis and fungus hæmatodes, and the whole of the urethra is in a melanoid state." But nothing is said of deposit in the lymphatics, or in any other organ.—*Idem*, p. 330.

Three or four other cases are reported in the above catalogue as similar to the foregoing; but the accounts given are not sufficiently complete to place their nature beyond doubt, and they are, therefore, not included here.

minal viscera affected. Other cavities not examined.—*Adams on the Prostate*, 2nd edit. p. 149.

This case formed the substance of a communication to the Med.-Chir. Soc., April 12, 1853, at which Mr. Adams stated that "the tumour had been examined by an experienced microscopist, who had pronounced it to be true schirrus in every particular."—*Lancet*, 1853, vol. i. p. 394.

Case No. XII.

A gentleman, aged 67. Severe vesical irritation for a considerable time; no retention: urine occasionally bloody; pain excruciating. The left side of the prostate presented tumour of an irregular knotty hardness. Indurated glands appeared in the groin. The whole pelvis became filled with cancerous growths, which after death were found to have pushed up the bladder towards the umbilicus. The left iliac vein was obstructed by pressure, the left ureter also. The disease was nearly limited to the parts named.

A case related to the Hunterian Society by Mr. Cock, and quoted by Mr. Adams.—*Op. cit.* pp. 147-49.

Case No. XIII.

C. F., aged 75, under the care of Mr. Fergusson. Duration of symptoms four years. He was admitted in a sinking condition, and died in the course of three days. P.M. The prostate was of the size of an orange; and both kidneys exhibited the presence of the same deposit: which in each instance was encephaloid.—*Lancet*, 1853, vol. i. p. 473.

Case No. XIV.

A gentleman, aged 59, under the care of Mr. Haynes Walton. Duration of symptoms between eight and nine months only. Hæmorrhage occurred several times. The tumour was so large at a late period in its course "that the finger could not be passed between it and the sacrum."

P.M. "The true pelvis was entirely filled by a tumour" of an encephaloid character. There was a cavity in the diseased mass, " with irregularly-ulcerated walls, containing unhealthy pus; there was no trace of urethra, the bladder opened directly into it." The adjacent lymphatic glands were affected.—*Path. Trans.*, vol. ii. p. 287.

Cases No. XV. and XVI.

B., aged 41, was an out-patient of St. Thomas's Hospital for some weeks with symptoms of stricture, and occasional retention of urine. Much disposition to bleed when catheters were passed. Subsequently, complete retention occurring, he was, on March 19, admitted to the hospital, under the care of Mr. Simon. Very sallow, not emaciated, bladder irritable, urine passed with effort; testicles swollen and tender. Rectal examination showed an enlarged prostate. He suffered much pain, but was relieved by occasional catheterism and opiates. In April he died from sloughing of the urethral aspect of the tumour, and consequent extravasation of urine.

P.M. "showed the prostate converted into an encephaloid mass the size of a small orange."

Mr. Simon adds, "The nature of the tumour had been fully recognized during life by the presence in the urine of granules and flocculi of animal matter, which under the microscope showed the large coherent nucleated cells of encephaloid cancer."—Lecture by Mr. Simon. *Lancet*, 1850, vol. i. p. 291.

Mr. Simon illustrates this case with a second. A man, aged 63, sallow and anxious. The chief symptom was irritability of bladder, which had existed more or less for about three years, with occasional bleeding during two years, and twice retention. No tumour in rectum. Flocculent matters passed, indicating the disease to be encephaloid. A few weeks after he died.

P.M. showed "an abundant growth of soft cancerous vegetations from the mucous membrane of the prostate." In both cases the neighbouring glands were enlarged.

Case No. XVII.

A man, aged 54, admitted under Civiale's care for calculus, 1837. Urine had long been thick, fœtid, and loaded with mucus: and the passing of it painful. Symptoms aggravated during last five months. On sounding a stone was detected, but his condition forbade any operative measures, and he soon died.

P.M. Large calculus in the bladder. A considerable tumour at the neck of the bladder reaching to the membranous portion of the urethra, where was a large excavation extending backwards from one and a half to two inches. The prostate was disorganized and softened at this part, whence by pressure escaped a cerebriform substance.

Reported as an illustration of cancer of the prostate.—*Civiale, Traité Pratique, deux⁰ partie, Paris*, 1850, p. 503.

Case No. XVIII.

A man, aged 70, at Hôtel Dieu, May, 1839. Urinary symptoms had existed about 18 months. Urine had been thick, loaded, and lately darkened by blood. Attempts at catheterism appear to have been unsuccessful. He gradually sank and died on the 13th day after admission.

P.M. Bladder contained urine, dark from blood; walls thick; mucous membrane slate-coloured. The prostate was the size of an ostrich's egg, and presented throughout, the tissue of softening encephaloid, with blood clots in it, a cavity containing one of these opened into the bladder. Encephaloid existed also at the lesser curve of the stomach.— By L. Aug. Mercier, *Recherches Anat. Path. &c., Paris*, 1841, p. 169.

Case No. XIX.

ENCEPHALOID DISEASE OF PROSTATE IN A CHILD.

W. M., aged 5 years. Mr. Stafford was called to see this child in the Marylebone Infirmary, on account of abdominal tumour, November, 1838. It reached two inches above the umbilicus, and proved to be a distended bladder, 25 ounces of urine being drawn off by the catheter. No pain was complained of; and no abdominal tenderness appears to have existed. Prostatic disease was not suspected, therefore no examination by rectum was made. Eight days afterwards he died.

P.M. All the viscera appeared to be healthy excepting the bladder and prostate; the kidneys were large. The mucous membrane of the bladder was thickened, in other respects the organ was sound. The prostate was equal in size to the largest walnut, was somewhat globular in form, and from its back part projected a tumour of the size of a small hazel-nut. Examination proved the mass to possess the "colour, consistence, and texture" of encephaloid; melanotic matter appearing to be intermixed.—*Trans. Med.-Chir. Soc.*, vol. xxii. p. 218, 1839. By R. A. Stafford, Esq. Now forms Prep. No. 17, series xxix. *Mus. Barthol. Hosp.*

Case No. XX.

A boy, aged, 8, "had experienced for a considerable length of time great difficulty in voiding urine." He was sounded for suspected stone, after which there was much hæmorrhage; and subsequently swelling in the perineum, supposed to be abscess, which was punctured, but nothing escaped but blood, and soon after a fungous growth. At the post-mortem the prostate was found to "be converted into a medullary sarcoma." There were also "several fungoid tumours in the liver."—*Catalogue of Preps. in the Museum of Geo. Langstaff, Esq.* London, 1842, p. 357.

Case No. XXI.

A child, aged 3, having been ill for some time, was observed to have difficulty in passing water, and subsequently retention. It was then brought to the London Hospital, and both catheterism and suprapubic puncture were resorted to, but without success. Death occurred shortly after, and the post-mortem revealed a "bladder distended with a large cancerous mass which had originated in the prostate gland and made its way into the bladder."—*Adams on the Prostate Gland*, 2nd ed. 1853, p. 145.

Case No. XXII.

A child, aged 3. Symptoms of vesical disease during six months; then retention, relieved with difficulty by the catheter, which was frequently employed afterwards, until the obstruction rendered it impossible. Puncture of the bladder was performed above the pubes. About three or four weeks afterwards, the child died from exhaustion. On dissection the prostate was "enlarged to the size of a hen's egg, and completely transformed into medullary matter."—*By Professor Bush, in Gross on the Urinary Organs*, 2nd ed. p. 719.

Case No. XXIII.

A child aged 3, under the care of Mr. Solly. Among the earlier symptoms was retention of urine, which was relieved with some difficulty, nearly one and a half pint being withdrawn. The presence of stone was suspected. The child died of peritonitis three months after the first accession of the symptoms.

P.M. The prostate was enlarged by encephaloid deposit to the size of a hen's egg.—*Path. Trans.*, 1850-51, p. 130.

Case No. XXIV.

A. C., aged 9 months, was placed under the care of Mr. Bree, of Stowmarket, March 9, 1843, for retention of urine of 24 hours' duration. A pint and a half of urine was removed by the catheter. A tumour was felt per rectum in the situation of the prostate. Obstruction increased, and ulceration took place through the sphincter ani in a few weeks. Death occurred April 22.

P.M. The bladder was highly inflamed and thickened, and the tumour proved to be a very large encephaloid growth from the prostate. Viscera generally healthy.—*Prov. Med. and Surg. Journal*, 1846, p. 76.

TABLE OF FOREGOING CASES OF MALIGNANT DISEASE OF PROSTATE.

ADULTS.

Case No.		Age.	Duration.	Character.	Other Organs.
7	Mr. Henry Thompson	60	2 years	Encephaloid	Spine and adjacent lymphatics.
8	Dr. Armitage	65	1½ "	Ditto	Lymphatics only.
9	Mr. Langstaff	68	5 "	Ditto	Lungs and liver.
10	Ditto	45	5 "	Ditto with matter resembling melanotic deposit	Lymphatics and vessels adjacent; stomach also.
11	Mr. Adams	59	3¼ "	Reported schirrus	Adjacent lymphatics only.
12	Mr. Cock	67	Not stated	Probably encephaloid	Adjacent lymphatics.
13	Mr. Fergusson	75	4 years	Encephaloid	Both kidneys.
14	Mr. H. Walton	59	¾ "	Ditto	Adjacent lymphatics.
15	Mr. Simon	41	Some months	Ditto	Ditto.
16	Ditto	63	Rather more than 3 years	Ditto	Ditto.
17	M. Civiale	54	Not stated	Ditto	Not stated.
18	M. Mercier	70	1½ year	Ditto	Stomach.

CHILDREN.

Case No.		Age.	Duration.	Character.	Other Organs.
19	Mr. Stafford	5	Few months	Encephaloid and melanotic	No other viscera.
20	Mr. Langstaff	8	"A considerable length of time"	Encephaloid	In liver also.
21	Mr. Adams	3	Not stated	Ditto	Not stated.
22	Prof. Bush	3	7 months	Ditto	Ditto.
23	Mr. Solly	3	3 "	Ditto	Other organs not examined.
24	Mr. Bree	9 ms.	About 3 months	Ditto	"Viscera generally healthy."

** Other cases of malignant disease involving the prostate are reported in the journals and elsewhere, but are excluded for want of sufficient particulars by which to form a fair conclusion respecting them. At least 30 cases instead of 18 might have been adduced had I not been careful to reject those in which the evidence was incomplete.

CHAPTER XV.

TUBERCULAR DISEASE OF THE PROSTATE.

TUBERCLE.—Rare in Prostate; almost invariably associated with Tuberculous Kidney or Testicle.—Morbid Anatomy.—Symptoms.—Diagnosis.—Treatment.—Tabular View of eighteen Cases.

TUBERCLE OF THE PROSTATE.—The prostate is very rarely the seat of the tubercular deposit, and when it is so, appears generally to be somewhat increased in size, until the later stages of the complaint are reached, when, after suppuration and discharge, its volume may become smaller than natural.

It would appear that at no period of the disease is the prostate affected alone, some other part of the genito-urinary track being the primary seat of the affection. In most cases the deposit appears to take place first in the kidney, or, at all events, to be present there in an early stage. The organ next in order of liability to the disease, among the genito-urinary group, is the testicle. Thus in 18 cases collected by myself, and forming a table at the end of this Chapter, in which the results of post-mortem inspections have been recorded, tuberculosis of the kidney is reported in 13, and of the testicle in 7. The state of the lungs has, I suspect, not always been recorded, but in 10 of these cases they are stated to have been diseased.

MORBID ANATOMY.—The form which the deposited tubercle assumes in the prostate is, at first, that of minute yellowish points, like millet seeds. These become larger, and numerous rounded collections of cheesy or curd-like consistence may be found, distributed throughout the substance of the organ. The processes of deposition and aggregation continue until the mass may reach the size of a chestnut; commonly, not larger than that of a full-sized marble. It is then generally surrounded by a thin limiting fibrous membrane, being isolated by it from the surrounding prostatic tissue. Among this mass may be found portions of thin cellular tissue intermingling, the remains, probably, of prostatic structures, which have given way before the accumulating tubercle. Central softening takes place at an earlier or later period; and if the patient survives, discharge of the detritus takes place by urethra. A space remains; and several such coalesce and form a cavity of larger size; pus is formed from the lining membrane, and a considerable portion of the prostatic substance is destroyed. But there is no reason to suppose, when such cavities are found in the prostate after death, that their origin has been tuberculous, unless the presence of such action or deposit can be distinctly verified in some other part of the urinary or genital organs.

Tubercular deposit appears to be more prone to affect the outer than the central parts of the prostate, occurring in the lateral lobes chiefly. By microscopic examination it may be inferred to commence in the glandular structures of the organ and not in the fibrous stroma.

SYMPTOMS.—From all these circumstances it follows that there are no symptoms which are, strictly speaking, proper to this affection of the prostate. Undue frequency and pain in making water, occasionally blood in the urine, and at times the signs of cystitis, are commonly experienced. Pains in the back and pelvis, and in the region of the bladder and urethra, are complained of; while wasting and extreme debility slowly show themselves in the system at large. The local complaint is to be regarded but as a part of the development of tubercular disease existing in other portions of the genito-urinary organs, and generally in other regions of the body also. It is no part of the design of this work to embrace a consideration of the symptoms and diagnosis of tuberculous kidney; a subject, nevertheless, of great interest to the surgeon, to whom the disease is perhaps quite as often presented as to the physician, on account of the increased frequency of passing water, and other symptoms of stone or obstruction which accompany it. The presence of pus in the urine, of occasional hæmaturia, of pains in the loins, perineum, and penis, give rise to suspicion of calculus, to be resolved sometimes only by careful sounding. The state of nutrition of the patient, his history, and the condition of the lungs, are among the main points to be considered in connection with the urinary derangements, which have probably more especially attracted his attention.

TREATMENT.—Nothing need be said of the constitutional treatment of tubercular disease, and little in relation to the local manifestation in the prostate. Mechanical interference is to be avoided, and every kind of irritating application. If suppuration takes the form of external abscess, it must be treated as other perineal or ischiorectal abscesses. But more commonly the discharge of purulent and tubercular matter takes place into the urethra. The improvement of the health, by all those numerous means which regulation of the diet, regimen, exercise, climate, and medicine enable us commonly to achieve in tubercular patients, constitutes almost the whole of the treatment to be employed in the affection, when involving the urinary or genital organs. The diagnosis once established, it is of great importance that the patient should be kept free from all instrumental treatment, which, in such cases, provokes irritation, and aggravates the disease, without conferring upon him any benefit whatever.

Tubercle of the prostate is a very rare affection. It is extremely uncommon to meet with it in the deadhouse. The preserved examples are few. There is no preparation of it in the Royal College of Surgeons, London, although there are one or two specimens of abscesses in the prostate, which are supposed to be of a scrofulous character or origin. There are two examples in the Museum of St. Bartholomew's Hospital, one in Guy's, one in St. George's, one in that of the College of Surgeons, Edinburgh, and one in a small collection at the Royal Naval Hospital, Greenwich.

The following cases, eighteen in number, have been condensed into a tabular form. They constitute the greater portion of the data on which our knowledge of this disease is founded. The age of the patient, the condition of the prostate itself, and that of other organs of the body also tuberculous are presented. The sources from which they are obtained are also added in a note.

CASES OF TUBERCLE IN THE PROSTATE.

Patient's Case.	Reported by	Age.	Condition of Prostate.	Condition of other Organs.
1	Mr. Lloyd.	23	Enlarged cavity containing an ounce of scrofulous matter.	"The bladder and kidneys were tolerably healthy;" tuberculous cavities in the lungs.
2	Mr. Adams.	26	Disappeared by suppuration.	Tubercle in left kidney, ureter, and testicle.
3	Mr. Hudson.	—	Several small points of softened tubercle, and superficial ulceration.	Kidney largely tuberculous; ureter, lungs, bones, lymphatic glands.
4	Dr. Basham.	29	Granular deposits.	Right kidney and bladder chiefly affected.
5	Dr. Gross.	27	Eight small masses, each about the size of a pea.	One kidney, ureter, and vesicula semin.; spine and lymphatic glands (not in lungs).
6	Vidal de Cassis.	19	Tuberculous cavity.	Testicles and vesicula semin.; brain and lungs.
7	Ditto.	50	Large tuberculous masses.	Kidneys, lungs, &c.
8	Lallemand.	55	Thirty small abscesses and tubercles.	Both kidneys.
9	Ricord.	58	Large abscess in prostate.	Miliary tubercles throughout ureter; testicle previously removed for tubercular disease.
10	Guy's Hospital Museum.	23	Tubercular deposits, "size of pin's head to that of a nut."	Ulceration of bladder; much congestion of kidneys; tubercle in both lungs, and in lumbar vertebræ.
11	Bartholomew's Museum.	Young Man.	Small circumscribed masses.	Tubercle in kidneys, lungs, and other organs; bladder ulcerated.
12	Ditto.	Do.	Large mass of tubercles in *left* lobe of prostate.	Tubercle in *left* kidney, *left* testicle; in lungs and other organs.
13	St. George's Hospital Museum.	35	Large tubercular cavity in prostate, capable of holding two ounces.	Kidneys extensively diseased; scrofulous abscesses in testicle, secondary to the prostate symptoms.

CASES OF TUBERCLE IN THE PROSTATE—*continued*.

Patient's Case.	Reported by	Age.	Condition of Prostate.	Condition of other Organs.
14	College of Surgeons, Edinburgh.	11	Prostate destroyed by ulceration.	Abscesses in perineum; ulceration of right ureter; both kidneys full of tuberculous matter.
15	Royal Naval Hosp., Greenwich.	76	Large masses of tubercle deposited in outer portions of the organ.	Tubercle deposited in mucous membrane of bladder; in vesic. sem., in right kidney, and in lungs.
16		32	Large deposits of tubercle in the prostate.	Tubercular meningitis; tubercle of lungs, and mesenteric glands; in the testicle.
17	Dr. Guerlain.	43	Large mass of tubercle in left lobe; smaller ones in right.	Tubercle of the lungs; and in bone.
18	Mr. Simon mentions a case in which the entire genito-urinary tract was more or less affected with tubercular deposit, from the testicle to the prostate, and from the kidney to the same point.			

1. Lloyd on Scrofula. p. 110. 1821.
2. Adams on the Prostate. 2nd ed. p. 127.
3. Trans. Path. Soc. Vol. i. p. 120.
4. Lancet, 1855. Vol. ii. p. 642.
5. Urinary Organs. 2nd ed. p. 721.
6. Annales de Chirurgie. 1845.
7. L'Union Médicale. 1850.
8. Des Pertes Séminales. 1836. Vol. i.
9. L'Union Médicale. 1819.
10. Guy's Hospital Museum. No. 2393, 75.
11 & 12. Barth. Museum. Series xxix.: 19 and 20.
13. St. George's Hosp. Museum. Series x.
14. Catalogue. p. 260.
15. Med.-Chir. Trans. Vol. xliii. p. 157.
16. Recueil de Mém. de Méd. de Chir., &c. 1860.
17. Bull. de la Soc. Anat. 1860. p. 133.
18. Lancet, vol. i. 1850. p. 290.

CHAPTER XVI.

CYSTS, OR CAVITIES, IN THE PROSTATE.

True Cysts unknown.—Cavities are of three kinds—Dilated Follicles; Purulent Depots; and containing Concretions.—Hydatid Cysts between the Prostate and the Rectum.

SMALL cavities, which have received from many authors the name of cysts, are frequently to be seen in the prostate, especially in those of aged subjects.

If by the term "cyst" it is designed to describe a thin-walled, cell-like body or closed sac, spheroidal, or nearly so, in form, containing fluid, a structure of new formation apparently, not a mere dilatation of some already-existing cavity, then I know of no such affection of the prostate either from personal examination or on the report of others. Such cysts are found in many parts of the body, containing hydatids, jelly-like material, serous fluid, &c. But in the prostate we find no such isolated cysts, as for example, in the kidney, nor any formation indeed which can be regarded as analogous to that, which may be considered as the type of simple fluid cysts. Although I have, in conformity with the practice of other authors, referred to "cystic disease" of the prostate, it does not appear that the use of the term is warranted by the phenomena presented; and if retained, it must be held to signify a formation of a wholly different kind from that which is indicated by it in the breast or kidney; neither is any species of proliferous cyst ever met with.

Cavities are frequently found in the prostate, but these manifestly belong to one of the three following orders:—

(a.) Mere dilatation of gland-follicles.

(b.) Cavities containing pus; abscess.

(c.) Cavities containing concretions or calculi, the walls of which become denser and firmer in proportion to their magnitude, which corresponds with that of the contained body.

Of the first kind, viz. dilated gland-follicles, examples may be found in most prostates from elderly subjects; as has been before seen, some of the gland-follicles generally exhibit a tendency to enlarge or dilate as age increases. The cavity is often filled with yellowish, semi-fluid material, or with yellow, semi-transparent concretions, of so small a size as only to be discernible with the microscope. And it appears that the increase in size of the contained and of the containing parts, takes place *pari passu*.

The second kind, viz. cavities containing pus, offers no subject for remark here. These have been fully examined and described in Chapter IV. on the results of inflammation, under the heads of Acute and Chronic Abscess.

The third kind of cavities consists of those which contain solid formations, whether concretions or calculi. It is common, when making sections of the prostate from an elderly subject, whether it is hypertrophied or not, to find cavities, of a somewhat irregular form, in its substance, having all the appearance of being dilated follicles of the glandular structure, which contain numerous little dark-coloured concretions. I have seen from thirty to fifty of these bodies occupying a cavity about the size of a grain of wheat or of a small pea.

But larger concretions, that is, of the size of pearl barley, may occupy each a separate recess of its own; and on removing the foreign body, a spherical, thin, and smooth-walled cavity is displayed. Sometimes hundreds of such small cavities may be found in one prostate,—but this is a very rare circumstance. A good example may be found in the museum of the College of Surgeons, Prep. No. 2519. See Chap. XIII. for numerous others.

The cavities referred to do not attain a sufficient size, nor, as far as I know, do they give rise to any symptom whatever, to render a knowledge of their presence possible during life. Generally speaking, they are capable of holding not more than a few minims of fluid. Mr. Coulson refers to one instance only in which he had seen such a cavity containing as much as half an ounce; and he regarded this as a dilated duct.* In relation to practice the diagnosis is unimportant, as no indication for treatment would be presented by the fact of their existence, were it ascertained.

HYDATIDS OF THE PROSTATE.—It is doubtful if hydatid cysts have ever been met with in the prostate. The only case on record which may have been an example is one which occurred in the Sussex County Hospital, and is recorded by Mr. Lowdell in the 29th vol. of the Med.-Chir. Trans. The author expresses a doubtful opinion on the subject, but appears to incline to a belief that the prostate was the seat of the hydatid formation. Six other cases may be referred to in which retention of urine and distension of the bladder occurred as a result of a hydatid cyst *between the bladder and rectum,* near to the neck of the former; but in which the prostate was not affected except by pressure. Prostatic enlargement was very closely simulated, certainly in three of them, and in two the prostatic catheter was employed under the belief of its existence.

The first was reported by John Hunter, in the Transactions of a Society for the Improvement of Medical and Surgical Knowledge, vol. i. p. 34. Here the retention caused death, and between four and five pints of urine were found in the bladder at the postmortem examination. The viscus was pushed up into the abdomen by the pressure of the cyst below.

The second, by Mr. Curling, appears as an appendix to Mr. Lowdell's paper in the volume referred to above, page 356. Here the same appearances were observed, but in a less-marked degree, relief having been given to the retention during life.

The third case occurred to the late Mr. Callaway, at Guy's Hospital, and is referred to in the *Medical Times* of Feb. 17,

* Op. cit. 5th ed. p. 587.

1855. Hydatids were removed when the catheter was passed. After death a large hydatid tumour was found between the bladder and rectum, pressing on the neck of the former.

The fourth, a man aged 40, was admitted into Guy's Hospital with retention of urine: no catheter could be passed, and he died. At the post-mortem, a large tumour occupied the pelvis and hypogastric region, the anterior and upper part of which was formed by the bladder, pushed out of its proper place. The tumour consisted of a cyst containing three pints of hydatids. The preparation is No. 2104[52]. Another preparation is preserved there very similar to the preceding.

The fifth occurred in a man, aged 59, admitted to the Westminster Hospital with retention of urine, under the care of Mr. White. A catheter could not be made to reach the bladder, which was therefore punctured through the perineum, a pint of urine escaping. He died next day, and a large hydatid tumour was found just above the prostate, pressing against the back of the bladder, so as to divide it into two portions, of which the upper still contained two pints of urine, the lower, which held one, having been evacuated by the puncture.

I have recently become acquainted with a sixth case, that of a lad, nine years old, and suffering from retention of urine, in whom a large and fluctuating tumour, occupying the position of the bladder, was felt within the rectum. No instrument could be passed, and the rectal puncture was performed, the result being, not the removal of urine, but of the contents of a hydatid cyst. This relieved the retention, and the boy recovered.

In the Museum of St. Bartholomew's there is a good example of a large hydatid cyst occupying a position between the bladder and rectum. In this case the prostate does not appear to have been affected. It is preparation No. 15, series xxix.

It does not appear unlikely that Mr. Lowdell's case may have belonged to the same category, the prostate being more or less absorbed by pressure from an external cyst, so that the latter came at length to occupy the situation proper of that organ. The case, is, however, given here in an abridged form.

CASE No. XXV.

HYDATID DISEASE OF PROSTATE. (?)

J. I., aged 64, in Sussex County Hospital, in July, 1844, under the care of Mr. John Lawrence, jun. During three or four years had experienced difficulty in making water and frequent micturition; and of late, almost complete retention. The bladder was now emptied by catheter, after great difficulty, and three pints withdrawn. Much pus and mucus passed afterwards. He died in a few days.

P.M. Bladder very much thickened, and in the situation of the prostate was a tumour larger than a fœtal head, which when cut open, proved to be a hydatid cyst, closely packed, the true substance of the prostate being lost in it. Hydatid tumours were also found in the omentum.

Whether the hydatid cyst was formed in the prostate itself, or external to the organ, destroying it by pressure alone, is stated to have been a matter of doubt. Appearances led Mr. Lowdell, who reports the case, to the former view. The facts of hydatid disease of the prostate being unrecorded, together with the existence of other tumours in the omentum, inclined him to believe that he "should be scarcely warranted in maintaining that opinion without question."— *Trans. Med.-Chir. Soc.*, vol. xxix. p. 253. 1846. By George Lowdell, Esq.

CHAPTER XVII.

THE BAR AT THE NECK OF THE BLADDER.

Close relation between this subject and Prostatic Enlargement.—Most organic obstacles at the Neck of the Bladder are Prostatic.—A few Cases which are exceptional.—Mr. Guthrie's recognition of them.—His views defined.—Views of Civiale, Mercier, Gross, Leroy.—A Bar may be due to repeated Contractions of the Bladder from any cause whatever, if long continued.—Shown to consist, in such cases, of Muscular Hypertrophy. — Examples.—CONCLUSIONS on the whole subject.—Rarity of any Affection meriting the appellation of Bar in absence of Enlarged Prostate.—DIAGNOSIS.—TREATMENT.—When due to Muscular Hypertrophy, as in Stone or Stricture, it will disappear on removal of the exciting cause.—Mr. Guthrie's proposal to divide Obstructions at the Neck of the Bladder.—Mercier's Modes and Instruments.—Results of Operations.—Consideration of these Proposals.

THIS is an affection so closely related to enlarged prostate, by identity of anatomical situation and of the symptoms resulting, that it is impossible to treat of one without also considering the other. As already seen, in the examination of the anatomy of the first-named affection presented in Chapter the fifth, a bar at the neck of the bladder is very frequently due solely to enlargement of some part of the prostate; but it is not less a fact that a somewhat similar obstruction is sometimes, though not very commonly, present, when that organ is not the subject of disease. It is to this latter condition that the term, as designating a distinct affection, has usually been applied in this country.

Although numerous forms of obstruction at the neck of the bladder have been frequently described, at some length, by the well-known French writers of the present century on urinary diseases, under the names of "bourrelets," "barrières urétro-vésicales," "brides," and "valvules," no specific distinction was recognized between the form of obstacle about to be described, and that which consists in a hypertrophy of the prostate, until the late Mr. Guthrie called attention to the subject in his Lectures at the Royal College of Surgeons in 1830. He examined it with care, and arrived at more precise views respecting it than any previous writer had done, pointing out the distinctive characters of the two affections, the prostatic and the non-prostatic; and the term which he employed to designate the latter is retained here, in the sense which he originally intended it to convey. The views which he entertained respecting the entirely-distinct character of the two affections, are summed up briefly by himself, and may be given here in his own words. He concludes:—

" 1. That an elastic structure exists at the neck of the bladder, and may be diseased without any necessary connection with the prostate gland.

" 2. That the prostate may be diseased without any necessary connection with the elastic structure."

He quotes two cases, one in which, " without any affection of the prostate, and particularly of the third lobe, the patient passed his water with great difficulty, in consequence of the barrier formed by this unyielding structure, and died ultimately of the disease after much suffering." Another, in which, as a consequence of unequally-enlarged lateral lobes of the prostate, the right being most so, the mucous membrane of the neck of the bladder had been drawn up "so as to form a bar across its under part. This bar," he adds, "is quite membranous, and does not include the elastic structure which is not diseased, *neither is that part called the third lobe*, nor is there any projection into the bladder, save the bar or valve formed by its mucous membrane at the very meatus." *

Of these two cases, as the author observes, one was exactly the reverse of the other. Each is, indeed, typical of two perfectly-

---

* On the Anatomy and Diseases of the Urinary and Sexual Organs; being the first part of the Lectures delivered in the Theatre of the Royal College of Surgeons, in 1830 By J. G. Guthrie. London, 1836. pp. 23 and 25.

distinct classes of abnormal conditions affecting the neck of the bladder. In the latter example there is presented merely a natural result of certain forms of prostatic enlargement which are occasionally met with, and some examples of which exist in our museums, in which the growth upward of the lateral lobes alone has the effect of drawing up the mucous membrane from parts below, and sometimes with it some subjacent fibrous and muscular structures, but in which there is little or no enlargement of the median portion. In the former case, there is an unnatural elevation of certain structures which underlie the mucous membrane at the posterior or vesical limit of the urethra, but which is unaccompanied by, and totally unconnected with, any enlargement of the prostate whatever.

It is particularly necessary to draw the distinction clearly between the affection, to which Mr. Guthrie thus applied the name of " Bar at the Neck of the Bladder," and that obstruction which is constituted solely by an enlarged median portion of the prostate itself. Views, differing very much from those which Mr. Guthrie held, have been frequently promulgated as his respecting it. Thus the bar has been described by more than one author as an eminence situated just *behind* an enlarged middle lobe of the prostate. Now this is clearly not what was intended by Mr. Guthrie; nor can such an eminence be said to have for its locality " the neck of the bladder " at all, inasmuch as it must necessarily lie considerably posterior to it. As we shall see hereafter, the eminence so indicated is formed by a hypertrophied condition of those muscular bands which intervene between the two orifices of the ureters, and which are generally known as " the muscles of the ureters."

Civiale devotes a chapter to the consideration of this subject. He describes the "vesico-urethral barrier" as formed in some cases by " a simple fold of membrane, smooth and thin, and almost transparent, extending from one lobe of the prostate to the opposite;" and in others by a fold, thicker, in the form of a rounded cord, which contains some fibrous or muscular tissue.* In others it partakes more or less of prostatic substance. And, he adds, that a considerable barrier may be found after death, when no symptoms have existed during life, while in other cases a train of very painful disorders of the urinary apparatus may result from an inconsiderable membranous fold, giving rise, however, to very decided obstruction at the vesical neck.

He also reviews the alleged causes of these valvular obstructions, that is, when not due to enlarged prostate; viz. spasmodic muscular contractions, rheumatism, &c., and expresses his opinion that we have no certain knowledge of the etiology of the affection.

Mercier, who is familiar with Mr. Guthrie's views on this subject, but does not altogether coincide with them, describes, as the usual form of urethro-vesical barrier, a "semi-annular eminence, very like the pyloric valve of the stomach, if it existed only on the lower half of the orifice." He recognizes two distinct kinds: one produced by spasm of the muscular fibres which close the neck of the bladder; these fibres having become permanently contracted in some individuals after long-continued and repeated attacks of the spasmodic affection, so that a permanent bar is the consequence. The other kind being simply the result of hypertrophy of the median portion of the prostate. Regarding the first, or muscular bar, he states that the seat of the affection is the tissue at the apex of the trigone, between the mucous membrane and the prostate. The occasion of it he believes may be the presence of stricture, calculus, or any other excitant of continued spasmodic action in the muscular fibres referred to. It may take its rise also in an attack of inflammation of the neck of the bladder, and following generally gonorrhœa, although sometimes due to other causes. Consequently, examples of it, he says, may occur in men much younger than those who are the subjects of the prostatic bar, and this will aid the diagnosis. But the angular

* Traité Pratique, vol. ii. p. 244.

sound will generally determine the difference, since in the muscular bar, the beak on being introduced can be rotated freely, but in the prostatic bar, the rounded eminence which it forms prevents this act of rotation.*

Dr. Gross, of Louisville, has devoted a few pages to a consideration of the "bar-like ridge of the neck of the bladder." From his observations, and from a drawing which he gives of the bar, it is obvious that he does not identify by that term the pathological condition described by Mr. Guthrie.† For indeed not only does he refer it to a different locality, but enumerates, as one of its most frequent causes, hypertrophy of the prostate. Whereas, as has been already shown, the term was assigned by Mr. Guthrie to an obstruction situated at the neck of the bladder, across the urethro-vesical meatus, and only when enlargement of the median portion of the prostate (middle lobe) is not present.

M. Leroy D'Etiolles appears not to recognize clearly the occurrence of valvular obstruction at the vesical neck, except when produced by prostatic enlargement. To this very common form of disease he has given considerable attention, treating of it in various papers in the French journals, and in his own published works, during nearly thirty years past.

The second form of bar described by Mr. Guthrie, viz., that which is formed by the drawing up of the mucous and sub-mucous tissues by the upward development of the lateral lobes of the prostate, can, of course, only be seen in those cases where enlarge-

---

* Recherches sur le Traitment des Maladies des Organes Urinaires, &c. Paris, 1856. Mémoire 6ième, p. 209 et seq.

† In the drawing referred to, the orifices of the ureters are situated one on each extremity of *the bar itself*, which is evidently the muscular ridge (muscles of the ureters, *Bell*), so often seen, extending between those orifices when the bladder is hypertrophied under circumstances of difficult micturition. If additional proof of this were wanting, it would be found in the description which follows the drawing, from which it appears that great prostatic enlargement was present; and that "the third lobe of the prostate, which is itself singularly enlarged and disfigured," * * * "formed a rounded prominent mass, which projected into the interior of the bladder, and *overhung the bar*."—*The Diseases, Injuries, and Malformations of the Urinary Bladder, &c.* By S. D. Gross, M.D. 2nd ed. Philadelphia, 1855, p. 236.

ment affects these, but not the median portion, or, at all events, not to a considerable extent. Examples of this kind are by no means common, for a tendency to the formation of outgrowth is usually manifested in the median portion, when any hypertrophy of the prostate exists; consequently, it cannot be regarded as a common, but, on the contrary, only as an unfrequent, affection. A few specimens exist in the Museum of the Royal College of Surgeons, in which such a barrier exists; those, for example, numbered 2488, 2489, and 2490—the first forming the upper figure in Plate X. of this work; the lower figure represents a very similar specimen from the Museum of University College.* But that condition in which a fold of mucous and sub-mucous tissues obstructs the posterior border of the neck of the bladder, no prostatic enlargement, stricture, or vesical disease being present, is still more rare.† Nevertheless, although I have occasionally seen an example of the kind here described, it has been so unfrequently, that I cannot do otherwise than regard it as extremely exceptional, notwithstanding that Mr. Guthrie has expressed himself to the contrary. I am persuaded, after careful and repeated examinations, that unnatural elevations at the neck of the bladder are more frequently caused by some enlargement or outgrowth, however small (and they may be seen in all degrees of development), springing from the posterior part of the prostate; and that such are not mere elevations of the mucous and sub-mucous structures only, but are commonly formed by genuine prostatic enlargements, united with the main body of the organ by proper ducts, not difficult to be traced.

In such cases the small prominence rising in the centre of the floor of the vesico-urethral orifice elevates a little crescentic

* No. 2489 is represented by Sir E. Home, in his work on the Prostate, Vol. I., Plate V. He describes it as "a transverse fold of the membrane of the bladder between the middle and lateral lobes," and speaks of it as "a part which merits particular attention, since it increases the obstruction to the passing of the urine, by preventing it from getting round the sides of the protuberance," p. 251.

† There is only one example of this in the College of Surgeons' Museum. It bears Mr. Guthrie's name, but is not numbered in the catalogue. It is one on which he set great value as demonstrating the existence of this affection. See op. cit. p. 23.

fold of membrane and underlying tissues on each side, and often, whatever magnitude the glandular growth subsequently attains, these folds still rise with it. It is through such a fold that the point of the catheter has been sometimes forced in performing perforation to relieve retention of urine, in which case, of course, the minimum of injury is inflicted on the parts; and could such a result be always anticipated from that operation, it would be the simplest and least-dangerous proceeding for relieving prostatic retention, beyond all question.*

But while such is the character of the commoner forms which obstructions at this spot assume, there is, undoubtedly, another, by no means unfrequently met with, and wholly distinct from it, which is probably that described by Mercier in the paragraph above quoted, as due to muscular spasm. In observing a specimen of the kind referred to, that which meets the eye is as follows:— the uvula vesicæ is unduly elevated and developed into a transverse ridge of varied size; there is a considerable decrease in the antero-posterior diameter of the floor of the bladder, or trigone, and the muscular bands which define that space are hypertrophied as is also the structure of the bladder generally. Further examination shows an absence of enlargement of the prostate, but usually either brings to light a stricture of the urethra, or it is ascertained that the patient was the subject of calculus or of some other cause of long-existing irritability in that viscus. In short, in a number of preparations of the urinary organs in which urethral obstruction has existed many years during life, the floor of the bladder presents the appearance of having been shortened or compressed from before backwards; the orifices of the ureters have approached very much nearer to the neck of the bladder than their normal situation, and the muscular emineuce, which unites them, is unduly developed; while the uvula, or the structures

* Preparation No. 2513, in the Museum of the Royal College of Surgeons shows well a case in which the perforation was done with precisely the result described. The patient lived five years afterwards. The Museum contains six other examples of this operation.

occupying its situation, appear also to have become more salient, and to project unnaturally, so as to form a marked prominence across the posterior part of the neck of the bladder, very different in character from a thin or merely membranous bar. On making a vertical section of the neck of the bladder and prostate, this elevation is seen to be due to an augmentation of the natural constituents composing the uvula, that is, of fibrous and muscular elements, while the prostate, as before said, is in no way increased in size. Now this condition of parts was examined by Sir Charles Bell, and formed the subject of a paper published in the Transactions of the Medical and Chirurgical Society, vol. iii. 1812, entitled "An Account of the Muscles of the Ureters, and their Effects in Irritable States of the Bladder."

He there demonstrated his views of the arrangement of those large bundles of muscular fibre which appear to be connected with the occluding of the orifices of the ureters, and with the opening of the neck of the bladder. These are greatly hypertrophied when increased action of the bladder has been long kept up by some source of irritation, and the result is that an approximation of the vesical neck to the orifices of the ureters takes place. But Sir Charles Bell stated that it was the "middle lobe of the prostate," to which this muscular apparatus was attached, and he suggested that the undue development of that lobe was, in all cases, mainly due to their mechanical action, drawing upon it, bringing it into its unnatural position, as it were, by repeated efforts; and by the irritation in it thus produced causing its hypertrophy. This theory gave too much prominence to the mere mechanical action described, which is obviously inadequate to give rise to the phenomena it was presumed to account for. At the same time, it made no advance towards the discovery of a primary cause for the prostatic enlargement, inasmuch as no explanation was offered of that which determined the muscles to act in this abnormal manner upon some prostates, and not upon others. But further, an examination of the preparations in our museums demonstrates that the development of the "muscles of the ureters" does not by any means necessarily co-exist with enlargement of the median portion of the prostate; but shows that it occurs when extraordinary contractions of the bladder have been long habitually exerted, *whatever* may have been the nature of the obstruction which has called them into play; and that the appearance of the prostatic enlargement itself, rising, as it often does, into the bladder in a polypoid form, with a pedicle at its base, does not indicate that it has been drawn into its position, or in any way produced by muscular action. Furthermore, the best examples of the hypertrophied muscular apparatus in question, and of the elevation at the neck of the bladder associated with it, are found in connection with long-standing stricture of the urethra, and not in cases of prostatic enlargement.

But the muscular bundles in question, it is now well understood, are not inserted, in the manner described, into any portion of the prostate; but are disposed in the following manner. Having united with the submucous layer of the bladder, they continue onwards to the vesical orifice, those of each side crossing each other there, and forming the "uvula;" they then become continuous with the longitudinal muscular coat of the urethra. This is clearly shown by Professor Ellis of University College, who has elaborately dissected and described the arrangement of muscular fibres throughout the genito-urinary apparatus.* A hypertrophied condition of these muscular bands, which associate the ureters with the urethra, and constitute the uvula, results from frequent and long-continued expulsive efforts of the bladder; they appear to be very closely identified with the action of the outlet, and their enlargement, thickening, and undue prominence, is the natural consequence of hyper-activity. At the same time it is probable, that repeated spasmodic actions of these muscular fibres, situated at the neck of the bladder, produce not only

* An Account of the Arrangement of the Muscular Substance in the Urinary and certain of the Generative Organs of the Human Body. By G. V. Ellis, Professor of Anatomy in University College, London.—*Medico-Chirurgical Transactions*, vol. xxxix. p. 327.

hypertrophy but inflammation; so that they lose (like the fibres of the iris in similar circumstances) the power of relaxing, and become permanently contracted, as Mercier has suggested. Such I believe to be the true pathology of an obstruction at the neck of the bladder, when it is not formed by development of the median portion of the prostate itself; nor by a simple fold of mucous and submucous tissue drawn up by enlarged lateral lobes; and when the median portion is not at all or only slightly hyper- -trophied.

It should be added that this appearance may be the only thing found sometimes after death, when no cause for urinary symptoms, which may have been severe during life, has been discovered. It is not necessary to suppose, as Mr. Guthrie suggested, that the bar had, even in such a case, been the source of those symptoms. Its existence simply proves that there has long been an undue amount of expulsive effort on the part of the bladder. It is itself but the result of that activity expressed in the form of hypertrophy; and the cause of the undue action which produced the bar has still to be sought. That there are cases of "irritable bladder," as it is termed, in which neither stricture, enlarged prostate, calculus, disease of the kidney, or of the rectum, or other satisfactorily-ascertained cause are present, is a matter of experience to most surgeons. In such, if the symptoms have been severe and long continued, we shall, undoubtedly, find, *as their result*, a state of the muscular coats of the bladder, which, in a degree more or less marked, presents the projecting bar at its neck, but we must search more deeply for the cause which produced the irritable bladder, and look upon the anatomical change as its consequence. Doubtless long-existing idiopathic chronic cystitis, which is not occasioned by any of the causes above named, or other such local sources of irritation, will likewise produce it, through the extreme irritability of the bladder so occasioned.

It is the presence of this bar in old cases of stricture, which, besides the enlarged lacunæ in the dilated urethra behind it, appears to be sometimes the cause of obstruction at the neck of the bladder, long after the instrument has passed the stricture; and which may occasionally cause some difficulty in drawing off the urine by the catheter. It appears also to explain the statement, often made, that a patient with strictured urethra has enlarged prostate also, a concurrence of circumstances less common than might otherwise be supposed. And, lastly, it accounts for some, although not for all, of those cases, in which difficulty of micturition, and other symptoms of obstruction persist in those who are the subjects of urethral stricture, even after the urethra has been satisfactorily and fully dilated. The bar remains at the neck of the bladder and is the cause of existing symptoms.

As this subject has been regarded in a manner which differs in some respects from that which has been followed by other writers, some examples of this form of bar will be referred to below, consisting in hypertrophy of the tissues at the neck of the bladder, arising from long-continued hyper-activity of the expulsive function, and generally as the result of stricture of the urethra.*

In order to conduce to the object of affording a clear exposition of the facts brought forward in relation to this subject, and of the views which have been founded upon them, I shall sum up with a brief statement in the form of conclusions.

It appears:

First.—That in the majority of cases in which there exists an organic obstruction, having more or less the form of a ridge

---

* ROYAL COLLEGE OF SURGEONS' MUSEUM.—Examples of thickening of the structures at the neck of the bladder, forming a bar, apparently due to hypertrophy of muscles controlling this orifice, are found in Preparations Nos. 2545, 2550, 2572, and 2567, the last-named being the most striking one.

UNIVERSITY COLLEGE MUSEUM. (No. 482.)—A marked example of bar; prostatic urethra dilated behind a stricture at the bulb. See Plate X., fig. 2.

In GUY'S HOSPITAL MUSEUM a good example may be seen (No. 2399) of valvular fold of mucous membrane only, forming a thin, sharp, and prominent bar at the neck of the bladder. The urethra passing through the prostate is dilated behind a stricture at the bulb. Nos. 2405 and 2406 exhibit bars at the neck of the bladder, co-existent with stricture.

In none of the foregoing is the bar prostatic in its nature.

or barrier, situated at the posterior border of the neck of the bladder, this unnatural elevation is constituted by an outgrowth arising from the median portion of the prostate.

Secondly.—That an organic obstruction may exist at the neck of the bladder, when there is no enlargement of the median portion of the prostate.

Thirdly.—That in this case it is most commonly due to hypertrophy of the muscular fibres at the neck of the bladder, produced by long-continued irritability of the viscus, and generally occasioned by stricture of the urethra, inflammation, or calculus of the bladder; and occasionally appearing without any of these conditions.

Fourthly.—That much less commonly it consists of a fold of mucous membrane and submucous tissue drawn upwards by enlarged lateral lobes of the prostate, the median portion being slightly, sometimes not at all, affected.

The two distinct conditions, described in the last two paragraphs, are those to which Mr. Guthrie applied the term "bar at the neck of the bladder," and which he employed in order to distinguish them from the well-known elevation there caused by enlarged "middle lobe," from which they are perfectly distinct, and with which it is obvious they cannot co-exist.

The foregoing conclusions show how rarely we have to encounter any affection meriting the appellation of Bar at the neck of the bladder, as distinguished from prostatic enlargement affecting that part, or from the condition described as muscular hypertrophy and generally resulting from stricture, calculus, or other similar cause of constant irritation.

DIAGNOSIS.—The subject of diagnosis of obstructions at the neck of the bladder has been fully considered in Chap. X., devoted to the subject. (See page 169 for the mode of recognizing the difference between prostatic tumour and non-prostatic bar at the neck of the bladder.) At the same time a knowledge of the age at which the symptoms appeared, will aid the inquiry as well as of the fact of absence or presence of prostatic swelling from rectal examination. We may be quite satisfied that prostatic obstruction does not exist before fifty-five or fifty-six years of age, and rarely at so early a period.

TREATMENT OF THE BAR AT THE NECK OF THE BLADDER.—It is obvious that when obstruction at the neck of the bladder consists only in a hypertrophied condition of the muscular apparatus described, produced by the irritation accompanying stone, stricture, or other source of continued, frequent, and painful micturition, no other treatment is necessary, or will be of any service, which does not remove the exciting cause. In most cases there is, indeed, no utility in regarding it separately from that cause; any more than exists for separately treating a hypertrophied or a dilated bladder, or an enlarged ureter, or most other consequences of urethral obstruction or vesical irritation. Remove the cause, and all these secondary evils will generally diminish or disappear. But the case is different when the cause being removed the symptoms are still present, or are but very slightly ameliorated. We must then seek the occasion of the continued distress either in chronic inflammation of the organs, in atony and inertia of the bladder, for the treatment of which see p. 181 ; or in still-existing obstruction, probably at the vesical neck. If on examination we find that a mechanical barrier does exist there, constituted either by the very common enlargement of the median portion of the prostate, or by the less-common one, a non-prostatic bar; and that this barrier, whatever its structure, is the cause of grave difficulties of micturition, the absence of calculus, stricture, vesical tumour, or other such source, being absolutely ascertained; it becomes a matter of consideration whether any relief, and if so what, can be afforded to the patient.

The late Mr. Guthrie, to whose experience, as the first to call attention to the subject, we naturally turn, says, that the treatment by simple dilatation and by permitting the catheter to remain permanently in the bladder, although often useful, does not always succeed; in which case, he adds, " the bar, or dam, at the neck of the bladder must be divided, and the question is, how is it to be done with the greatest safety?"*

The same conclusion has been arrived at by the French surgeons. Thus, the tying-in a large catheter for a few days in some cases, and in others attempts to depress the salient portion by mechanical means, as by sounds of large curve, &c., have been resorted to with some benefit in the most simple cases, but with none whatever in those which are more confirmed.†

After discussing the propriety of following the example of Sir Wm. Blizard, who, in a few instances, practised incisions resembling those made in lateral lithotomy for the purpose of affording relief in very confirmed and advanced enlargement of the prostate, Mr. Guthrie suggests that a simple bar may easily be divided, and mentions two instances in which he had so acted with advantage. In these he employed a modification of one of Mr. Stafford's stricture instruments, consisting of a prostatic catheter, containing a blade which was easily projected from the side of the instrument at its extremity, after this had been passed into the bladder, so as to make an incision in the act of withdrawing it; or two incisions might be made, one during the act of entering, and the second as just described. He also intimates his belief that in some cases much relief would be obtained by making an incision in the median line of the perineum into the membranous part of the urethra, and thence dividing the prostate, together with the bar or any other form of obstruction which might exist. This, Mr. Guthrie adds, he has not yet put in practice. The operations thus initiated seem to have fallen into disuse—or rather perhaps it would be more correct to say that they never came into general practice in this country; but similar proceedings were adopted in Paris about the same time; M. Leroy D'Etiolles having addressed the Academy of Sciences

* Op. cit. p. 274.
† Traité pratique. Civiale. 1858. T. 2ᵉ. p. 157.—Also Mercier and others.

(1832-3) for the purpose of advocating scarifications and incisions of obstructions situated at the neck of the bladder. The instrument which he subsequently employed, which was exhibited to the Academy in 1837, and was figured, I believe, for the first time in 1840, resembled very much that just described, excepting that the blade could be projected from either the convexity or the concavity of the curve, this latter being extremely small, like that of the exploring sound exhibited at page 167. Besides this, however, there was another blade, which was intended to act against the original one, like the blades of a pair of scissors, and to serve the purpose of excising a small outgrowth when necessary.*

M. Civiale states that he has frequently practised moderate incisions of the obstructing bar under certain circumstances. After describing his method he adds, in terms which will commend themselves to the minds of English surgeons:—" But I cannot too much insist that we should thus operate only in cases which are perfectly understood, after having recognized with precision the spot on which the instrument is to act, after having placed the patient in the most favourable condition, and after becoming absolutely certain that all other means of cure have been exhausted." †

M. Mercier, who has paid great attention to this particular subject, has employed the practice of incising freely, and also, in some cases, of excising portions from the obstructing bar. The annexed figure of an instrument for incision is copied from his work. ‡

Fig. 20.

* Compte-Rendus des Seances, vol. iv. p. 551.
† Traité pratique. 1858. T. 2e. p. 171.
‡ Recherches, &c. Par A. Mercier. Paris, 1856. p. 216. M. Mercier has stated, in an extended analysis of my works, which he has done me the honour

It consists of a silver canula, having the form of his exploring instrument, containing a blade, cutting from either the convex or concave aspect of the curve of the canula, as already described. When the blade is sheathed, the instrument is employed as a sound, and being introduced into the bladder, previously injected with a few ounces of water, the beak is turned downwards, as in sounding behind the prostate. By means of the screw at v' an adjustment is made which regulates the distance to which the blade is permitted to slide out of the beak and along the shaft; in the figure that distance is indicated by the letter L. By drawing the circular handle R towards himself, the operator, therefore, makes an incision of the bar against which the beak rests, dividing the tissues more or less perfectly between it and the point L. To ensure complete division, the blade is pushed back again into the beak, and this process may be repeated if necessary. The instrument may also be used in front of the prostate, or an incomplete incision made from within the bladder, in the manner just described, may be completed by placing the beak, directed upwards, in front of the bar, and making the knife project from the canula on its convex aspect, as seen in the figure, and indicated by the asterisk. In this case the screw v' is first turned in order to regulate the movement. In this manner Incision is performed; the method recommended by M. Mercier, where the bar is narrow, or thin and prominent.

If the bar is rounded and wide, as most prostatic enlargements in this situation are, he practises Excision by means of an in-

to publish (Étude sur Divers Points, &c. Paris, 1860 ; and Gazette Hebdom., 1860, several papers), that, in the first edition of this work I did him some injustice in my notice of his views and proceedings, in describing his instrument and operation as a modification of those of Leroy d'Etiolles. After a re-examination of the subject, I come to the conclusion that the difference is essential, and that although there is an obvious resemblance in the form of the instruments employed, the proceeding of M. Mercier is original and distinct as compared with the scarifications made by Leroy. I exceedingly regret that I should have erred in not attributing the utmost value to the labours of one for whose original and sagacious observations I entertain the highest respect.

strument somewhat similar to a lithotrite (fig. 21), by which, the beak being turned downwards, he seizes a portion between the blades of the instrument, fixes the yielding tissue by means of an arrow-headed needle, seen in the figure, which renders it immovable, then closes the blades, and in so doing excises the portion contained. This, being fixed by the needle aforesaid, is removed when the instrument is drawn out. Frequently some bleeding, generally a good deal, appears to follow this process; sometimes it has occurred to a dangerous extent. This is treated by small injections of cold water, by elevating the pelvis above the level of the head, and by internal astringents. After the operation, the patient is desired, in the surgeon's presence, to empty the bladder of the water injected previously. It is important, says the author, that he should do this while lying on his side, and without effort. If it does not flow, a full-sized catheter, with large eyes, should be passed. A catheter should not be left lying in the canal, as it provokes spasm, and repeated desires to pass water, and so favours bleeding. It is better to pass an instrument from time to time when the patient feels the desire to urinate. After some extended remarks upon the subject of hæmorrhage, M. Mercier says, " It is at present the only serious accident which I have observed to result. I repeat, that by the aid of the means indicated, I have never failed to overcome it, and my experience rests on a sufficiently extensive base, since my operations have reached the number of 300, some of my patients having been operated upon several times."* More recently he has stated that there is less to be feared on this score from Excision than from Incision, because the vessels are more crushed by the former process, while in the latter they are more cleanly cut. About six days after the operation he passes a flex-

Fig. 21.

* Op. cit. p. 240.

ible catheter with great care lest hæmorrhage be excited, and, according to circumstances, but generally about the 10th or 11th day, having introduced the same catheter, he passes into it an elastic steel stilet, in order to make pressure with it upon the seat of the wound by way of preventing union, increasing the force employed from the second to the third week following the operation. He speaks in very decided terms of the benefits which have followed the practice in numerous cases, citing the reports of 15 in illustration. The names and residences of these patients are appended; in most instances the operation was performed in presence of some one or more of the most distinguished surgeons of Paris; and in five cases the patients were examined by the Commission of the Argenteuil prize. The reports are therefore open to correction if any undue partiality for the proceeding has been manifested by the author. Of these, two died; one of dysentery, one of "fever." Both appear to have been in extremely bad health, and their antecedents would, in London, have been held by most surgeons to contra-indicate the performance of any operation not absolutely necessary. In five cases, in which the urinary difficulties, such as habitual retention and dysury had been considerable, complete relief was afforded, and had been verified as continuing in one case during four years, in another two years. Seven were improved in various degrees, some very considerably; these were reported by the author as cases of cure. In one, no alteration was produced by the operation, in the power of passing water, but he can pass a catheter, which previously he was unable to do. Such are the results, judging from a perusal of these fifteen histories,[*] and they may be taken as a fair example of the practice. M. Mercier

[*] M. Blandin, lecturing on a case of retention of urine from bar at the neck of the bladder, at the Hôtel Dieu, speaks highly of Mercier's method by incision, which he prefers to excision, as more certain and equally efficacious.—*Gaz. des Hôpitaux.* Jan. 23, 1849.

M. Demarquay, of the "Maison Municipale de Santé," has performed excision 10 or 12 times. He is reported to have lost one case by hæmorrhage. Two cases are given, in one of which the operation was followed by orchitis; both were ultimately successful.—*Gaz. des Hôpitaux.* March 14, 1861.

regards them as belonging to the most serious and obstinate class of cases, the milder ones being remediable, either by mechanical pressure, or by incision. He nevertheless will shortly publish another series, of which the proportion of deaths is still smaller, and he appeals to the advantages gained by his practice, as amply compensating for the risk.*

This is a question for very cautious consideration. So much can the condition of almost all patients suffering with obstruction be ameliorated by care and management, and by emptying the bladder once or twice a-day, when necessary, that the cases must be rare and exceptional in which the proposal to employ this treatment can be entertained, in this country.

At the same time, with care to avoid constitutions which are obviously bad, or the subjects of advanced renal disease, I can conceive that there are cases, as Mr. Guthrie observed, in which the operation might be advantageously practised, and I have been the more induced to enter upon its consideration on account of the recognition by that distinguished surgeon of the existing necessity for some mode of overcoming the obstructions in question. Such a proceeding should be employed by none but those who have been thoroughly familiarized by the use of instruments in the urethra and bladder, and then much caution and judgment must be exercised in the selection of suitable cases. For my own part I should desire to be fully assured that all means had been exhausted without success, and that the patient was in a condition which urgently required relief, before I consented to apply this or any other similar operation to the prostate. All experienced surgeons, however, must sometimes meet with such cases, and feel the necessity for some more decided measures than are commonly employed here. Hence we cannot but observe with interest the adoption and the results of any method of affording relief which emanates from practised hands, even although the proposals may at first sight appear hazardous, compared with the advantage to be obtained.

* Étude sur Divers Points, &c. Paris, 1860. p. 32.

CHAPTER XVIII.

PROSTATIC CONCRETIONS AND CALCULI.

PROSTATIC CONCRETIONS.—Distinct from Calculi.—Physical Characters to naked Eye and Microscope.—Chemical Characters.—Analysis.—Are they Natural or Abnormal Formations?—Mode of Formation, and History.—May become Nuclei for Calculous Matter.—Analogy between Prostatic and Biliary Concretions—and other Concretions.—Views of Dr. Jones and Mr. Quekett; Virchow, Wedl.—PROSTATIC CALCULI.—Different kinds, sizes, &c.—Analyses.—Situation. —Numerous Examples.—Operative Measures for Removal.

IT is exceedingly common, when a urethra has been laid open, to observe in its prostatic part, lying around the verumontanum, and in the orifices of the prostatic ducts especially, numerous little brownish or blackish bodies, the largest of which ordinarily met with are about the size of poppy-seeds. They do not lie free in the canal, but generally just within some of the orifices alluded to, sometimes beneath the epithelial layer of the mucous membrane of the urethra, barely visible over them. Again, when a section of the prostatic substance is made, especially if the knife be directed backwards and outwards on either side of the verumontanum, many of the same bodies may be frequently seen. Indeed they may be dispersed through all parts of the organ, although they rarely attain the size named in situations very remote from the urethra.

These bodies have received the name of Prostatic Concretions. They have in their origin no relation to urinary calculi; nor should they be confounded with the hard, white, porcelain-like masses which are sometimes met with in the prostate, and from which they differ also. It is the more necessary to draw the distinction, as the concretions have not unfrequently been spoken of as urinary calculi, and still oftener are viewed as having a similar

PHYSICAL CHARACTERS OF THE PROSTATIC CONCRETIONS. 313

constitution, and as not being generically distinct. They will therefore be considered here separately.

PROSTATIC CONCRETIONS.—The bodies described above under this term are not generally observed before middle or advanced age. But, although, as a rule, to which there are occasional exceptions, no examples visible to the naked eye can be detected during the periods of youth and early manhood, the microscope reveals them of very small size at all ages, except that before puberty. In the first series of fifty prostates examined by myself, they were present *in every instance*.

In the early stages of formation, they present very beautiful microscopic objects, varying in size, for the most part, from about the one-thousandth to the one-hundredth of an inch; generally oval in form, sometimes angular, and then apparently from the result of mutual pressure. They have a yellowish hue, varying between the faintest tint and a deep orange, evidently acquiring intensity of colour with age or increase in size. They have a well-defined outline, as if limited by some kind of cell-wall, and exhibit numerous concentric rings in their anterior, which, although the object is more or less translucent, suggests strongly to the observer the appearance of a uric acid calculus when cut (Plate XII.); a resemblance which no doubt originally suggested a relationship with formations of that class. The central part, or nucleus, as it has been termed, does not usually exhibit these rings, but has a cellular appearance, as if constituted by a conglomeration of corpuscles partially fused together (Plate XII. ƒ). This central mass varies in size, but is usually present; and around this it is that the concentric circles are placed. Often two, three, or more, of very small concretions are seen lying closely together, each possessing more or less angularity, and forming the nucleus of a larger formation (Plate XII. *e h*): occasionally lines are seen radiating from the nucleus to the circumference (*d*): sometimes cleavage of the mass takes place in the direction of these lines. Sometimes the concretion appears to consist almost entirely of the agglomerated corpuscles; at others these exist in very small relative proportion to the concentric rings, in which latter no trace of corpuscular arrangement can be detected. It is easy, when examining a considerable number, to trace a series of these small semi-transparent bodies, as it were in different stages, until they gradually become the dark, almost opaque, forms at first described. Even in the latter some remains of the concentric arrangement may be occasionally observed, but, generally speaking, they are too dark and opaque to transmit light, and only appear as dark-brown masses of spheroidal form, with enough of translucency to enable the eye to detect easily at their edges that they have really a deep-orange or red colour. In consistency they have also considerably changed. In the early stages these bodies are soft, and readily cleave or divide as a soft body does under superincumbent pressure, even when as large as the one-hundredth of an inch, by the application of a very little weight to the thin glass which covers the object under the microscope. But when they have arrived at the size and colour above described, they have acquired also considerable solidity, are firm, and even brittle when force is applied to break them.

Their chemical reactions are peculiar. The small, soft, pale, and yellow bodies are not acted on by acetic, nitric, hydrochloric, or sulphuric acids cold: nor by sulphuric ether, liquor-potassæ, or ammoniæ. The larger, hard, and brownish concretions I have found to be unaltered by the addition of alkalies, except that they occasionally become rather more translucent. By hydrochloric and nitric acids they are sometimes influenced, giving off a few bubbles of gas, and slightly diminishing in size. Sulphuric acid liberates gas more rapidly, and sometimes after they have ceased to be affected by the former acids. Occasionally they become soft, and disintegrate into amorphous matter, losing very little of their colour or bulk, by immersion in sulphuric acid. On the other hand, some specimens appear to be very little affected by any reagent.

In order to obtain an exact qualitative and quantitative analysis of these bodies, I submitted about 200 of the hard dark-coloured concretions to my friend, Dr. William S. Squire, who, after a careful investigation, has furnished me with the following report.

"Immersed in acetic acid the concretion resists its action, but if broken before this agent is applied, it becomes slightly softened and swollen. Nitric acid, cold, has no effect; when hot it dissolves the concretion entirely, producing a very slight yellow tint. A portion of the concretion treated with a solution of iodine, does not change colour. Sulphuric acid and sugar do not produce any effect characteristic of protein. When treated with a solution of potash, hot or cold, there is no change, and on adding an acid, no precipitate is formed in the alkaline liquid. When the concretion is heated, a strong ammoniacal odour is evolved.

"From these reactions, I conclude that the organic constituent of the concretion is not a true protein body, but most probably belongs to that class of nitrogenized substances, sometimes termed protein derivatives, of which fibroin, gelatin, and chitin, are examples.

"The following results were obtained by incineration:—
  Concretions ·0244 gramme.
  Yielded of Ash ·0112, which equal 49·5 per cent. of residue. This consisted chiefly of phosphate of lime, with a small quantity of the carbonate."

Is the presence of these concretions in the prostate to be regarded as a natural or as an abnormal circumstance? This question has been variously answered. I have examined not less than a hundred prostates, at all ages over 20 years, and have detected them in every one. I have found them, also, in one specimen at 14 years of age; and have failed to find them in childhood. There is much difference, however, as to their number in different cases. In some—in most after 50 years of age—they are obvious enough to the naked eye in the urethra around the verumontanum as soon as the canal is laid open; in others, it is necessary to make section of a lateral lobe, to scrape the surface, and place the milky, semi-transparent fluid under the microscope, when they may be seen, sometimes in small, sometimes in large, numbers. At other times it may be necessary to make several such sections before finding any. They are smaller, paler in colour, or even almost destitute of it, in young subjects; and *vice versâ* in aged : but there is not an unvarying relation in regard of number, size, &c., between their development and the age of the subject. In No. 15 of the table, from a man aged 66 years, I found more than in any I have ever examined, including subjects at 90. I estimated the number of dark-coloured concretions visible to the naked eye, in this case, as amounting to several thousands.

Generally speaking, they have been considered abnormal. One of the most recent writers on the subject, Wedl, regards them as the product of *enlarged* prostate.* From the facts just brought forward, I cannot, however, but conclude, that their existence is a necessary result of the performance of natural functions on the part of the prostate, although there does not appear to be any evidence to show that they ultimately disintegrate and yield up "some elements to the natural secretion of the gland," as has been suggested by an observer who has studied them closely.

My examination of the preparations referred to has led me to the following conclusions respecting these bodies. It may, however, first be stated, in what manner the inquiry has been prosecuted. The urethral surface of each prostate has been exposed, and some concretions, if present, removed for examination. These have been submitted to the chemical tests enumerated, of which the results have been already given. Next, several sections have been made with Valentin's knife, and examined *in situ*. The best of these I have mounted in preservative fluid, of which I possess not less than a hundred examples, and from some of these the drawings (Plates XII. and XIII.), most admirably and truthfully

* Rudiments of Pathological Histology. By Carl Wedl. Translated by G. Busk, F.R.S. London, 1855. p. 269.

executed by Mr. Tuffen West, illustrating this part of the work, have been taken. The fluid, also, which exudes from the prostatic ducts on pressure has been separately examined.

By such means, I have observed that in addition to the corpuscles, obviously epithelial or glandular, which are found abundantly in the prostatic fluid, however obtained, there are always present also, and in considerable numbers, some small yellowish bodies, in appearance sometimes granular, sometimes homogeneous, about the size of red blood corpuscles, but not so uniform, being from about $\frac{1}{5000}$ to $\frac{1}{3500}$ of an inch in diameter (Plate XIII. fig. 5); they possess considerable refractive power, nearly so much as to give them a resemblance to oil globules. They are not acted upon by ether, however, nor by the other reagents mentioned. They are not only found in the prostatic fluid, but may be washed from sections of the organ, and may be found lying in clusters, aggregated and adhering in various parts of it. Further, it is common to see small ducts and cœcal pouches stuffed with these bodies, as well as with yellow granules of smaller size, several marked examples of which are represented (Plate XII. *i i*, & Plate XIII. figs. 1, 2). In the larger masses, obviously occupying crypts, or follicles, of the glandular structure, these may be seen, not only cohering, but fused together, and then an appearance is presented identical with that which may be seen occupying the centre of the larger concretions. Judging from these appearances and the frequency of their occurrence, I cannot but conclude that the coalescence of these yellow bodies, or granules; their partial fusion into a mass, more or less homogeneous; the stratification, perhaps, in part, of this mass itself, or more probably the deposit upon its surface of fresh layers of fluid matter, similar to that which originally constituted the yellow bodies; and, finally, some additions of earthy matter to it, either by infiltration or accretion, are the steps by which the formation of a prostatic concretion is very frequently accomplished. Most, if not all, of the appearances thus described are very accurately shown in the drawings referred to. Those concretions which are not found lying at the orifices of the ducts, as at first described, occupy, generally, the larger ducts and follicles of the secreting portion of the organ. In examining sections under a power of 200 or 250 diameters, it is easy to see the circular arrangement of fibres around the concretion, when it has been preserved *in situ*, and in other places around the openings from which these bodies have escaped (Plate XIII. fig. 4).

Perhaps all these yellow bodies are originally composed, that is in their earliest stages of formation, of purely-organic matter, and that matter a product of the secreting structures of the prostate. At a very early period, however, they seem to be impregnated with the earthy constituent, in some form or proportion, which does not much impair the translucency of the object. Most appear to remain in this condition, although liable to considerable increase in size after their formation, which, probably, at first, depends on that of the cavity in which the aggregation occurs.

When circumstances give rise to an addition of *opaque* earthy matters to the concretion, their size increases not only by such addition, but by that also of the organic matters with which such deposits are associated. It is then that they acquire density, more or less opacity, and give proof of the presence in large, that is, predominating quantity, of mineral constituents to chemical reagents.

The acquisition of this opaque earthy mattter by a concretion may be thus explained. One of these bodies having formed within a follicle, and the size having greatly increased by fresh layers of secretion, it becomes, sooner or later, a kind of foreign body, and as such creates a certain degree of irritation. Now it is well known that secreting membranes, throughout every part of the body, are prone to deposit opaque earthy matter under certain forms or degrees of irritation, the product in all cases consisting chiefly of the phosphate, with a little of the carbonate of lime; and in this manner the secreting membrane of a prostatic follicle appears to produce a fluid from which earthy salts are precipitated upon the nucleus, until its size has increased from that of a microscopic object, under favourable circumstances, it may be, even to that of a grain of pearl barley, or of a pea, or even larger. Together with this earthy matter, mucus, gland cells, &c., in varying proportions, will be also intermixed. Thus at various stages of their formation the chemical analysis of a prostatic concretion will exhibit a different result, being found to contain a larger proportion of organic matter in the early period, when only visible to the eye as dark points, and more of inorganic, when as large as those of the extreme sizes just described. Analyses of the latter have frequently been made, and the inorganic matters have amounted to 85 per cent., instead of $49\frac{1}{2}$, as found in the smaller kinds examined here. Arrived at this stage, the walls of the follicle originally containing it, have become absorbed and have disappeared; other calculi from neighbouring crypts have come into contact with it by a similar process; and now, perchance, a number of these bodies will be found occupying together a single sac, in which the work of deposition ceases, and then they may lose their spheroidal form, and acquire facets by attrition or juxtaposition. There is, therefore, nothing more of a urinary character about these concretions than belongs to salivary, biliary, or other glandular concretions. But the case is somewhat different with some of the calculi which are found in diseased prostates, and which appear to be the products of bladder derangement as well as of prostatic secretion; these will be considered separately hereafter. Meantime, it will be desirable very briefly to illustrate the history here given of the structure of the prostatic concretion, and of the steps by which it becomes approximated in constitution to a true calculous body, by an allusion to analogous processes which occur in mucous membranes of other parts of the system. As stated above, wherever there is a mucous membrane, or a simple follicular membrane exercising a secreting function, there exists the requisite agency for the production of earthy matter when certain kinds of irritation are applied to it. Thus is formed the phosphatic calculus of the urinary bladder, resulting from the irritation of chronic cystitis; so different in origin from the uric acid stone which descends from the kidney; and thus also the coating of phosphates, which the latter so frequently acquires after its arrival. So we find the salivary glands containing phosphatic calculus, a little spicula, as from the husk of some grain taken as food, having originally supplied both the irritation and the nucleus. The tonsil is the seat, although rarely of a similar occurrence. Still more rarely do the nasal passages, the pharynx and œsophagus, furnish like forms, and doubtless through the agency of similar exciting causes. Throughout the intestinal canal, phosphatic formations have frequently been observed; in the north especially, the hard husk of the oat, which forms a staple article of diet, having there usually furnished the nucleus. With biliary calculi all are familiar. The histological elements of the liver secretion, inspissated by removal of water, enter largely into their composition, at first. Ultimately phosphatic salts, generally as in other concretions, the phosphate of lime, are added, and form a constituent part of the biliary calculus. In many respects there appears to be a considerable analogy between this product and the prostatic concretion, not only in manner of formation, as originating from an organic basis, and in the ultimate relative proportion of the mineral constituent, but in their frequency of occurrence. Besides these, I shall but allude to the similarity which obtains in the character and mode of formation of calculi in the lacrymal gland, in the frontal sinuses, in the mouth, in the pancreas, in the vesiculæ seminales, in the mammary ducts, and elsewhere.*

The views entertained here differ, it should be said, in some respects from those which have been held by some who have at a former period paid much attention to the subject. Dr. Handfield Jones published, in 1847, a very interesting paper describing the results of his researches,† which many of the observations

* For an interesting and very complete account of these formations, see an article in the Cyclopædia of Anatomy and Physiology, under the title of "Adventitious Products," by Dr. W. H. Walshe, Professor of Medicine in University College.

† The Medical Gazette, Aug. 20, 1847, and Report of Pathological Society, 1846-7, p. 129.

here recorded have but fully confirmed. He, however, then believed the concretions to originate in a simple vesicle, containing granular matter, and generally a nucleus; and that they increase mainly by endogenous growth; regarding their origin and constitution to be entirely organic in the earlier stages of existence. Mr. Quekett, on the other hand, describes them as "commencing by a deposit of earthy matter in the secreting cells of the gland, increasing either by aggregation or by deposition in the form of concentric layers."*

With the first-named observer I am disposed to agree as to the organic composition of the concretions at first, a fact which must be regarded as proved by their behaviour with chemical reagents. As to the vesicular origin described, favoured as it seems to be by the appearance of some of the smaller formations, I feel, nevertheless, some difficulty in admitting the probability of such a mode of production, especially as we know of no process in regard of any other cell-formation, normal or abnormal, seen in the human body, which can be said to be at all analogous with it. As to the larger formations, I have little doubt, judging from the appearances already described, that they are often, if not always, produced by the aggregation in follicles, ducts, and in interstices of adjacent tissue, of the small yellow bodies which have been referred to.

What the "yellow bodies" are, I am not able satisfactorily to explain. Whether they are themselves altered secreting cells, from the gland follicles, or whether only the product of secretion by these structures, does not at present clearly appear. Thus much I have observed, that the small fusiform and spheroidal epithelium lining the prostatic ducts is often loaded with yellowish granular matter, appearing in all respects to be identical with that seen in a free state elsewhere, which seems to favour the view that it is a natural product of gland secretion (Plate XIII. figs. 3, 4).

* The Anatomy and Diseases of the Prostate Gland. By John Adams. 2nd ed. 1853. p. 153.

It may be added that I have constantly found similar yellow granules in the fluid of the vesiculæ seminales also, in which, at all events in some elderly persons, they appear to be more constant than are spermatozoa (Plate XIII. fig. 6). In this situation, bodies which are evidently the same, have been described in a recent article on the vesiculæ seminales, by Mr. Pittard, appearing in the "Cyclopædia of Anatomy and Physiology," as "very numerous insoluble globules, which have a great tendency to coalesce, and appear very much like oil; their refractive power is, however, less, and there are reasons for doubting that they are really globules of oil." Besides these, the same observer finds "suspended little conglomerated masses of transparent solid, just visible to the naked eye," which have, under the microscope, the appearance of "a nodulated mulberry-like surface, as if composed of smaller balls;" these, he thinks, are made up by a coalescing of the minute globules before mentioned.*

Virchow believes the concretions to be derived from a peculiar insoluble protein-substance mixed with the semen. Wedl, before referred to, regards them as identical with certain bodies found abundantly in the enlarged thyroid gland, and with those met with in the brain and spinal cord, especially in elderly subjects, and named "amyloid bodies" by Kölliker and Virchow. He believes all these to be the result of a pathological exudation frequently occurring in various organs in advanced age, and which he terms "colloid matter," on account of its resemblance in physical characters to liquid glue, and he proposes to call the concretions in question "concentric colloid corpuscles." He regards them as "principally composed of an organic substance, and consequently the names of 'stones' and 'concretions' to be inappropriate;" although he has occasionally observed that the colloid matter is deposited upon some of the rounded or nodular forms of calcareous salts, which are found scattered in the parenchyma of

* Cyclopædia of Anatomy and Physiology, p. 1433.

the prostate, and which thus form the nuclei of some few of the concretions.* A further study of this subject, since the above appeared in the first edition, confirms me in my opinion that evidence is wanting at present to show that these bodies are examples of amyloid degeneration; or indeed that they are the products of any morbid change at all.

Reviewing, however, all that has been ascertained of the mode of formation and constitution of these bodies, I see no valid objection to the use of the term concretions, at all events for the small formations which have hitherto been described. Nor, perhaps, at present, is it easy to find a better, since it is one which involves no theory, except the simple one that the mode of aggregation of their component elements is mechanical, rather than organic in its essential character. When they have arrived at a size sufficient to occasion, as sources of irritation, the deposit of dense, opaque, earthy matter in the manner above referred to, they may be regarded as belonging to the category of calculus, rather than of concretion.* The inorganic component now becomes predominant, the body increases in size; and although there is no exact period at which it can be said to cease to be a concretion and to become a calculus, yet there can be no hesitation as to which of the two terms should be applied to most examples met with of either kind.

PROSTATIC CALCULI.—Prostatic calculi exist in very various sizes and forms. The smaller examples, which are most frequently met with, are rounded or ovoid; the larger are irregular, often elongated, sometimes branched, and commonly consist of several fragments uniting to form a mass. These fragments fit almost accurately, one to the other at their adjacent surfaces, but, nevertheless, appear to be separate and distinct calculi which have become adapted in form one to another, by close proximity. The small isolated formations are about the size of grains of pearl barley, rarely as large as peas; and these form the purest speci-

* Rudiments of Pathological Histology. By C. Wedl. Translated for the Sydenham Society by G. Busk, F.R.S. 1855. pp. 38 and 271.

mens of prostatic calculus. The masses formed by coalescence are of all sizes, but have been seen reaching the length of four or five inches in very rare instances. In the latter case they extend into and along the urethra, and even into the bladder. Still in these circumstances chemical analysis shows them to be mainly composed of phosphate of lime, and to have but a small admixture of the ordinary vesical or urinary product, the triple phosphate of ammonia and magnesia. They are in consistence hard, and so close in texture as to bear some resemblance to porcelain. They are white, fawn, or pale brown in colour, the surface being usually of a darker tint than the interior.

Analyses of several of the small round variety have been made, and the composition has pretty generally corresponded with the result of Dr. Wollaston's examination, who first pointed out their chemical character, and showed that they were not urinary products. Dr. Wollaston described them as composed of the neutral phosphate of lime, tinged with the secretion of the prostate.* Among modern observations, that of Lassaigne, which is generally quoted, may be adduced as follows:—

Phosphate of Lime	84·5
Carbonate of Lime	0·5
Animal matters, &c.	15·0
	100·0

We have already seen, however, that the proportions of these constituents may vary considerably. It appears that as we approach the earliest stages of their formation, the mineral constituent is found in diminished relative proportion to the animal. Thus the small concretions described at pages 313-15, were composed of about equal parts of organic and inorganic matters.

These small prostatic calculi are often found lying, each in a separate space for itself, or hollow in the substance of the organ,

* Phil. Trans., 1797, p. 397.

corresponding with the size of the calculus itself.* At other times several occupy a larger space or cavity, in which they are movable; and thus their spheroidal form gives place to a more or less angular one, from their mutual pressure or attrition. In this state they may sometimes be felt by the finger introduced into the rectum, and the grating, from their movements one upon another when pressure is made, plainly perceived. At the same time a similar sensation is communicated to the hand by the catheter passed along the urethra, when the instrument traverses the prostatic part. I recently exhibited at the Pathological Society, some good examples, which fell under my own observation. They were taken from a patient who died at the age of 89 years. The prostate was enlarged, although not greatly so, but occupying a cavity in each lateral lobe, and also immediately beneath the verumontanum, were numerous dark-coloured calculi, very hard in texture with polished surfaces; each calculus having several irregular facets, varying from the size of a grain of pearl barley to that of a large pea. These characters distinguished them completely from calculi of renal or of vesical origin.†

The larger masses formed by coalescence, while generally consisting mainly of the phosphate of lime, have usually enough of the triple phosphate in their composition to relate them more or less closely to the class called fusible calculus; this term being understood to embrace many varieties in regard of the relative proportions of the two phosphatic salts. These formations often occupy large spaces in the prostatic substance and among the adjacent tissues; irregular cavities which enlarge with the increasing bulk of the calculous formation. It is worthy of observation that they are most frequently met with in young men.

* Good examples of these small calculi, embedded in the substance of the prostate, may be seen as under:—
　　　Royal College of Surgeons, Nos. 2519 and 2520.
　　　University College Museum, Nos. 1640 and 3844.
Some examples of encysted calculi of the prostate are well represented in Mr. Crosse's Treatise on Calculus. London, 1835. Pl. xi.
† Transactions of Path. Soc. Vol. xii. 1861.

One of the most complete descriptions on record of a case in which an extremely large calculus of this kind existed, is that by Dr. T. Herbert Barker of Bedford. This gentleman successfully removed a mass formed by 29 portions, weighing three ounces four drachms and a grain, from a patient 26 years of age. He describes them as of "a whitish colour, and porcelainous lustre and hardness; indeed the latter character is so well marked that it is with some difficulty that any impression can be made upon them with a knife."* Dr. Golding Bird found it to "consist of phosphate of lime (like salivary and bronchial calculi), with a rather larger proportion than usual of the ammoniaco-magnesian phosphates." When the stone was restored by the adjustment of the fragments it measured $4\frac{5}{8}$ inches in length.

A case very closely resembling this is recorded by the late Mr. Benjamin Gooch of Norwich. The calculus consisted of 16 fragments, which when applied to each other formed a mass nearly six inches in length. They are described as being "like alabaster in colour, and of as fine or rather a finer polish." A drawing is appended representing the natural size.†

Numerous cases have been detailed varying but little from the foregoing;‡ except in the much-smaller size of the calculus met

* Trans. Prov. Med. and Surg. Ass., 1847. Illustrated by an excellent drawing of the stone.
† Cases and Practical Remarks on Surgery. Norwich, 1777. Vol. ii. p. 174.
‡ Cases of calculi formed in the posterior part of the urethra (not merely lodged there), all being probably of prostatic origin:—

Mr. Jos. Warner, of Guy's Hospital, removed, from a man aged 20, two hard and polished calculi, weighing 350 grains, from the perineum, where they could be felt by the finger before the operation.—*Phil. Trans.*, vol. li. p. 304, with plate.

A second case, in which the calculi were very much larger, is reported by the same surgeon: patient aged 22.—*Phil. Trans.*, vol. lii. p. 258, with plate.

Dr. Livingston, of Aberdeen, two cases.—*Edinburgh Essays and Observations*, vol. iii. p. 546. 1771.

Dr. Cheston, of Gloucester, one case.—*Medical Records and Researches*, p. 163. 1798.

Mr. Wickham, of Winchester, a post-mortem case.—*Medical Facts*, vol. viii. p. 126. 1800.

Dr. Marcet relates a case in which 100 calculi were found.—Drawn, Pl. ix. *Essay on Calculous Disorders*. 2nd ed. London, 1819.

with. Many surgeons of the last century refer to them, and to operative measures for their removal.*

The operative proceeding by which large prostatic calculi have been removed is usually an incision in the perineum carried into the urethra upon a grooved staff, in the manner and situation of lateral lithotomy. Occasionally, the opening has been made in the median line, *i. e.* in the raphe of the perineum. Undoubtedly this situation is the best and safest for the incision, inasmuch as the median opening gives a more complete command of the position occupied by the stone; and is also a nearer and less-hazardous route to the neck of the bladder under these circumstances. The operation is far less dangerous than that of ordinary lithotomy, as the bladder remains untouched, supposing there is no vesical calculus also, a point which must be carefully investi-

Sir Astley Cooper relates three cases.—*Surgical Lectures*, 1825. Vol. ii. pp. 295, 296.

Sir B. Brodie, a case in which the calculi were lodged in a sac, from which he removed some by urethral forceps.—*Urinary Organs*, 4th ed. p. 362.

Mr. Crosse, of Norwich, several cases.—*Treatise on Calculus*. London 1835. p. 26, *et seq.*

Mr. Liston removed one, of characteristic appearance, in several fragments by a scoop, through the urethra.—Drawn, *Lancet*, Oct. 28, 1843.

Mr. Fergusson of London, a case of thirty fragments, forming a mass as large as a walnut.—*Lancet*, 1848, vol. i. p. 91. Another large specimen.— *Lancet*, 1849, vol. ii. p. 552.

Mr. Erichsen, of London, a case of prostatic associated with vesical calculus in a youth.—*Lancet*, 1850, vol. ii. p. 575.

Dr. B. Jones, a post-mortem case, 10 fragments.—*Described and engraved in Path. Trans.* vol. vi. p. 254.

Cases on record in which calculi of this kind have escaped externally through abscess in the perineum.

Dupuytren, after dilating with the knife some perineal fistulæ, removed 12 calculi with articulating surfaces, from, as he believed, the prostate. Thénard analyzed them, and found 86 per cent. of phosphate of lime, 13 of animal matter, and traces of carbonate of lime.—*Journal Univ. des Sciences Méd.*, Aug. 1820.

Lenoir and Nelaton, a case each; made up of several fragments; removed by simple pressure, by lithotrity and by cutting.—*Gaz. des Hôpitaux*, 1846.

Good examples are preserved in the Museum of the College of Surgeons: the best are those numbered H. 13, 15, and 23. The first, which is the largest, weighs one ounce and ninety-five grains.

* Dionis. Oper. de Chir. par La Faye, p. 221. Deschamps, sur la Taille, tom. iv. p. 161, *et seq.*, 1796. Sabatier, Méd. Opératoire, tom. iii. p. 136. 1810.

gated beforehand. Especial care must be taken at the time of operation to remove all the fragments which are lodged in the prostate so as not to leave nuclei for fresh deposit.

Sir B. Brodie relates a case in which he removed small prostatic calculi with the long urethral forceps; but some of these escaped also into the bladder, and had subsequently to be removed from that situation.

The existence of these bodies when small, and embedded in the prostate, is not revealed by symptoms during life. When by their increased size irritation is set up, abscess may be formed, or obstruction to the flow of urine be occasioned; the latter may take place also from the escape of small calculi into the urethra. The treatment of these consists in their removal if possible by means of the forceps, or long cuvette, in the same manner as advised for the removal of fragments after the operation of lithotrity, in the succeeding chapter. But when marked symptoms are present, and a foreign body can be ascertained by the sound, or catheter, to be embedded anterior to the neck of the bladder, or when it can be recognized from the rectum or perineum, an incision from the latter spot will offer a simple and efficient means of removing it.

CHAPTER XIX.

ON THE RELATION BETWEEN HYPERTROPHIED PROSTATE AND STONE IN THE BLADDER.

Vesical Calculus a frequent result of Enlarged Prostate.—How this may be accounted for.—Calculus often overlooked.—Best means of discovering it, by Sounding, &c.—Difficulties in removing it.—Lithotomy and Lithotrity.—Objections to each considered.—General applicability of Lithotrity.—Experience of various Surgeons.—Removal of Fragments by Scoop-lithotrite; by evacuating Catheter; Sir P. Crampton's Apparatus.—Position of the Patient.—Injection of Solvents.—Dr. Hoskin's proposal.—Decomponents.—Impaction of Fragments.—Treatment.—Course to be pursued when Bladder is extremely irritable.—Value of Treatment.—Preparatory Measures.—Palliatives.

So closely connected is enlargement of the prostate, through the resulting chronic cystitis, with the formation of earthy deposits in the bladder, that it is almost impossible to avoid a consideration in these pages of that great subject, at all events under one of its several aspects, and that certainly neither the least difficult, nor the least important one. It would be almost as easy or as consistent to decline treating of the best methods of affording relief to complete urinary retention arising from the same enlargement, as to overlook a result, no less important although somewhat less frequent. For it is impossible that any man can have the care of many cases of enlarged prostate, without meeting also several of calculous formation in the bladder, although it is not less true that he may occasionally, perhaps more frequently than has been supposed, overlook its existence, so much are the symptoms of the one malady masked by those of the other. Of this I have witnessed not a few instances. And it is not a matter to which the surgeon may be indifferent, even should the distress produced by the calculus in any given case of the prostatic affection be comparatively slight in degree. Because, as we have seen in a previous chapter, any source of vesical irritation—and there are few more potent than the presence of such a foreign body—tends to augment the difficulties attendant upon the already-enlarged organ, it may be to increase its rate of development, and to hasten the catastrophe which all our treatment is, or should be, directed to avert. The importance of the calculous complication is therefore to be estimated, not altogether according to the marked character of the symptoms by which its presence is rendered obvious, nor by the actual degree of suffering which it causes to the patient himself.

That calculous disease and prostatic enlargement frequently coexist is a fact which our museums testify in unmistakable language, and which experience corroborates.

Not only does our observation of the living point to the same fact, but the grounds of the relation are so obvious as to render it almost impossible that the result should be otherwise than it is. The calculus which is met with under these circumstances is generally, although not invariably, a vesical product, that is, one owing its existence mainly to the bladder itself. From the altered condition of urine depending upon constant or long-continued retention within the viscus, of a certain quantity which cannot be expelled by the efforts of the patient, owing to the existence of obstruction at its neck, irritation of the mucous lining is set up, and much viscid secretion is often poured out. This action having long continued, it will frequently be observed that some whitish soft or gritty matter, a phosphatic deposit, is evolved from the same source; at first, perhaps, only in inconsiderable quantity. This may pass off entirely with the mucus, in which streaky portions are seen to be entangled: and no more than this may occur. The formation may take place in small quantities, and may possess no very great cohesive power, in which case the bladder can be maintained tolerably clear by occasionally injecting it with warm water, either unmixed, or to which a minute quantity of mineral acid has been added. On the other hand, the calculous deposit may assume a more solid consistence, a nucleus may be formed, and aggregation taking place, a phosphatic stone may not slowly result. This condition is very much favoured by the state of the urine itself under circumstances of retention, as instead of being acid, and so affording a menstruum favourable to the solution of a phosphatic formation, it becomes alkaline, and not only aids in giving rise to irritation of the mucous membrane, but also in maintaining the existence of the earthy formation when produced. In the same manner, also, if a solid body be introduced into the bladder while the urine remains in this unhealthy condition, it is almost certain rapidly to acquire a coating of this same deposit. And so it happens that if a small renal calculus which consists of uric acid or urates, or of oxalic acid, descend at this time through the ureter, a large phosphatic stone will probably at no very long time be formed upon it as a nucleus. But this descent of the renal product is no mere contingency under the circumstances, no mere unlikely coincidence with the vesical state; there is very little reason to doubt that, in some cases, the formation results from abnormal action of the kidney set up by irritation propagated upwards from the bladder. Perhaps the vesical origin of renal calculus is not always sufficiently recognized; and it may be a question whether we are not rather too prone to attribute its existence to a calculous tendency in the system,—to a uric acid, an oxalic, or a phosphatic diathesis. Far be it from me to ignore the constitutional tendencies which undoubtedly give rise to calculous formation in the human constitution; I only believe that the phenomena presented by calculous patients must lead us in some cases, but especially among those who are the subjects of enlarged prostate, to regard their complaints as of local rather than of constitutional origin.

The relation which prostatic enlargement bears to these formations may be explained more fully. Two circumstances commonly concur to play a chief part in the production of calculous matter. These act and react on each other, and intensify the state which favours such production.

First,—There is the altered condition of the urine itself, resulting from its retention within the bladder by obstruction at the neck. This change consists in its alkalinity, and in its consequent tendency to deposit the earthy phosphates in the form of an insoluble precipitate. The alkalinity may be attributed primarily to the following source; viz., to the production of carbonate of ammonia from the decomposition of urea, favoured by the presence of some organic matter (probably mucus); a process which takes place in the urine of a healthy person, if permitted to stand in the air for a day or two after its removal from the body. It now deposits the phosphate of lime and magnesia, which in small proportion are normal constituents of healthy urine, but which require its normally-acid condition in order to remain in their natural state of solution. As soon as it becomes alkaline, these tend to precipitate, and doing so in presence of the carbonate of ammonia just referred to, there results the formation of a triple phosphate of ammonia and magnesia, with some phosphate of lime, and a very small quantity of carbonate of lime. Such are the constituents of the deposits so frequently met with in these circumstances, and the same are found entering largely into the composition of a very considerable proportion of the calculi formed at all ages, but particularly of those which occur at advanced periods of life.

Secondly,—There is the unhealthy state of the mucous membrane lining the bladder, which results from the altered condition of the urine, and augments the tendency both to alkalinity and to deposit.

The highly-irritating salt carbonate of ammonia, being habitually produced in the manner described, unnatural vascular excitement in the mucous membrane is set up, and an unusual quantity of its secretion is poured out which is naturally alkaline; this added to the urine, even when the latter is in its normally-acid condition, is sufficient to render it alkaline. But when the urine is already decomposed from retention, the action of the irritated mucous membrane considerably intensifies the morbid quality. But again, this mucus, or muco-pus, which is so familiarly known by the tenacious and adhesive character which it presents when removed from the body and cooled, contains itself also earthy phosphates, chiefly the phosphate of lime, with a trace of the carbonate, and often to a large amount; these being of course, insoluble in the alkaline secretion, are also precipitated in addition to those derived from the urinary secretion proper. Thus the inorganic constituents of a phosphatic calculus are abundantly supplied, and in circumstances particularly favourable to its formation, viz., in a surrounding menstruum in which solution of the earthy precipitate cannot be effected; contained in a cavity from which, both on account of its form and impaired vital powers, the contents are with difficulty expelled, and in which, consequently, aggregation and concretion are promoted; while, lastly, all this takes place in presence of an adhesive organic material, well adapted to form a binding cement for the saline particles of calculous matter. As might be expected under such circumstances, the resulting formation most commonly met with is the fusible calculus, composed of the phosphate of ammonia and magnesia, intermixed with the phosphate of lime in greater or less abundance; the proportions doubtless depending on the relative preponderating influence of either of the two sources of the deposit pointed out.

The calculous contents of the bladder resulting from, and frequently associated with, enlarged prostate, assume two forms as regards their physical condition, viz., spheroidal or ovoid masses of moderately-firm consistence, and semi-solid matter, well compared to mortar, to which it bears a great resemblance both in appearance and texture. Of these, the former are more commonly met with than the latter.

Whenever the symptoms accompanying enlarged prostate, especially pain and involuntary straining, are in any case more severe than those which are ordinarily encountered; whenever the occasional appearance of blood in the urine has been noted, unassociated with the use of the catheter, and occurring especially after moderate exercise; inquiry should be made for calculus.

We have already seen how much the ordinary signs of its presence are frequently masked by the enlargement, partly from the fact that the foreign body is less liable to come in contact with the neck of the bladder, and partly because the viscus itself is often unable to contract altogether upon its contents, and so the pain at the end of the act of micturition is but slight, or may be absent altogether. Nevertheless, it is not the less important to verify the fact of its existence or the contrary, and this occasionally requires a mode of search more rigorous in some respects than that adopted in ordinary cases, and one which is specially adapted to the circumstances resulting from this form of complaint. The ordinary catheter will probably fail to encounter the stone, and thus it is that its presence is often never suspected, although it may have existed for years; the daily use of such an instrument having been deemed incompatible with its non-discovery. Generally, the foreign body lies in a depression behind the enlarged prostate, below the level of the urethro-vesical orifice, and hence will rarely be detected except by the use of the short-curved or angular sound or catheter (See fig. 22, on the next page), the beak of which, if sufficiently short, that is, less than one inch in length, can, after introduction into the bladder, be turned downwards with perfect ease behind the base of the prostate into this depression. In some cases even this movement fails to reach the stone, and other means must be adopted. Thus the finger of the left hand being introduced into the rectum, sometimes suffices to elevate the base of the bladder or to displace the stone, and permit contact between it and the sound to be made. A more certain mode perhaps than any other is a change in the position of the body of the patient. The pelvis may be raised considerably above the level of the shoulders, on a large, hard, and well-stuffed cushion, so that the upper part of the bladder becomes the depending part, into which the stone may, especially by a sudden slight movement, be made to roll. Sometimes this has been achieved by suddenly lowering the head and shoulders of the patient, a position provided for by Baron Heurteloup in his rectangular bed; and much adopted by the late Mr. Aston Key, who simplified that apparatus for his own use, and with a special view to the discovery of a stone which might be concealed by enlarged prostate.* Or, as Sir B. Brodie remarks, " the same purpose will be answered sufficiently well, if the patient be placed on a light sofa, the end of which may be raised by an assistant."†

The sound to be employed should not only possess the form which has been indicated, but should be hollow, so as to admit of the passage of a stream of water through it either inwards or outwards as may be required. Such an instrument saves the preliminary passing of a catheter to determine the quantity of urine, or the capacity of the bladder, and serves for the purpose of injecting fresh fluid; and the urethra is thus traversed once only, instead of two or three times, while all these objects are accomplished. The sound should be made of steel, that it may possess weight and solidity; characters in which Heurteloup's catheter, made of silver in the same form, and employed for the same purposes, was deficient. It should also be plated, that the channel, which is necessarily smaller than that of a middle-sized catheter, may not be blocked up by rust, an accident which is otherwise very apt to happen. There should be also a plug or stopcock, at the handle, the opening in the latter being adapted to fit the nozzle of the injecting instrument. The annexed woodcut exhibits an instrument which possesses these advantages.

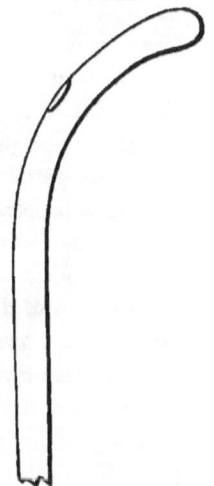

Fig. 22.

The sound having been introduced into the bladder, when it is neither distended

* Mr. Key's Chair is engraved and described in the Guy's Hospital Reports, Vol. iv. pp. 45-56.
† Med.-Chir. Trans., Vol. xxxviii. p. 174.

nor empty, but holding perhaps about four to eight ounces of urine, the search may be conducted by passing the beak laterally from the prostate to the upper part of the bladder, first on one side, then on the other. The instrument having been directed through the upper fundus and back part is now conducted to the depression behind the prostate. Withdrawing it until the beak arrives as near to the neck of the bladder as the tumour permits, the handle is slowly depressed between the patient's thighs, so that it can be fairly rotated, and its beak turned downwards towards the rectum. It is then moved gently to the right and left, backwards and forwards, during which it will probably be felt to glide lightly over some projecting muscular columns, if the bladder be fasciculated. Before this last-named movement is executed, it may be sometimes necessary to inject two or three aditional ounces of water, provided it is easily retained, in order to distend the bladder a little more, and so permit perfect freedom of motion to the beak of the sound. At this point, if nothing is found, the patient's pelvis may be elevated in order to dislodge the suspected calculus from behind the prostate, and move it into the body, or towards the upper part of the bladder. Nothing being still discovered, the patient may be placed in the upright posture, or he may be partially seated, while the water is permitted to flow through the sound ; during which it sometimes happens that the stone is brought down against the sound, or that the relations of parts are in some way altered, so as to permit a stone to be felt which had hitherto escaped detection. In cases of doubt, a lithotrite of moderate size offers some advantages as a sound. The ability to open the blades a little, presents a fresh mode of searching, and enables the operator to sweep a larger field in the distended bladder, adding to our chances especially when seeking to detect small stones or fragments. Civiale often employs it thus, and in order to avail himself of the advantages possessed by the hollow sound, he has designed a lithotrite, of which the male blade contains a channel, so as to permit water either to be injected through it, or to escape, during its employment in the bladder, whether for sounding or for crushing the stone.

The presence of calculus having been verified, the mode of removing it comes next under consideration.

Enlargement of the prostate is doubtless a source of difficulty in the performance both of lithotomy and lithotrity.

In lithotomy it prevents the finger from reaching the bladder, and determining the situation of the stone, renders the application of the forceps much more uncertain, and the seizing and extraction of the stone much more difficult. The distance from the surface of the perineum to the vesical cavity is increased in proportion to the degree of enlargement. The prostatic urethra, as we have seen, may be lengthened from one and a half inch to three inches, or even a little more, and beyond this there may be a tumour projecting, over which any instrument must be carried before the calculus, which lies deeply behind it, can be reached. In any case of notable enlargement, yet far short of an extreme size, the finger can barely reach the bladder, it certainly cannot touch the stone, much less verify or influence its position. Again, the depth of the wound limits the motions of the forceps considerably; it is not merely necessary to seize the object without the assistance of the finger, but the range of movements possible to the instrument is circumscribed—indeed, in some cases it would scarcely be practicable to obtain contact with the calculus unless the blades were specially curved to enable the operator to search behind the prostate for it. On the same ground, the extraction is considerably more difficult; an increased length of passage requires a proportionate augmentation of diameter, in other words a more extended use of the knife, to permit a facility for extraction equal to that enjoyed in ordinary cases.

In lithotrity, the enlarged prostate affords an obstacle more or less considerable to the easy introduction of the instrument; causes a little more difficulty sometimes in seizing the stone; and opposes a bar to the free discharge of the fragments.

The first two sources of difficulty are, however, comparatively of small consequence; the last is by far the most serious.

A careful and judicious manipulation of the lithotrite will not fail in overcoming the difficulties met with in traversing the urethra, even when considerably deviating from the normal direction and length. Directions for its management in these cases have been given in chapter the twelfth in connection with the subject of Catheterism in enlarged prostate; and the remarks just made in reference to the operation of sounding will still further aid in illustrating the method to be adopted. The lithotrite having been introduced and brought to bear in the process of search upon the rectal aspect of the bladder, the blades may in this position be opened, and the stone picked up, usually without much difficulty. The instrument may then be carefully turned round, so as to direct the beak upwards, before the screw is applied and the crushing effected; or, without making this movement, pressure may be made by the hand alone, or by the screw, in the reversed position. I believe it will usually be difficult, sometimes impossible, to seize a stone lodged in a deep or depressed basfond below the level of the vesico-urethral orifice, without searching for it by reversing the blades; carrying the instrument, in short, to the stone, not waiting for the latter to fall into the instrument. It is true that the stone may sometimes be dislodged and removed to another situation, as has been already stated, by elevating the pelvis of the patient very considerably, but in this case it becomes equally an object of search; whereas in the condition described it may almost certainly be found by proceeding in the manner specified. It is not sufficient simply to introduce the lithotrite and open the blades, for assuredly in these cases the stone will not fall between them, as often happens in the bladder which retains its normal form and capacity, or nearly so.

There can be no doubt that the main objections to lithotrity in cases of enlarged prostate, do not lie on the ground of any particular difficulty in performing the operation. A far more formidable obstacle is a contracted irritable bladder, which will not easily retain the requisite quantity of fluid; in which there is neither space enough to work, nor ability to bear the necessary manipulation. Neither of these conditions commonly exist in association with enlarged prostate, for, on the contrary, the bladder is generally dilated, and the urethra often well accustomed to instrumental contact. The objection on the ground of the difficulty of expelling the fragments is certainly more considerable. The power of the bladder being impaired, at all events an obstruction existing at the outlet, it does happen that the detritus is not always got rid of with facility. This, however, is by no means an insuperable difficulty, and will be specially considered hereafter.

In consequence of these circumstances, lithotrity has been held by some authorities to be contraindicated in presence of an enlarged prostate. Sir B. Brodie's experience leads him, however, to speak in favour of its adoption. Thus, he says, in relation to the introduction of the lithotrite, in "cases of considerable enlargement of the prostate gland, I have never met with an instance in which the difficulty was not overcome by a cautious and gentle manipulation, nor with any in which any injury was done to the neck of the bladder in this part of the operation."* The increased difficulty in seizing the stone, which he regards as "the most important part of the operation," he represents as readily overcome by ordinary management. And in relation to the removal of the fragments, he expresses himself as follows:—

"In an elderly person, in whom there is usually more or less enlargement of the prostate gland, the fragments do not come away so readily as in those who are younger. This especially happens where the patient has lost the power of emptying the bladder by his own efforts. It would be a mistake, however, to suppose that the incapability of expelling the whole of the urine prevents them from coming away altogether; still the process is

* Notes on Lithotrity. By Sir B. Brodie, Bart.—*Trans. Med. and Chir. Soc.*, 1855, p. 172.

more tedious, and requires some assistance beyond that which is required under ordinary circumstances. Firstly, the patient should be directed to void his urine stooping forward, or even in the recumbent posture, lying with his face downwards. Secondly, tepid water should be injected by means of a syringe, or elastic gum bottle, through a large silver catheter, having a wide aperture near its extremity on the concave side, by which means fragments below a certain size may be washed out of the bladder. This may be done daily, the injection being repeated on each occasion three or four times. In one instance, in which a complete retention of urine followed the crushing of a large calculus, in the course of two or three weeks the whole of the fragments were thus brought away, the patient regaining the power of emptying the bladder afterwards."*

On these points Mr. Coulson holds nearly similar opinions. He considers a paralyzed condition of the bladder by no means to contraindicate lithotrity.† The late Sir Philip Crampton also, whose experience was considerable, believed that this "state furnishes no objection to the application of lithotrity, provided sufficient means be employed to rid the bladder of the fragments of the broken calculus," adducing the case of a gentleman, aged 71, from whom he removed, by six applications of the lithotrite, a stone two inches in diameter; in this case the "bladder was so completely paralyzed that, for several years, he had been obliged to draw off the urine by the catheter four or five times a day."‡

---

* Notes on Lithotrity. By Sir B. Brodie, Bart.—*Trans. Med. and Chir. Society*, 1855, pp. 178, 179. In this extremely valuable paper are presented the results of the author's practice in 115 operations of lithotrity, performed upon upwards of 100 adult patients. Of these 115 cases, the sequel was unfavourable in 9; but death could be only attributed directly to the operation in 5 instances. In 4 it resulted from old-standing disease brought into activity by the shock of the operation. The success of the operation was thus shown to have been as somewhat more than 12½ to 1. For further particulars, see also analysis of the paper, *Lancet*, 1855, vol. i. p. 316.

† Diseases of Bladder and Prostate Gland. 5th ed. London, 1857. Pp. 470 and 485, *inter alia*.

‡ Lecture on Lithotrity.—*Dublin Quarterly Journal*, 1846, vol. i. p. 52. Sir P. Crampton relates the particulars of 20 operations by lithotrity on 17 adults.

Civiale, the great master of the art, treats at length of this subject in his various writings, and adduces his experience respecting it. The sum of this forms the basis of an opinion in favour of applying lithotrity to ordinary, but not to extreme or extraordinary cases of enlarged prostate. For such, he would prefer, an operation being imperative, the operation of hypogastric lithotomy. With respect to the instrumental difficulties of the proceeding, he was accustomed to overcome them even with the straight lithotrites which he originally employed, and he regards the adoption of the curved instruments now in use, as removing the objections made on the score of difficulty of introduction in presence of a large prostatic tumour. He recommends the stone to be seized in the manner already described—a method of manipulation absolutely necessary, in his opinion, to detect the residual fragments which are certain to lodge behind the prostate. He recognizes the influence of obstruction at the vesical neck in occasioning retention of detritus, but believes it may be overcome in most cases by the use of full and repeated injections after each application of the lithotrite. He appends, in his "Traité de la Lithotritie," p. 163, brief particulars of fifteen cases in which he successfully crushed calculi complicated with prostatic enlargement. The treatment was in some of these more difficult, painful, and prolonged than usual, but otherwise no particular deviation from the ordinary course of things is remarked. The ages of the patients varied from 60 to 81 years, the majority being about 70.

Of the objections which have been made to the employment of lithotrity in enlarged prostate, one of the principal is that the

There was no death. All were successful, with the exception of one who discontinued treatment from causes unconnected with the operation, and one who died of rupture of the stomach from excessive drinking during his treatment. Of the successful cases, one was a patient aged 65, who submitted to two operations, with an interval of some years, and lived in perfect freedom from the complaint four years after the last. Another, aged 71, submitted to two operations with six or seven months' interval, and lived ten years with no return of symptoms. Another, aged 71 (the case alluded to in the text), with "immense prostate," was completely freed from the stone, and lived three years without return.

cavity of the bladder may be so diminished by the prostatic outgrowth, that sufficient room is not left for the proper manipulation of the lithotrite. In reply to this, without denying the existence of the state described, I am in a position to affirm that it is extremely rare. Dilatation of the bladder is almost invariably the sequence of prostatic enlargement, and that to an extent proportionate with the amount of the obstruction. This is undoubtedly a law almost universal in its application. Hence it follows, and a large field of observation proclaims the fact, that abundant compensation in the matter of space usually exists in dilatation of the viscus, for any encroachment on its cavity arising from the growth of a prostatic tumour into it.

A valuable source of evidence is afforded us in a paper by Dr. Ivanchich of Vienna, embodying a statistical account of one hundred cases of lithotrity treated by himself; the great majority of patients being males of 50 years old and upwards.

With very few exceptions, indeed, Dr. Ivanchich gives the name at length, the residence, and occupation of each patient; also the age, the peculiar complication present, if any, the composition of the calculus, the weight of the detritus removed, the number of sittings required, the number of days occupied by the treatment, and the final results. From this we learn that 81 male patients were operated upon by lithotrity, of the following ages:—

    31 were between 50 and 60 years,
    34 ,, ,, 60 and 70 ,,
    16 ,, ,, 70 and upwards.
    ——
    81

In 53 the calculus was uric acid; in 24 phosphatic; in two mixed; and in two the composition is not stated.

Of the 81 patients, eight died within a short period after the operation from fever, shock, &c.; but in one the disease was complicated with stricture of the urethra and renal calculi; in another with large renal calculus. Most were above 70 years of age. One, aged 75, exhibited "a remarkable example of valvular prostatic disease." It would not be fair to state that in every one of these eight instances the cause of death was altogether attributable to the operation. Besides the eight, one died two months, and another six months after the operation of diseased kidneys; and a third died of cancer at the end of a month. Of the 70 remaining cases, four or five are reported incomplete, and the remaining 65 as successful. In some instances the sittings were numerous, and the treatment protracted; but these cases were few and exceptional.

But of the 81 cases, eight had notable hypertrophy of the prostate: two of these had been operated on three years before. In one the results were "incomplete;" all the others were successfully treated; one only, alluded to above, dying with great enlargement at the age of 75. In five of these the stone was phosphatic; and in three, of uric acid.*

Taking the most unfavourable view of 81 cases of patients above 50 years of age here recorded, the deaths amount to not more than 1 in 10; a result which, it is needless to say, is vastly superior to our experience of lithotomy in patients of corresponding age. The experience of our metropolitan hospitals, at this period of life, records a fatal result in about two out of every five cases subjected to the latter operation.

We now come to a consideration of the method to be adopted in the management of that which I have heretofore characterized as constituting, in some of these cases, the most troublesome part of the process, viz. the withdrawal of the calculous fragments from the bladder.

There are four methods of proceeding, each one of which may be brought to bear in cases where any difficulty is apprehended, or encountered. All may be employed under certain circumstances conjointly, and with advantage.

These are—the employment of the scoop lithotrite; repeated injections of water, returned through a full-sized catheter with a large opening near its extremity, with or without a special exhaust-

* Wien Wochenschrift, 1856. Beilage, No. 51.

344 MODE OF REMOVING THE CALCULOUS FRAGMENTS.

ing apparatus attached; the position of the patient during subsequent acts of micturition; and the injection of solvents into the bladder. Each of these we may now examine in detail.

1. The application of the scoop lithotrite.

The use of the scoop lithotrite is so obvious, and its characters so well known, that it is necessary to do little more than mention it here. By its aid we may in some cases remove a good deal of fragmentary or of semi-solid calculous matter, without injury to the narrow passage through which it is to be withdrawn. It is, however, a slow process: involves a great deal of passing and repassing along the urethra, and may do much mischief when rough spiculæ are caught between its jaws, and project a little beyond their borders; an occurrence which will sometimes happen and is not always absolutely to be guarded against; although it may be in a great measure prevented by invariably screwing home the male blade completely, before attempting to withdraw the instrument. The mechanical action of hard bodies within the bladder,

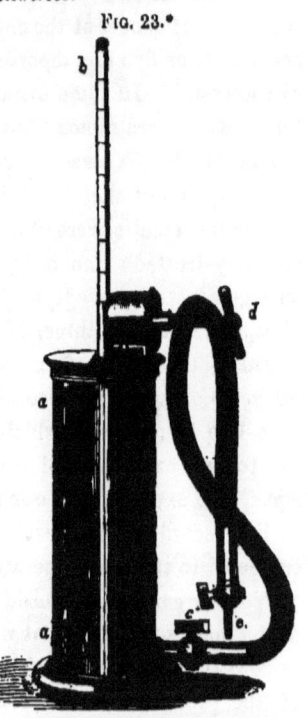

Fig. 23.*

a a. Cylinder containing 16 oz. of water.
b. The piston-rod, graduated. The descent of each mark, when the instrument is in action, indicates that two oz. have been injected.
c. The regulating stopcock, and elastic tube for adaptation to the catheter at the stopcock, *e.*
d. Key for winding up the spring, which gives the motive-power.

---

* This apparatus for injecting the bladder with water or other fluids not exerting a chemical action on metal is advantageous, because the fluid is propelled by a spring, at the will of the operator, and independently of any effort on his part, except for the purpose of regulating the force of the current, which, by simply placing a finger on the stopcock, is effected in the most perfect manner.

however, is always to be avoided when not strictly necessary. In the management of stone in the bladder, no axiom is better established, or more religiously to be followed, than this; viz. the smaller the mechanical power expended in the attainment of any given effect, the more successful will be the final result. The lithotrite is to be employed for no purpose which can be attained by milder and better agencies. We may, perhaps, find means of a mechanical nature which are superior even to the lithotrite under certain circumstances, in the attainment of the object desired; while in others we may have a safer and a better resource in the more subtle dynamics of chemical action.

2. Injections of water and the evacuating catheter.

The catheter usually employed immediately after a calculus has been crushed is one not smaller than No. 12, if it can be easily introduced; it may be even larger if the urethra will readily admit it, which is not unfrequently the case. An oval aperture, about three-quarters of an inch in length, is made in its concavity, through which fragments of moderate size may make their escape. This instrument being introduced when the lithotrite is withdrawn, the patient may stand upright or lean forwards a little, when a few ounces of tepid water are quickly injected, and permitted instantly to flow out, before the débris, stirred up by the act of injecting, can subside; when some of it usually escapes, although commonly less passes than one might at first thought suppose. The process should, however, in order to prove successful, be rapidly repeated three or four times, if not productive of uneasiness to the patient. In this manner much débris may be removed, and many small fragments, if the stone has been well crushed, and not merely broken. The catheter is then withdrawn, care being taken, in commencing to remove it, to recognize the occurrence of any degree of obstruction, while its extremity is passing through the neck of the bladder, as a fragment may be lodged in the opening described, but with a rough or sharp angle protruding beyond it; and great pain, if not some mischief to the neck of the bladder, may result if the possibility of this contingency be not remembered and provided against. If, therefore, on beginning to withdraw the catheter, anything like obstruction is felt, or a sharp pain is complained of, it is better to inject again, which will, probably, displace the fragment, and enable us to remove the catheter with ease. If this fails, however, a flexible but strong stilet, of a size sufficient to fill the catheter, with which it should always be provided, may be passed down to the end of the instrument; this will always succeed in getting rid of the obstacle, and the removal of the catheter follows without difficulty. The catheter of Heurteloup is, perhaps, preferable to that just described. This instrument, made of steel, possesses two long oval openings, situated near to the extremity. Each should be about three-quarters of an inch long; and, placed *laterally*, one rather nearer to the point than the other. It is represented in connection with other instruments by fig. 26, on the opposite page.

Sir Philip Crampton has applied with success an exhausting apparatus to the extremity of the catheter, for the purpose of removing detritus from a distended and atonied bladder. One of the cases reported in his valuable paper on lithotomy and lithotrity before referred to, was that of a Mr. Rodger, aged 71, who had "paralysis of the bladder, immense prostate, urine mucopurulent and bloody, and a stone two inches in diameter." This gentleman "had totally lost the expulsive power of the bladder" for several years before.* In this case the whole of the detritus was removed in this manner. Six sittings were sufficient to free the patient from every vestige of the stone, and he lived three years after with no return of the calculous affection.

The apparatus is thus described in the paper referred to. It consists "of a strong glass vessel of an oval form, and six or eight inches in length, by three in diameter, capable of holding about a pint and a half of water; to this vessel is attached a tube of about half an inch bore, furnished with a stopcock. The air being exhausted by means of an exhausting syringe, and one

* Dublin Quarterly Journal, Feb. 1846, pp. 22 and 43.

of Heurteloup's wide-eyed steel evacuating catheters being introduced into the bladder, it is next attached to the exhausted vessel; the stopcock is then turned, and a communication being thus established between the bladder and the glass, the pressure of the atmosphere is by this means brought to bear on the bladder, and supplies an expulsive power, which may be increased to any required amount."*

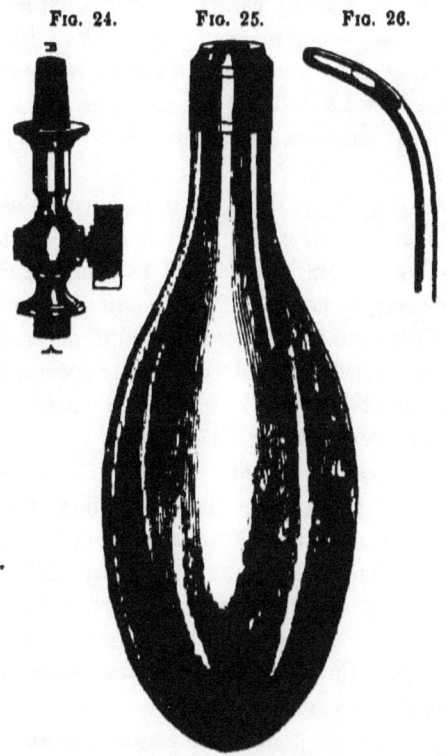

Fig. 24.   Fig. 25.   Fig. 26.

Figs. 24 and 25 represent Sir P. Crampton's apparatus—Fig. 24 is the piece which unites the exhausted receiver and the catheter. A is the end which screws to the receiver; and B that which fits the catheter.
Fig. 26 represents the lower end of Heurteloup's steel catheter.

The only precaution necessary in its use that must be observed,

* Op. cit. p. 22.

is to proportion the degree of exhaustion in the glass receiver to the quantity of water previously injected into the bladder. By a few preliminary trials its power is ascertained, the operator learning in this manner that so many strokes of the exhausting syringe will produce a vacuum equal to so many ounces of water. I am indebted to the kindness of the late Sir Philip Crampton for the apparatus represented at figs. 24 and 25. He also favoured me with the latest particulars of his experience in its employment, which confirms the belief in its utility which he expressed more than ten years ago. It is necessary to mention, that he advised that it is better to use the apparatus a day or two subsequent to the crushing, rather than at the time of the operation itself.

Especially when the position of the patient's body can be easily commanded, a Heurteloup's evacuating catheter, or one which has a large opening in the *convexity*, instead of the concavity of the curve, appears to afford the best chances of success. When the patient leans forward, or is placed in the prone position, the openings are much more advantageously situated for the reception of débris suspended in the injected fluid, and are thus more readily removed from the vesical cavity.

3. The position of the patient.

The position of the body during the outflow of the injected fluid is an important element in relation to success in withdrawing calculous detritus. The best unassisted attitude of the body is doubtless that of standing, and with a gentle inclination forwards of the trunk. A better position is obtained by desiring the patient to place his hands, and rest his weight, against the wall of the apartment, or upon some projection, as the mantelpiece, while his feet are gradually removed backwards; he may thus with ease maintain for a few minutes, a position of body in which its axis makes with the ground an angle of about 60°. By means of a couch, so contrived that a large opening can be made in its centre when required, the prone position may be insured without disturbing the patient, or even requiring him

THE POSITION OF THE PATIENT.

to rise from the recumbent position after the operation. In this case he lies in the usual manner on the couch, the pelvis being raised more or less on a cushion, which may be elevated to any degree at the will of the operator. After crushing has been performed the catheter is introduced, a cushion is drawn from under him, and with it a portion of the entire thickness of the couch to which it is attached. See fig. 27, which represents a couch designed and employed by myself. The patient

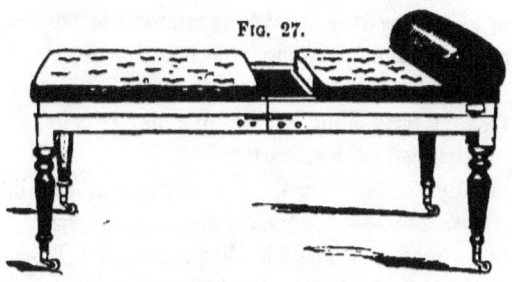

Fig. 27. The couch with pelvic cushion removed, showing the central aperture. There are two cushions—the ordinary one, level with the rest of the surface, as seen *in situ* at fig. 28; the other, much thicker, for elevating the pelvis when necessary.

is requested simply to turn on his face, when the neck of the bladder becomes the most depending part, and the catheter appears through the opening resulting from the removal of the cushion. The end of the couch corresponding with the feet may be raised if necessary, and the patient placed in any angle re-

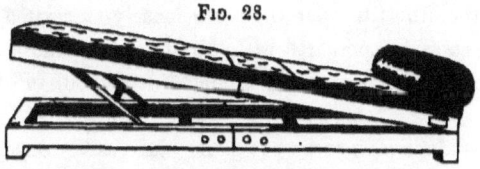

Fig. 28 shows the couch with the top raised. The entire couch is contrived so as to pack in a space of 3 feet by 2½, for the sake of portability.

quired for sounding or operating, without any exertion on his own part. Injection is made into the bladder by means of a flexible tube fastened to the catheter, and the outfall takes place into a vessel below.

By the means thus described a much freer exit is obtained, and greater success follows in the removal of fragments at the time of the operation, than is insured by the methods ordinarily followed. The advantages of such a result are too manifest to require any additional remark or illustration. There are few cases, even among those in which the bladder has lost all power of contracting, in which the influence of well managed currents and position will not be adequate to remove the detritus resulting from an operation by lithotrity. When, on the other hand, as mostly happens, we prefer to make but little attempt to remove the detritus of the stone immediately after crushing, the injecting in the position suggested should be employed subsequently ; say, usually after an interval of about forty-eight hours. We may then reap the full advantage of the method described.

It is especially for those cases which exhibit unusual difficulty, as well as for those where the pain which results from the use of instruments, is more considerable than usual, that I believe we may obtain assistance from the following method, viz. :—

4. The injection of solvents into the bladder.

We may call to mind here the fact already stated, that the majority of formations met with in the bladder, in connection with enlarged prostate, are phosphatic in their character. And with this we may associate another fact, viz. that these, of all others, are most susceptible to the influence of chemical agents, not only on account of their own chemical constitution, but on the ground of the state of physical aggregation which they usually affect. The action of the lithotrite upon the more loosely-associated elements of these masses, as compared with the uric and oxalic compounds, produces more of powdery or granulated detritus, than of angular fragments.

The finer deposit, however, is often less readily expelled from the bladder than the smaller fragments, especially when adhesive mucus is present. But at the same time it possesses an amount of surface in the state of minute division, which offers unusual advantage for the action of fluids endowed with solvent power.

Such circumstances, then, usually associated with that condition of the bladder in which it is incapable of contracting sufficiently or at all, present one of the contingencies in which the employment of solvents may be useful. The chemical character of the calculus being known, as it generally is in such cases, or if not, the scoop lithotrite will obtain sufficient to determine the question, the nature of the solvent to be employed is readily decided upon.

Much has been attempted, and some positive results have been attained, by means of chemical solutions for the destruction of calculus, wholly apart from mechanical interference. At present, however, there are perhaps few cases in which the practice can be entertained as adequate, of itself, to accomplish the desired end. But it does not follow that they should therefore be altogether thrown aside. The function which these agents should now be called on to perform, and one which would, I believe, be attended with material advantage, is that of aiding us in the exercise of the mechanical power. If ever they are to play a very important part in the surgical, as well as the medical treatment of calculus, as I cannot but believe that they will, it will probably not be until the initiatory step has been taken of employing them first as adjuncts to the present method. Granted that they may exercise some influence upon a calculus in its integral state, as has often been proved, how much greater should be the result of their action when exerted upon the detritus which results from crushing. When we find, after two or three applications of the lithotrite, in which considerable progress has been made, that we are compelled to wait for a season, owing to the increased sensibility of the bladder that has been occasioned, the use of appropriate solvents might at all events continue the work begun, if not in some circumstances complete it altogether.

I have derived some aid by applying this principle in two cases of lithotrity; one for phosphatic, the other for uric acid calculus. In this latter instance, my patient, who was 77 years of age, had the advantage of medical care from my friend Dr. Webster, of Dulwich, and we both considered the result, which was perfectly successful, to have been promoted by the use of lithia taken internally, reversing as it did the reaction of the urine which had been intensely acid. In the former I combined with good effect the use of weak acetate of lead injections into the bladder, as hereafter described.

From a very early period the alkalies have enjoyed a high reputation for their powers to dissolve calculus; and perhaps they might, in absolute ignorance of the nature of the formation, be rightly regarded as more useful in the majority of all cases than any other agent. The predominance of uric acid formations, whether alone or associated with bases, is a fact of itself sufficient to support this statement.* But it is not improbable that even some

---

* In the early history of the use of solvent remedies against stone, the alkalies appear to have formed the basis of the agents employed.—In the time of Arctæus and Paulus Ægineta (10th century) lime in water, or with vegetable infusions of a diuretic kind, was given with the view of dissolving the stone. The latter mentions the stones found in sponge, dried goat's blood (an exceedingly popular remedy with all ancient authors), and numerous decoctions, as those of maidenhair, parsley, and couchgrass. Avicenna, Rhazes, and the Arabian school, recommended solutions of carbonate of soda and potash for the same purpose.

Coming down to later times, we find that Crollius, in his "Basilica Chymica," Frankfort, 1608, recommends salt of tartar in infusion of parsley, also mixtures in which lime was the principal ingredient; although for others the sulphuric and hydrochloric acids.—*Vide* pp. 117-166, 220, 247. Daniel Sennertus, in his "Praxis Med.," not only mentions the internal use of the alkalies, but advises them to be injected into the bladder by a catheter.—Lib. iii. Part viii. § 1, cap. ii. 1650. Other writers at this time also refer to them. Boerhaave tells us that Val. Basil, in the fifteenth century, advised for the solution of stone the internal use of an alkaline salt made from vine cuttings.—*Elem. Chimiæ*, vol ii. p. 53. 1732.

But a new impetus was given to inquiry by the fame of Mrs. Joanna Stephens' practice, the secret of which was purchased by Parliament in 1739 for £5000. The agents in this case were alkaline, usually salts of potash and lime. Great interest in the subject was excited, and numerous trials made, in which enormous quantities of alkali, chiefly uncombined potash largely diluted, were administered both by mouth and by injection into the bladder. In rare instances, the calculus appears to have been removed; in numerous others, the effects were palliative to a very considerable extent. Among the numerous authors who subsequently wrote on this subject, the following are some of those that give the

phosphatic calculi have been disintegrated by this class of agents, probably by their action on the animal matter which cements together the calcareous particles. Nevertheless, from their influence in determining the precipitation of phosphatic salts from the urine, they must be regarded as contraindicated in these cases. Indeed it is not impossible to induce the formation of a phosphatic deposit upon a pre-existing uric acid calculus, by employing too freely, or too continuously, the use of alkalies, for its solution. And there is little doubt, during the extensive employment of soap and lime-water for this purpose, which was made in this country about a century ago, that not only numerous patients, but medical men also, sometimes mistook the phosphatic salts which were precipitated from the urine by the remedy, for the result of its disintegrating action on the calculus.*

But, as has before been shown, the calculus associated with enlarged prostate is, in most instances, phosphatic. The examination of a fragment after the first crushing will, however, determine the question. Supposing it to be so, the agent to be selected may be one of the mineral acids which combine with the

best and most accurate information:—Hartley, who modified and improved Mrs. Stephens' plan, 1739; Hales, 1740; Butter, Edinburgh, 1754; Whytt, Edinburgh, 1755. See also Dr. Rutty's paper at the Royal Society, 1741; two practical essays by Alex. Blackrie, London, 1766-71; also, N. Hulme, 1788.

Dr. Lobb, in 1739, had prescribed lemon-juice largely, and vegetable diet; Dr. M. Dobson, 1779, "fixed air" and alkalies. Spallanzani proposed gastric juice as a solvent.—"Expériences sur la Digestion." Paris, 1783. So also did Dr. Darwin and others. Subsequently Drs. Physick and Dorsey made numerous experiments on calculi with this agent: the latter tried it in a case of stone in the bladder for a short time, with a partial success, but does not state why he did not persevere.—See Dorsey's work on the lithontriptic virtues of the gastric liquor. Phil. 1802.

In France, at an early date, the alkaline remedies were strongly recommended; by Darcet, "Annales de Chimie," in 1720; and Pierre Desault, 1736. By Fourcroy and Vanquelin; and more recently by C. Petit, 1834. The first and last-named employed the Vichy waters. See also a report of the Royal Academy of Medicine of Paris upon these agents for numerous cases.—Bulletin d'Académie; tome iii. 1839. P. Desault recommended injections and baths of Barèges water for the same purpose.

* This was discovered by later writers than the above-named. See Dr. Austin's "Gulstonian Lectures," 1791; Murray Forbes, 1793, and others, who point out this source of error.

A A

calcareous bases, and form soluble salts. Of these the nitric has hitherto been esteemed the best.

Sir Benjamin Brodie's case, in which two phosphatic calculi were removed from a diseased bladder by this means, is well known. He employed from two to two and a half minims of strong nitric acid to the ounce of distilled water, which solution was allowed very slowly to pass through the bladder by means of a double-current catheter, for a period of fifteen to thirty minutes every two or three days. From this case he draws the following conclusions:—

"1. That a calculus, composed externally of the phosphates, may be acted on by this injection so as to become gradually reduced in size, while it is still in the bladder of a living person.

"2. That there is reason to believe that small calculi, composed throughout of the mixed phosphates, such as are met with in some cases of diseased prostate gland and bladder, are capable of being entirely dissolved under this mode of treatment." *

Such being the power of chemical solvents on the unbroken stone, it cannot be doubted that their efficacy would be considerably increased when applied after the stone had been reduced to fragments. Hence their employment must be regarded as particularly applicable to those cases now under consideration, presenting, as they so frequently do, an atonied state of bladder inimical to the due expulsion of its foreign contents; cases in which a return of the complaint is more to be dreaded, and the more carefully to be guarded against.

The agents which have been referred to, act by their elective affinity for the base of the salt which forms the calculus. From their very nature, they tend also to act as irritants upon the mucous membrane of the bladder. Hence the necessity to sacrifice power by greatly diluting them, and thus to render the chemical action much slighter than it would be, could they be

* Op. cit. pp. 306-311.

used in a more concentrated form. Attempts have consequently been made to discover an agent which should exert no injurious effect on the bladder, but at the same time possessed of considerable power to disintegrate the stone. Dr. S. E. Hoskins of Guernsey has endeavoured to effect this by securing a double chemical decomposition in the bladder, by means of an agent sufficiently unirritating to be retained there almost for an indefinite period of time. The first published account of his experiments is to be found in a paper which was presented to the Royal Society in 1843.*

The distinction which he there draws between SOLUTION and DECOMPOSITION is important. The *solvent* acts on the calculus, and on the tissues; the *decomponent* only on the calculus. The explanation of these actions may be given in the author's own words: "the active agent of the decomponent is liberated gradually, and neutralized by the earthy bases of the calculus, before it can come in contact with the living tissues; * * * The *base* of the decomponent unites with the *acid* of the calculus, while the acids of the former combine, and form soluble salts, with the bases of the other. The combined acids are set free in definite proportions, neutralized in their nascent state, and removed from the sphere of action before any stimulating effect can be exerted on the bladder."†

The agent employed by Dr. Hoskins in several cases reported, was the nitro-saccharate of lead, which he recommended to be employed in the following manner. To each fluid ounce of water is to be added, one grain of the salt previously dissolved in five minims of strong acetic acid. The mixture to be heated to the boiling point, and subsequently used at 100° Fahrenheit. A few ounces are to be injected, permitted to remain 10 or 15 minutes, and renewed two or three times if thought proper, at each application.

* Phil. Trans. 1834, p. 7.
† Decomposition of Phosphatic Calculi.—Lond. Jour. Med. 1851, vol. iii. p. 891. By S. Elliott Hoskins, M.D., F.R.S.

Subsequently Dr. Hoskins informed me that further experience has, in some measure, modified his plan. He writes me as follows: "The salt which I have, in some more recent cases, had recourse to, is the pure acetate of lead (one grain to the ounce of water), with the smallest possible quantity of acetic acid, and no more, that is, sufficient to secure solution, and render the liquid transparent." In this liquid, which is perfectly unirritating, very rapid decomposition of a phosphatic calculus will take place, as may be easily demonstrated by experiment. If a fragment be suspended in a small quantity of the solution, a dense white and very fine precipitate of phosphate of lead immediately occurs, and an acetate of the base or bases is formed in solution. I have injected this liquid in the presence of phosphatic stone and of highly-charged phosphatic urine, in several instances, without producing irritation, and the expelled fluid has deposited the insoluble phosphate of lead. The frequency with which the injection is used must depend upon the degree of sensitiveness manifested by the bladder. It may be used either by permitting a few ounces to remain in the bladder as long as it can be retained, by passing a stream of it through a double current catheter for some minutes, or by a still more prolonged current after the manner which Dr. Willis several years ago designed. This consisted of a "reservoir for the fluid, raised a foot or two above the bed or sofa on which the patient is laid, to be connected by means of a flexible tube guarded by a stopcock, with a double-current catheter. In this way, a constant stream of the injection can be kept circulating through the bladder, and acting on the stone to the very best advantage. The reservoir should be a double vessel of tinned iron, the outer one being filled with water at from 95° to 98° Fahrenheit, and kept at this temperature by means of a small spirit lamp. There is no necessity for any very rapid current through the bladder."*

The impaction of fragments in the urethra may sometimes

* On the Treatment of Stone in the Bladder by medical and mechanical means. Lond. 1842, p. 179.

be a source of difficulty, although less commonly in these cases than in younger subjects, in whom the action of the bladder is vigorous; particularly in those who, whatever their age, exhibit irritability or spasm of the viscus after operation. The consideration of this subject at length is not within the scope of my design. When situated far back, as in the prostatic part, perhaps the best proceeding is to push them gently towards the bladder with a large wax bougie, or by means of a strong injection of water passed through a catheter, which has been carried down to the spot. Painful sensations about the neck of the bladder, and frequent desires to make water, appearing a day or two after a sitting, are not unfrequently caused by the lodgment of a small fragment in that situation, although it may be insufficient to interfere with the flow of urine. At all events, the passage of a *full-sized* instrument, the best for the purpose being a soft flexible bougie, will sometimes remove all these symptoms, and probably by carrying such a fragment back into the bladder. This is a point in the after treatment which Civiale recommends the operator never to forget.

Mr. Skey speaks in high terms of the use of a catheter of the largest size, "the lower end of which is cut off, and the rounded extremity supplied by a round knob or ball, connected by a wire to the handle. When," says he, "the instrument strikes the fragment, the ball is withdrawn, and by forcing on the catheter, little by little, the stone is pushed back into the bladder, the sharp circle of the catheter surrounding the fragment, while its large size fully distends the canal." *

The employment of the long urethral forceps, of the long cuvette, or of the small lithoclast, will usually suffice to remove fragments which are impacted in the anterior and middle portions of the passage.

There is one condition liable to be encountered in a case of prostatic enlargement complicated with calculus, which remains to be noticed, and the consideration of which may appropriately

* Clin. Lect. on Lithotrity.—*Lancet*, vol. i. 1855, p. 554.

close this chapter. It is the existence of extreme irritability on the part of the prostate, or of the bladder itself. This condition, if insuperable, certainly contraindicates the application of lithotrity. The necessary manipulation would not only be distressingly painful but fraught with danger to the patient. Cystitis would, probably, be produced, and might soon prove fatal, or extension of inflammation to the kidneys might follow, and death occur from disorganization of these organs already perhaps impaired by chronic disease. When, however, the irritability is purely local, that is, when it consists only in an irritated condition of the prostate and bladder, unassociated with renal disease, much may be done in the way of preparing the patient, and we may even then succeed in rendering him a fit subject for lithotrity. Preparatory measures are desirable in most cases; in many necessary, and in these they are especially effective.

We should commence with soothing means, enjoining rest, attention to the condition of the urine, the employment of acids or alkaline salts, combined with those infusions which the nature of the case indicates as most appropriate (Chap. XI.). The judicious use of hip baths, thorough friction to the surface of the skin, attention to the digestive functions and state of the bowels, with the regulation of the diet, and sedatives by the rectum or otherwise, will often exert a very beneficial influence on the state of the patient. Besides these, the condition of the bladder and its contents must be carefully watched; and either by ensuring from time to time, as often as is necessary, that the viscus is completely emptied, or by the use of soothing injections and other local treatment, which may be indicated, all sources of discomfort and irritation should, as far as possible, be removed.

If the patient has not been accustomed to the employment of instruments in the urethra, he should, when the irritability of the organs has been diminished, be gradually accustomed to the presence of a catheter in the canal, until a full-sized instrument passes with comparative ease, and its presence there for two or three minutes ceases to excite pain, or much irritation, or other unpleasant consequence. The time and attention of the surgeon is well bestowed in such preliminary treatment; and, although sometimes necessarily tedious in an obstinate case, is generally well repaid by the successful issue which it tends so much to ensure. Under such measures, large discharges of mucus may almost disappear, difficulty and exhausting suffering in the act of micturition be greatly mitigated, a disposition to hæmorrhage removed, and the general power of the patient wonderfully recruited. In those cases which appear at the outset most unpromising, we should, nevertheless, try the effect of the measures described, since, with whatever result, so far as the anticipation of lithotrity is concerned, they can scarcely fail to be, to some extent, beneficial; and we should only relinquish the hope of applying that operation on discovering that it is not within our power thus to effect a considerable amelioration of the patient's condition. It then becomes a question whether future treatment should be merely palliative, or whether he should be submitted to the alternative of lithotomy. If, after all our efforts, the symptoms are but little improved, especially if the preliminary use of instruments is followed by severe rigors or fever, and the signs of inflammation in the bladder are so produced or augmented, we may not run the hazard of attempting to crush a calculus. Better is it, under such circumstances, to choose a fitting time to rid the sufferer of his complaint, by one operation adequate to its complete removal, than to expose him to the tedious course of repeated and painful efforts, which such a state of bladder would render inevitable, granting even the possibility of the application.

In *all* cases of lithotrity we shall do well to adopt certain preliminary measures, and to attain for our patient as good a state of health generally, and as quiet a condition of the urinary organs particularly, as we can accomplish, before entering upon the operation properly so called. The use of instruments as a preparatory measure, so strongly recommended by Civiale, is undoubtedly important, as I have had some opportunities of observing. The susceptibility to constitutional irritation as the result of instrumental interference with the urethra and bladder, which some individuals exhibit in a marked degree, apparently as an inherent peculiarity, and to which all are, under certain circumstances, more or less liable, may be greatly diminished by gradually accustoming the passages to contact with foreign bodies, through the progressive steps of soft bougie, catheter, and lithotrite. Where this peculiar susceptibility exists, evinced by the occurrence of rigor, or other disturbance of the nervous system, after the use of solid instruments especially, no sign of organic disease being present, we cannot be too careful to adopt the precautions recommended. Where, on the other hand, unmistakable signs of renal disease are present, little else but harm will accrue to the patient, as a rule, from the employment of lithotrity.

Supposing, however, that, for the present at all events, the employment of operative measures is not to be entertained, much may be done by purely palliative treatment. Conjoined with this we should apply, so far as we may be able, and in any manner in which they can be borne, either locally or through the system, the agents calculated to produce a disintegration of the stone. Whatever may be the amount of the action so exerted upon it, there can be no question as to the relief to painful symptoms, which is afforded by a perseverance in the use of diluents, containing in solution the acid, saline, or alkaline agents which the case may require. We possess the best evidence, and in great abundance, from sources already referred to, in support of this opinion. I believe that such a mode of treatment has been of late too much neglected, a result probably of the great advance in the surgical or mechanical treatment which the last century has witnessed. And there appear to be good grounds for anticipating that its employment might still be exceeding beneficial, in the management of those cases occurring in elderly patients who are judged unpromising subjects for the performance of any operation.

INDEX.

	PAGE
Abscess of the prostate, acute	67
morbid anatomy of	67
symptoms	70
treatment	72
Abscess of the prostate, chronic	69, 73
cases of	74, 76
Acetate of lead injections	202
Acid injections for bladder	202
Acute prostatitis	50
Acute and chronic abscess	67
Adams, on course of urethra in prostate	25
on use of alkalies	196
on scirrhus of the prostate	265
Age, the, when hypertrophy appears	137
Alkalies, value of	196
Amyloid origin of concretions, doubtful	322
Analogy between prostatic and uterine tumours	113
analysis of 164 prostatic dissections	137
Angular catheter of Mercier	238
mode of passing	239
Atony or inertia of the bladder	154
diagnosis of	172
often mistaken for paralysis	154
treatment of	181
Atrophy of the prostate	256
forms of	257
symptoms and treatment	261
Balsams, the use of	191, 194
Bar at the neck of the bladder	293
conclusions respecting	303
diagnosis of	169, 304
treatment of	305
Baths, hot, in prostatitis	58
in cystitis	184
in retention	229
Benzoic acid, value of	198
Bladder to be emptied daily	176
when not to be emptied	181, 242
not to be emptied always	242
case of death from so doing	243
Blood in the urine	203

	PAGE
Brander's, Dr., puncture through symphysis	248
Brodie, on hypertrophy of prostate	125
on catheterism	181
on injecting the bladder	202
on position in lithotrity	335
on lithotrity with enlarged prostate	339
on value of solvents for stone	354
Buchu, properties of	184, 191, 194
Calculi, prostatic	319, 323
Calculus diagnosed from hypertrophy	171
with enlarged prostate	329
causes of	330
mode of sounding for	335
lithotrity for	338
solvents for	352
Cancer of the prostate	262
frequency of	263
morbid anatomy of	267
symptoms of	269
urine in	271
treatment of	271
cases of	272
Capsule of the prostate	39
Cases, 74, 76, 119, 121, 198, 250, 272-282, 286, 290, 292	
Cases of abscess of prostate	74, 76
Cases of prostatic tumour	119, 121
Catheters, gum elastic	178, 232, 235
Catheter, of tying in the	179, 243
Catheterism, when necessary	176
by the patient himself	177
in retention	230
mode of performing	234
Causes of hypertrophy	122
opinions respecting	123
exceedingly obscure	123
of prostatitis	51
Changes in urethra from hypertrophy	86
Characters of urine in hypertrophy	145, 332
Children, prostate in	40-42

	PAGE
Chronic prostatitis	59
Civiale, on causes of prostatic hypertrophy	126
on bar at the neck of the bladder	295
on operations for	307
on lithotrity with enlarged prostate	341
Compression	222
various modes	223
Conclusions respecting operations for retention	254
Concretions, prostatic	313
Congestion of hypertrophied prostate	208
treatment of	208
Couch for lithotrity	349
Coulson on lithotrity with enlarged prostate	340
Counter-irritation in cystitis	183
Course of urethra through prostate	24
Crampton, on lithotrity with enlarged prostate	340
apparatus of	347
Cruveilhier on prostatic tumours	103
Crypts or vesicles of prostate	33
Cystitis, chronic, diagnosed from hypertrophy	172
caused by hypertrophy	182
treatment of	183
Cysts or cavities in prostate	288
varieties of	289
Danger of catheterism in some cases	181, 242
Demulcents, the use of	191
modes of making	192
Diagnosis of hypertrophy	161
of tumour at neck of bladder	169
of bar at neck of bladder	169, 304
Dietetic rules in hypertrophied prostate	209
Different forms of hypertrophy	95
Dilatation of the bladder	159
Dissection to define prostate	3
Division of the prostate into lobes	17
Effects of enlarged prostate on micturition	151
prostatic obstruction	159-160
Ejaculatory ducts	29
Electricity	225
Ellis on anatomy of prostate and bladder	30
Encephaloid of prostate	264
Engorgement of the bladder	152, 155
Enlargement from hypertrophy	84
from prostatitis	65
Enormous prostatic calculi	326

	PAGE
Epithelium of prostate	35
Examination of the living prostate by rectum	162
by the urethra	165
External relations of prostate	12
False passages	240
mode of avoiding	241
Mercier's instrument for	242
Fœtal development of prostate	40
Follicles of the prostate	34
Form of sounds to be employed	167, 168
Frequency of hypertrophy	135
of atrophy	259
of prostatic tumours	101
Function of micturition in hypertrophy	151
Gland lobules of prostate	36
Glandular structures of prostate	33
Gross on size and weight of the prostate	10, 40
on causes of hypertrophy	126
on remedies for cystitis	190, 194, 195
Gum elastic catheters	178, 232, 235
Guthrie on bar at the neck of the bladder	294
on operations for	226, 306
Hæmaturia	203
Hæmorrhage into bladder	204
treatment of	205
Healthy prostate, measurements of	8, 10
Hemlock in hypertrophied prostate	213
History of solvents for stone	352
Hodgson, Dr., on size of the prostate	10
on course of urethra in	25
on hypertrophy of urethra	98
Home, "on the third lobe"	18
Hoskins, Dr., on solvents for stone	355
Hydatids near to the prostate	290
Hydrochlorate of ammonia	214
Hypertrophy of the prostate	79
Hypertrophy, morbid anatomy of	79
physical characters of	80
parts affected by	82
amount of enlargement from	84
changes in the urethra in	86
structural changes in	93
different forms of	95
not an effect of age	135
rarely causes incontinence	156
almost always causes retention	153
treatment of	174
Hunter "on the third lobe"	22

INDEX.

	PAGE
Incontinence, often used erroneously	151
of urine, treatment of	207
of urine is rare	156
Inertia of the bladder	154, 181
Inflammation of prostate, acute	50
causes	51
morbid anatomy	53
symptoms	55
treatment	56
of prostate, chronic	59
causes	60
morbid anatomy	61
symptoms	62
treatment	62
enlargement resulting from	65
Injections in lithotrity	345
Iodine in hypertrophied prostate	215
Irritability of bladder	200
suppositories for	201
Ivanchich on lithotrity	342
Jones Handfield on tissue of prostate	30
on hypertrophy	98
on prostatic concretions	320
Kölliker on anatomy of prostate	31
Kreuznach waters	217
Lawrie's, Dr., operation in retention	253
Leroy D'Etiolles, on bar at the neck of the bladder	297
his treatment of	306
Lithotomy, with hypertrophied prostate	337
Lithotrity, with hypertrophied prostate	338
Lithotrite scoop, use of	344
Lithotrity, injections in	345
position in	348
couch for	349
preparation necessary	358
aid of solvents in	350
Lobules of the prostate	34, 36
Lymphatics of the prostate	16
Lythrum salicaria, Prout on	189
Management, general, of patients with hypertrophy	209
Matico, properties of	189
Measurements of prostate	8, 9, 41
Melanosis of prostate	266
Mercier, on course of urethra in prostate	25
on paralysis of the bladder	153
on hypertrophy causing incontinence	156
his angular catheter	238
mode of avoiding false passages	242

	PAGE
Mercier on bar at neck of the bladder	296
his operations for	308
Mercury in hypertrophied prostate	214
Messer, Dr., on weight of prostate	11
his dissections	45, 85
on absence of symptoms	92
on tumours in prostate	101
Microscopical examination of hypertrophied prostate	96
history of concretions	316
Minute anatomy of prostate	30
Kölliker on	30
Ellis on	30
Mode of examining the minute structures	37
examining the prostate	162
Morbid anatomy, 53, 61, 67, 79, 267, 283, 303	
Mucus or muco-pus in the urine, 146, 332	
Muscular fibres of prostate, 5, 6, 31	
Numerical prevalence of hypertrophy	135
Operations for diminishing hypertrophy	226
Outgrowths from the prostate	111
Over-distention of the bladder	154
Overflow of urine	152, 155
Paget's, Mr., of Leicester, cases	249
Paget on prostatic tumours	106
Paralysis of the bladder diagnosed from hypertrophy	173
Pareira brava, properties of	184
Perforation of the prostate in retention	245
instrument for doing it	252
Phosphatic urine	202
injections, use of for	202
Preparation for lithotrity	358
Prevalence of hypertrophied prostate	135
Position in lithotrity	348
Prostate, an independent organ	2
form of	3, 7
in young subjects	40, 42
Prostatic calculi	323
analysis of	324
cases of	326
operations for	327
concretions	313
analysis of	315
ducts	26, 34
hypertrophy	79
ligaments	13
muscles	14

INDEX.

	PAGE
Prostatic outgrowths	111
tumours, simple	101
urethra	26
Puncture of the bladder, suprapubic	246
by rectum	247
through symphysis pubis	248
appreciation of methods	249
case of	250
Quain, Mr., on an artery supplying the prostate	15
Rare form of prostatic tumour	117
Rees, Dr., on use of alkalies	196
Regimen and diet	209
Result of treatment in hypertrophy	136
Retention of urine, treatment of	228
catheterism in	230
puncture of bladder in	246
Retention produced by prostatic tumours	91, 158
the common result of hypertrophy	156
Rokitansky on prostatic tumours	105
Sacculation of the bladder	159
Scirrhus of the prostate	265
Section of prostate in lithotomy	11
Senega, properties of	188
Senile atrophy of prostate	259
Skey on fragments of stone in urethra	357
Silver catheters, mode of passing	234
Simpson, Dr., on analogy between uterus and prostate	113
Sinus prostaticus	26
Solvents, use of	350
history of	352
Sounding for stone when prostate is enlarged	334
Sounds to be employed	167, 168, 335
Spence, Mr., on irregular vessel near prostate	16
Stafford's treatment of hypertrophied prostate	215
Statistics of hypertrophy	135
Stricture diagnosed from hypertrophy	171
Structure of prostatic glands	33
Structural changes in hypertrophy	93

	PAGE
Structures of prostate and bladder continuous	5
Suppositories	201
Symptoms of hypertrophied prostate	139
Table of diseases of prostate	49
of 194 prostates	43-48
Tanchou's cases of cancer	262
Treatment of hypertrophy of prostate	174
of its results	175
general	209
special	212
Triticum repens, the value of	188, 194
Tubercle of the prostate	283
morbid anatomy of	283
symptoms	284
treatment	285
cases of	286
Tumour of the bladder, diagnosis	171
Tumours, simple, of prostate	101
frequency of	101
Cruveilhier on	103
Rokitansky on	105
Paget on	106
Velpeau on	104
examples of	107
description of	109
Tumours of prostate and uterus compared	113
Tying in the catheter	179, 243
Ulceration of the prostate	77
Urethra, course of	24
relation of, to the gland	25
Urethral mucous membrane	27
Urine, alkaline, action of	145, 332
Urine, the, in hypertrophy of prostate	145, 332
Utricle, the	28
Uva ursi, properties of	187, 194
Uvula vesicæ	26
Veins of the prostate	16
Velpeau on prostatic tumours	104
Vesicles of the prostate	33
Vessels of the prostate	14
exceptional	15
Virchow on concretions	322
Weight of healthy prostate	11
Willis, Dr., on injecting the bladder	356

Woodfall and Kinder, Printers, Angel Court, Skinner Street, London.

London, New Burlington Street,
April, 1865.

MESSRS. CHURCHILL & SONS' Publications,

IN

MEDICINE

AND THE VARIOUS BRANCHES OF

NATURAL SCIENCE.

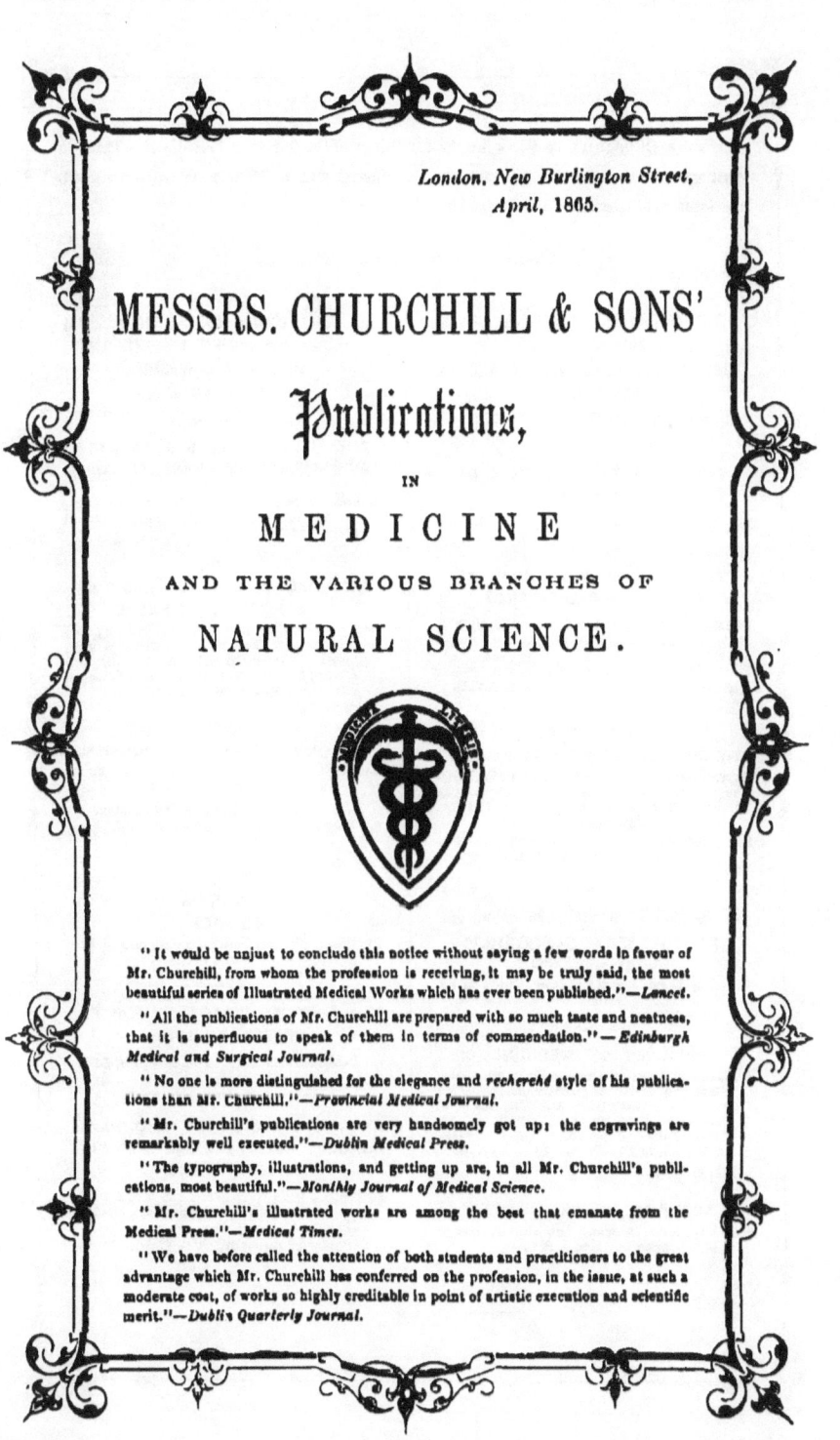

"It would be unjust to conclude this notice without saying a few words in favour of Mr. Churchill, from whom the profession is receiving, it may be truly said, the most beautiful series of Illustrated Medical Works which has ever been published."—*Lancet*.

"All the publications of Mr. Churchill are prepared with so much taste and neatness, that it is superfluous to speak of them in terms of commendation."—*Edinburgh Medical and Surgical Journal*.

"No one is more distinguished for the elegance and *recherché* style of his publications than Mr. Churchill."—*Provincial Medical Journal*.

"Mr. Churchill's publications are very handsomely got up: the engravings are remarkably well executed."—*Dublin Medical Press*.

"The typography, illustrations, and getting up are, in all Mr. Churchill's publications, most beautiful."—*Monthly Journal of Medical Science*.

"Mr. Churchill's illustrated works are among the best that emanate from the Medical Press."—*Medical Times*.

"We have before called the attention of both students and practitioners to the great advantage which Mr. Churchill has conferred on the profession, in the issue, at such a moderate cost, of works so highly creditable in point of artistic execution and scientific merit."—*Dublin Quarterly Journal*.

Messrs. Churchill & Sons are the Publishers of the following Periodicals, offering to Authors a wide extent of Literary Announcement, and a Medium of Advertisement, addressed to all Classes of the Profession.

THE BRITISH AND FOREIGN MEDICO-CHIRURGICAL REVIEW, AND QUARTERLY JOURNAL OF PRACTICAL MEDICINE AND SURGERY.
Price Six Shillings. Nos. I. to LXX.

THE QUARTERLY JOURNAL OF SCIENCE.
Price Five Shillings. Nos. I. to VI.

THE QUARTERLY JOURNAL OF MICROSCOPICAL SCIENCE,
INCLUDING THE TRANSACTIONS OF THE MICROSCOPICAL SOCIETY OF LONDON.
Edited by DR. LANKESTER, F.R.S., and GEORGE BUSK, F.R.S. Price 4s. Nos. I. to XVIII. *New Series.*

THE JOURNAL OF MENTAL SCIENCE.
By authority of the Association of Medical Officers of Asylums and Hospitals for the Insane. Edited by C. L. ROBERTSON, M.D., and HENRY MAUDSLEY, M.D.
Published Quarterly, price Half-a-Crown. *New Series.* Nos. I. to XVII.

THE JOURNAL OF BRITISH OPHTHALMOLOGY AND QUARTERLY REPORT OF OPHTHALMIC MEDICINE AND SURGERY.
Edited by JABEZ HOGG, Surgeon. Price 2s. 6d. No. I.

ARCHIVES OF MEDICINE:
A Record of Practical Observations and Anatomical and Chemical Researches, connected with the Investigation and Treatment of Disease. Edited by Dr. LIONEL S. BEALE, F.R.S. Published Quarterly; Nos. I. to VIII. 3s. 6d.; IX. to XII., 2s. 6d., XIII., XIV., 3s.

ARCHIVES OF DENTISTRY:
Edited by EDWIN TRUMAN. Published Quarterly, price 4s. Nos. I. & II.

THE ROYAL LONDON OPHTHALMIC HOSPITAL REPORTS, AND JOURNAL OF OPHTHALMIC MEDICINE AND SURGERY.
Vol. IV., Part 3, 2s. 6d.

THE MEDICAL TIMES & GAZETTE.
Published Weekly, price Sixpence, or Stamped, Sevenpence.
Annual Subscription, £1. 6s., or Stamped, £1. 10s. 4d., and regularly forwarded to all parts of the Kingdom.

THE HALF-YEARLY ABSTRACT OF THE MEDICAL SCIENCES.
Being a Digest of the Contents of the principal British and Continental Medical Works; together with a Critical Report of the Progress of Medicine and the Collateral Sciences. Post 8vo. cloth, 6s. 6d. Vols. I. to XL.

THE PHARMACEUTICAL JOURNAL,
CONTAINING THE TRANSACTIONS OF THE PHARMACEUTICAL SOCIETY.
Published Monthly, price One Shilling.
*** Vols. I. to XXII., bound in cloth, price 12s. 6d. each.

THE BRITISH JOURNAL OF DENTAL SCIENCE.
Published Monthly, price One Shilling. Nos. I. to CV.

THE MEDICAL DIRECTORY FOR THE UNITED KINGDOM.
Published Annually. 8vo. cloth, 10s. 6d.

ANNALS OF MILITARY AND NAVAL SURGERY AND TROPICAL MEDICINE AND HYGIENE,
Embracing the experience of the Medical Officers of Her Majesty's Armies and Fleets in all parts of the World.
Published Annually. Vol. I., price 7s.

A CLASSIFIED INDEX
TO
MESSRS. CHURCHILL & SONS' CATALOGUE.

ANATOMY.
	PAGE
Anatomical Remembrancer	3
Flower on Nerves	11
Hassall's Micros. Anatomy	14
Heale's Anatomy of the Lungs	14
Heath's Practical Anatomy	15
Holden's Human Osteology	15
Do. on Dissections	15
Huxley's Comparative Anatomy	16
Jones' and Sieveking's Pathological Anatomy	17
Maclise's Surgical Anatomy	19
St. Bartholomew's Hospital Catalogue	24
Sibson's Medical Anatomy	25
Waters' Anatomy of Lung	29
Wheeler's Anatomy for Artists	30
Wilson's Anatomy	31

CHEMISTRY.
Abel & Bloxam's Handbook	3
Bowman's Practical Chemistry	7
Do. Medical do.	7
Fownes' Manual of Chemistry	12
Do. Actonian Prize	12
Do. Qualitative Analysis	12
Fresenius' Chemical Analysis	12
Galloway's First Step	12
Do. Second Step	12
Do. Analysis	12
Do. Tables	12
Griffiths' Four Seasons	13
Horsley's Chem. Philosophy	16
Mulder on the Chemistry of Wine	20
Plattner & Muspratt on Blowpipe	22
Speer's Pathol. Chemistry	26
Sutton's Volumetric Analysis	27

CLIMATE.
Barker on Worthing	4
Bennet on Mentone	6
Dalrymple on Egypt	10
Francis on Change of Climate	12
Hall on Torquay	14
Haviland on Climate	14
Lee on Climate	18
Do. Watering Places of England	18
McClelland on Bengal	19
McNicoll on Southport	19
Martin on Tropical Climates	20
Moore's Diseases of India	20
Scoresby-Jackson's Climatology	24
Shapter on South Devon	25
Siordet on Mentone	25
Taylor on Pau and Pyrenees	27

DEFORMITIES, &c.
Adams on Spinal Curvature	3
Barwell on Clubfoot	4
Bigg on Deformities	6
Do. on Artificial Limbs	6
Bishop on Deformities	6
Do. Articulate Sounds	6
Brodhurst on Spine	7
Do. on Clubfoot	7
Godfrey on Spine	13
Hugman on Hip Joint	16
Tamplin on Spine	27

DISEASES OF WOMEN AND CHILDREN.
	PAGE
Ballard on Infants and Mothers	4
Bennet on Uterus	6
Do. on Uterine Pathology	6
Bird on Children	7
Bryant's Surgical Diseases of Children	7
Eyre's Practical Remarks	11
Harrison on Children	14
Hood on Scarlet Fever, &c.	16
Kiwisch (ed. by Clay) on Ovaries	9
Lee's Ovarian & Uterine Diseases	18
Do. on Diseases of Uterus	18
Do. on Speculum	18
Seymour on Ovaria	25
Smith on Leucorrhœa	26
Tilt on Uterine Inflammation	28
Do. Uterine Therapeutics	28
Do. on Change of Life	28
Underwood on Children	29
Wells on the Ovaries	30
West on Women	30
Do. (Uvedale) on Puerperal Diseases	30

GENERATIVE ORGANS, Diseases of, and SYPHILIS.
Acton on Reproductive Organs	3
Coote on Syphilis	10
Gant on Bladder	13
Hutchinson on Inherited Syphilis	16
Judd on Syphilis	17
Lee on Syphilis	18
Parker on Syphilis	21
Wilson on Syphilis	31

HYGIENE.
Armstrong on Naval Hygiene	4
Beale's Laws of Health	5
Do. Health and Disease	5
Bennet on Nutrition	6
Carter on Training	8
Chavasse's Advice to a Mother	9
Do. Advice to a Wife	9
Dobell's Germs and Vestiges of Disease	11
Do. Diet and Regimen	11
Granville on Vichy	13
Hartwig on Sea Bathing	14
Do. Physical Education	14
Hufeland's Art of prolonging Life	16
Lee's Baths of Germany	18
Moore's Health in Tropics	20
Parkes on Hygiene	21
Parkin on Disease	21
Pickford on Hygiene	21
Robertson on Diet	24
Routh on Infant Feeding	23
Rumsey's State Medicine	24
Tunstall's Bath Waters	28
Wells' Seamen's Medicine Chest	30
Wife's Domain	30
Wilson on Healthy Skin	31
Do. on Mineral Waters	31
Do. on Turkish Bath	31

MATERIA MEDICA and PHARMACY.
Bateman's Magnacopia	5
Beasley's Formulary	5
Do. Receipt Book	5
Do. Book of Prescriptions	5
Frazer's Materia Medica	12
Nevins' Analysis of Pharmacop.	20
Pereira's Selecta è Præscriptis	21

MATERIA MEDICA and PHARMACY—continued.
	PAGE
Pharmacopœia Londinensis	22
Prescriber's Pharmacopœia	22
Royle's Materia Medica	24
Squire's Hospital Pharmacopœias	26
Do. Companion to the Pharmacopœia	26
Steggall's First Lines for Chemists and Druggists	26
Stowe's Toxicological Chart	27
Taylor on Poisons	27
Waring's Therapeutics	29
Wittstein's Pharmacy	31

MEDICINE.
Adams on Rheumatic Gout	3
Addison on Cell Therapeutics	3
Do. on Healthy and Diseased Structure	3
Aldis's Hospital Practice	3
Anderson on Fever	4
Austin on Paralysis	4
Barclay on Medical Diagnosis	4
Barlow's Practice of Medicine	4
Basham on Dropsy	5
Brinton on Stomach	7
Do. on Ulcer of do.	7
Budd on the Liver	8
Do. on Stomach	8
Camplin on Diabetes	8
Chambers on Digestion	8
Do. Lectures	8
Davey's Ganglionic Nervous System	11
Eyre on Stomach	11
French on Cholera	12
Fuller on Rheumatism	12
Gairdner on Gout	12
Gibb on Throat	13
Granville on Sudden Death	13
Gully's Simple Treatment	13
Habershon on the Abdomen	13
Do. on Mercury	13
Hall (Marshall) on Apnœa	14
Do. Observations	14
Headland—Action of Medicines	14
Hooper's Physician's Vade-Mecum	15
Inman's New Theory	16
Do. Myalgia	16
James on Laryngoscope	17
Maclachlan on Advanced Life	19
Marcet on Chronic Alcoholism	19
Meryon on Paralysis	20
Pavy on Diabetes	21
Peacock on Influenza	21
Peet's Principles and Practice of Medicine	21
Richardson's Asclepiad	23
Roberts on Palsy	23
Robertson on Gout	24
Savory's Compendium	24
Semple on Cough	24
Seymour on Dropsy	25
Shaw's Remembrancer	25
Smee on Debility	25
Thomas' Practice of Physic	27
Thudichum on Gall Stones	28
Todd's Clinical Lectures	28
Tweedie on Continued Fevers	29
Walker on Diphtheria	29
Wells on Gout	30
What to Observe at the Bedside	19
Williams' Principles	30
Wright on Headaches	31

CLASSIFIED INDEX.

MICROSCOPE.

	PAGE
Beale on Microscope in Medicine	5
Carpenter on Microscope	8
Schacht on do.	24

MISCELLANEOUS.

Acton on Prostitution	3
Barclay's Medical Errors	4
Bascome on Epidemics	4
Bryce on Sebastopol	8
Cooley's Cyclopædia	9
Gordon on China	13
Graves' Physiology and Medicine	13
Guy's Hospital Reports	13
Harrison on Lead in Water	14
Illingeston's Topics of the Day	15
Lane's Hydropathy	18
Lee on Homœop. and Hydrop.	18
London Hospital Reports	19
Marcet on Food	19
Massy on Recruits	20
Mayne's Medical Vocabulary	20
Part's Case Book	21
Redwood's Supplement to Pharmacopœia	23
Ryan on Infanticide	24
Snow on Chloroform	26
Steggall's Medical Manual	26
Do. Gregory's Conspectus	26
Do. Celsus	26
Whitehead on Transmission	30

NERVOUS DISORDERS AND INDIGESTION.

Birch on Constipation	6
Carter on Hysteria	8
Downing on Neuralgia	11
Hunt on Heartburn	16
Jones (Handfield) on Functional Nervous Disorders	17
Leared on Imperfect Digestion	18
Lobb on Nervous Affections	19
Radcliffe on Epilepsy	22
Reynolds on the Brain	23
Do. on Epilepsy	23
Rowe on Nervous Diseases	24
Sieveking on Epilepsy	25
Turnbull on Stomach	28

OBSTETRICS.

Barnes on Placenta Prævia	4
Hodges on Puerperal Convulsions	15
Lee's Clinical Midwifery	18
Do. Consultations	18
Leishman's Mechanism of Parturition	18
Mackenzie on Phlegmasia Dolens	19
Pretty's Aids during Labour	22
Priestley on Gravid Uterus	22
Ramsbotham's Obstetrics	23
Do. Midwifery	23
Sinclair & Johnston's Midwifery	25
Smellie's Obstetric Plates	25
Smith's Manual of Obstetrics	26
Swayne's Aphorisms	27
Waller's Midwifery	29

OPHTHALMOLOGY.

Cooper on Injuries of Eye	9
Do. on Near Sight	9
Dalrymple on Eye	10
Dixon on the Eye	11
Hogg on Ophthalmoscope	15
Holthouse on Strabismus	15
Do. on Impaired Vision	15

OPHTHALMOLOGY—cont.

	PAGE
Hulke on the Ophthalmoscope	16
Jacob on Eye-ball	16
Jago on Entoptics	17
Jones' Ophthalmic Medicine	17
Do. Defects of Sight	17
Do. Eye and Ear	17
Nunneley on the Organs of Vision	21
Walton on the Eye	29
Wells on Spectacles	30

PHYSIOLOGY.

Carpenter's Human	8
Do. Comparative	8
Do. Manual	8
Heale on Vital Causes	14
O'Reilly on the Nervous System	21
Richardson on Coagulation	23
Shea's Animal Physiology	25
Virchow's (ed. by Chance) Cellular Pathology	8

PSYCHOLOGY.

Arlidge on the State of Lunacy	4
Bucknill and Tuke's Psychological Medicine	8
Conolly on Asylums	9
Davey on Nature of Insanity	11
Dunn's Physiological Psychology	11
Hood on Criminal Lunatics	16
Millingen on Treatment of Insane	20
Noble on Mind	21
Williams (J. H.) Unsoundness of Mind	30

PULMONARY and CHEST DISEASES, &c.

Allson on Pulmonary Consumption	3
Billing on Lungs and Heart	6
Bright on the Chest	7
Cotton on Consumption	10
Do. on Stethoscope	10
Davies on Lungs and Heart	10
Dobell on the Chest	11
Fenwick on Consumption	11
Fuller on Chest	12
Do. on Heart	12
Jones (Jas.) on Consumption	17
Laennec on Auscultation	18
Markham on Heart	20
Richardson on Consumption	23
Salter on Asthma	24
Skoda on Auscultation	20
Thompson on Consumption	27
Timms on Consumption	28
Turnbull on Consumption	28
Waters on Emphysema	29
Weber on Auscultation	29

RENAL and URINARY DISEASES.

Acton on Urinary Organs	3
Beale on Urine	5
Bird's Urinary Deposits	6
Coulson on Bladder	10
Hassall on Urine	14
Parkes on Urine	21
Thudichum on Urine	28
Todd on Urinary Organs	28

SCIENCE.

Baxter on Organic Polarity	5
Bentley's Manual of Botany	6
Bird's Natural Philosophy	6

SCIENCE—continued.

	PAGE
Craig on Electric Tension	10
Hardwich's Photography	14
Hinds' Harmonies	15
Howard on the Clouds	16
Jones on Vision	17
Do. on Body, Sense, and Mind	17
Mayne's Lexicon	20
Pratt's Genealogy of Creation	22
Do. Eccentric & Centric Force	22
Do. on Orbital Motion	22
Price's Photographic Manipulation	22
Rainey on Shells	23
Reymond's Animal Electricity	23
Taylor's Medical Jurisprudence	27
Unger's Botanical Letters	29
Vestiges of Creation	29

SURGERY.

Adams on Reparation of Tendons	3
Do. Subcutaneous Surgery	3
Anderson on the Skin	3
Ashton on Rectum	4
Barwell on Diseases of Joints	4
Brodhurst on Anchylosis	7
Bryant on Diseases of Joints	7
Callender on Rupture	8
Chapman on Ulcers	9
Do. Varicose Veins	9
Clark's Outlines of Surgery	9
Collis on Cancer	9
Cooper (Sir A.) on Testis	10
Do. (S.) Surg. Dictionary	10
Coulson on Lithotomy	10
Curling on Rectum	10
Do. on Testis	10
Druitt's Surgeon's Vade-Mecum	11
Fergusson's Surgery	11
Gant's Principles of Surgery	13
Heath's Minor Surgery and Bandaging	15
Higginbottom on Nitrate of Silver	15
Hodgson on Prostate	15
Holt on Stricture	15
James on Hernia	17
Jordan's Clinical Surgery	17
Lawrence's Surgery	18
Do. Ruptures	18
Liston's Surgery	18
Logan on Skin Diseases	19
Macleod's Surgical Diagnosis	19
Do. Surgery of the Crimea	19
Maclise on Fractures	19
Maunder's Operative Surgery	20
Nunneley on Erysipelas	21
Pirrie's Surgery	22
Savage's Female Pelvic Organs	24
Smith (H.) on Stricture	25
Do. on Hæmorrhoids	25
Do. (Dr. J.) Dental Anatomy and Surgery	26
Squire on Skin Diseases	26
Steggall's Surgical Manual	26
Teale on Amputation	27
Thompson on Stricture	27
Do. on Prostate	27
Do. Lithotomy and Lithotrity	27
Tomes' Dental Surgery	28
Toynbee on Ear	28
Wade on Stricture	29
Watson on the Larynx	29
Webb's Surgeon's Ready Rules	29
Williamson on Military Surgery	30
Do. on Gunshot Injuries	30
Wilson on Skin Diseases	31
Do. Portraits of Skin Diseases	31
Yearsley on Deafness	31
Do. on Throat	31

www.ingramcontent.com/pod-product-compliance
Lightning Source LLC
Chambersburg PA
CBHW030602300426
44111CB00009B/1069